T0133486

Cybernetics in C++

RIVER PUBLISHERS SERIES IN SOFTWARE ENGINEERING

Indexing: All books published in this series are submitted to the Web of Science Book Citation Index (BkCI), to CrossRef and to Google Scholar.

The "River Publishers Series in Software Engineering" is a series of comprehensive academic and professional books which focus on the theory and applications of Computer Science in general, and more specifically Programming Languages, Software Development and Software Engineering.

Books published in the series include research monographs, edited volumes, handbooks and textbooks. The books provide professionals, researchers, educators, and advanced students in the field with an invaluable insight into the latest research and developments.

Topics covered in the series include, but are by no means restricted to the following:

- Software Engineering
- Software Development
- Pogramming Languages
- Computer Science
- Automation Engineering
- Research Informatics
- Information Modelling
- Software Maintenance

For a list of other books in this series, visit www.riverpublishers.com

Cybernetics in C++

Avi Bhattacharyya

Shomiron Institute
UK

River Publishers

Published, sold and distributed by:
River Publishers
Alsbjergvej 10
9260 Gistrup
Denmark

River Publishers
Lange Geer 44
2611 PW Delft
The Netherlands

Tel.: +45369953197
www.riverpublishers.com

ISBN: 978-87-93609-45-7 (Hardback)
        978-87-93609-44-0 (Ebook)

©2018 River Publishers

CONTENTS AT A GLANCE (1)

CONTENTS AT A GLANCE (2)

CONTENTS

PREFACE

I have been working in computer science all of my adult life, and a significant proportion of my child life. Throughout this time, as both a student and instructor, I have been both fascinated, and sometimes disturbed, at the often unique challenges in learning computer science, especially programming. It is a subject where skill needs to be built, as well as knowledge. Building up confidence in computer science is an almost unique challenge, and seeing how easily it can be destroyed in most people is often quite shocking. The need for clear, lucid explanations from instructors is especially important.

In my experience, I have found that a major issue when learning is organisation – knowing how concepts fit together, how one category comes under another. As an instructor, I always ask myself the questions "What kind of teaching does my audience want?" and "What teaching do I wish I had been given, in retrospect?" then deliver this to my students. Several have remarked that the bullet point style, which I use in my course materials, is especially useful for them, since it makes it crystal clear from the start how ideas are organised. They have also discussed how they would like to see this in a textbook. This book, along with the others in this series, aims to do just that.

A final note, I have always believed that the joint most important part of an instructor's job is to generate enthusiasm in his/her students, inspire them to go further. Computer technology is my life's greatest passion as well as my career, and one of my aims with this book, is to ensure at least some of that rubs off on you.

The source code of this book can be downloaded from the River Publishers' website, http://www.riverpublishers.com

Avi Bhattacharyya

DEDICATION

To my Parents, Asim and Sima Bhattacharyya

LIST OF FIGURES

LIST OF PROGRAMS

INTRODUCTION

Computer Science, along with its subset Software Engineering, is an immeasurably beautiful subject and industry for several reasons including:

❖ Technology which is driving the modern world

➢ Ask the question "How is the world different from, and better than 100, 50 or even 10 years ago," most answers will concentrate largely if not exclusively on computer technology

➢ Along with computer hardware, software is what has brought about this phenomenon

➢ It is fair to say that in modern times, there is not a single business/ charity which does not heavily use computer technology

- The majority have at least one person, usually at least a team, dedicated to full time computing
- Consequently, whichever industry you are passionate about, you will be able to apply your new found software skills there

❖ The best of both worlds: Arts and the Sciences

➢ In your childhood you might have been asked, which do you prefer "The Arts or the Sciences?" You may well have had trouble choosing between the two

- Science and Mathematics are fascinating, but are limited in how much they allow self expression
- The Arts have abundant scope for self expression, but have problems that results can be subjective

➢ Computer Science has

- Not only all the fascination of Science and Mathematics but the further exhilaration of applying them to the real world
- Abundant scope for self expression
  - You really know you are getting somewhere when you see that you are developing your own style

1

❖ Enhancing abundant skill sets

➢ Learning software will deepen your understanding of IT

• e.g. Learning about Abstraction will show you how databases are implemented

➢ Since software puts several ideas in mathematics in practice, your understanding of these will be enhanced

➢ The skills you acquire in software are applicable in most other fields, many non-technical

• e.g. Object-Oriented Management is largely on the increase

❖ Logistics: Cheap to produce, high profit margin

➢ Very little apparatus is needed to program, only a computer, which you will need anyway

➢ Only further software is needed, and a lot of this is free

➢ Several of the world's most prosperous companies, such as Microsoft, Oracle and Google, started with a few people programming

Within the Software Engineering field, C++ is a leading language; is not only beautiful and fascinating language in its own right, but one that underpins several others. Several world-class programs are written in C++, examples being:

❖ Operating Systems:

➢ Microsoft Windows was written largely in C++, along with its predecessors C and Assembler,

➢ The KDE interface, very popular with Linux, is written in C++

➢ Apple Mac OS was reengineered, with C++ playing a major role in all post-1990 versions

➢ Mobile Phone/PDA:

• Google Android and Apple iOS are programmed largely in C++ and C

• Blackberry OS, Symbian and MS Windows Mobile are programmed in C++

❖ Databases

➢ MySQL

• one of the most popular web databases, used by Facebook, Twitter, Flickr, You Tube, Joomla, Wordpress, Drupal

❖ Computer Desktop Applications:

➢ Adobe's products, including their flagship Creative Suite which comprises Photoshop, Dreamweaver, InDesign, Illustrator, are written in C++

➢ Microsoft Office was written entirely in C++, today some parts are added using C# which is based on C++

➢ Mozilla Foundation's applications, including Firefox and Thunderbird, are written in C++

❖ Search Engines

➢ Google's file system, as well as several of Google's applications such as Google Chrome and Google Earth, are written in C++

❖ Graphics & Animation

➢ Maya, used for 3D modelling and animation is implemented in C++

❖ Finance

➢ Much of Bloomberg software is written in C++

In addition, C++ is extremely powerful since its supports several paradigms of programming, including:

❖ Imperative
❖ Procedural
❖ Memory & Resource Management
❖ Object Oriented
❖ Generic

The extra breadth in C++ does not mean a compromise in depth. In fact, by its powerful nature, C++ covers more depth than most languages which cover only some of these paradigms

❖ C++ covers all the of concepts of Imperative, Procedural and Memory & Resource Management Programming in C, and considerably more

❖ C++ covers all the of concepts of Object Oriented and Generic Programming in Java, Swift, C#, and other .NET languages and all web-programming languages, and considerably more

❖ A comment which is frequently made in Java and C# is "Primitive variables are placed on the stack, objects are placed on the heap,"

➢ Somebody with C++ experience will have a full understanding of this

This book covers each of these, and shows how. Note that this book aims to cover

❖ Principles of Programming, especially with regards to the various programming paradigms
❖ The syntax used in C++
❖ The culture of C++ programmers and the way we think, especially when it is different from programmers of other languages

The Order is as follows:
❖ Imperative Programming (1)

➢ This section looks at the Principles required by most modern day programming languages, including Data Types, Operators, and Control Statements
➢ Abstract Data Types are also covered

❖ Procedural Programming

➢ Dividing and managing a program into functions is covered, as is the concept of Recursion

❖ Data Structures & Algorithms

➢ The Array, the most commonly used data structure is discussed, along with its applications

❖ Memory & Resource Management Programming

➢ This section looks at how to get down to the "Grass Roots" level and manipulate individual hardware registers

❖ Object Oriented Programming

➢ This section looks at the sophisticated, modern day concepts of Abstraction, Encapsulation, Inheritance and Polymorphism

❖ Imperative Programming (2)

➢ This section looks at exchanging Exception Handling and dealing with data in External Files

❖ Generic Programming

➢ The powerful concept of the Template, and its applications, are discussed here
➢ Further concepts such as Specialisation and Overloading are also covered

❖ Getting It Together: Programming in C++

➢ This part covers a general summary

➤ Several concepts encountered on the course, such as duplicate keywords and symbols, are revised and further clarified here

The style of this book is different from most textbooks in that it uses bullet points rather than prose.

❖ In computer science, of all subjects, being able to categorise is of prime importance

❖ Readers need to understand how every detail fits into the wider picture,

➤ Then how that fits into each respective paradigm
➤ and so on with broader categories

❖ For example, consider the following:

➤ Smart Pointers are available in several open source C++ libraries, including the STL and Boost, the two best known and most widely used, as well as private vendors:
➤ STL

• Auto Pointer

➤ Boost

• Scoped Pointer
• Shared Pointer
• Weak Pointer
• Scoped Array Pointer
• Shared Array Pointer
• Intrusive Pointer

➤ In addition, the following are available

• Loki
• Yasper (Yet Another Smart Pointer Library)

➤ Note that several vendors provide their own equivalent of smart pointers:

• Microsoft: For use with their CLI (Common Language Infrastructure) Library, they provide

○ Handles
○ Pinning Pointers

■ These provide further safety when passing copies of handles
■ These are used with the "managed heap," a kind of special heap memory used in Microsoft .NET

- QT/Nokia: For use in the QT Environment, they provide
  - Guarded Pointers
    - These are set to 0 when the object they are pointing to is deleted
- ❖ At each point, this clarifies which category which smart pointer belongs to

Using prose, covering the same information with equal clarity would have still been possible but taken far more words

IMPERATIVE PROGRAMMING (1)

❖ The Imperative Programming Paradigm is the practice of giving the computer instructions using sequential statements in a structured program

➢ The bulk of modern day programming languages support this paradigm, including:

- C
- Java,
- .NET languages, including C# and Visual Basic
- Web Programming Languages, including JavaScript and PHP
- Python

➢ Imperative Programming is often used as an umbrella term

- Further paradigms, including
  - Procedural,
  - Memory & Resource Management,
  - Object Oriented and
  - Generic Programming,

  are all said to be subsets, since they comprise using statements in a program

➢ Languages which are not Imperative include

- Purely low level languages, such as Machine Code and Assembler[1]
- Scripting Languages, including
  - IT Scripting: Shell Scripting in Unix/Linux, Batch Files in DOS
  - SQL Structured Query Language
  - Markup Languages, including

    - HTML
    - XML

---

[1]Whether or not Assembler and Machine Code can be classified as Imperative is disputed. They are made up of statements. However, the overall structure is not considered modern Imperative programming

❖ C++ is often said to be "most powerful language available", since it has a high versatility, and supports several paradigms of programming.

❖ The principles covered here are applicable to all programming languages

  ➤ Some modern day languages attempt to hide some of these principles from the programmer but they still go on under the bonnet

   ● Most web programming languages, such as PHP and JavaScript, avoid determining data types altogether
   ● On one hand, this creates a more programmer friendly environment
   ● The downside is that the final program will be far from optimal
     ○ All data items must be of the most sophisticated type, which leads to inefficient usage of memory

❖ This part covers

  ➤ Getting started with the mandatory "Hello World" program, and the various stages of program development
  ➤ Looking at the telling the computer

   ● What should be done, using the built-in Data Types and Operators
   ● How it should be done with Control Statements, particularly Conditional Statements and Loops

1. GETTING STARTED

1.1 SETTING UP THE PROGRAMMING ENVIRONMENT

❖ In order to practice programming, you will primarily need

  ➤ A Text Editor
  ➤ A C++ Compiler and Linker

❖ Text Editor

  ➤ Your activity, when programming, is to write text which the computer will follow as instructions
  ➤ Any plain text editor will do

   ● A possibility is Microsoft Notepad
   ● A custom-built editor for programmers, which colours the code to aid you, is generally more desirable; examples include:
     ○ Eclipse www.eclipse.org
     ○ Emacs http://www.gnu.org/software/emacs/

- o TextWrangler `https://www.barebones.com/products/ textwrangler/`
- o Komodo Edit `https://www.activestate.com/komodo-ide/ downloads/edit`
- It is important to avoid using any editor built for another purpose, such as a word processor or website designer, since
  - o These include several control characters, hidden from the user, which may well be interpreted as errors

❖ Compiler and Linker

- ➢ The primary activity of the compiler and linker is to convert your text program into a program which the computer can run
- ➢ The GNU GCC compiler is a possibility `http://gcc.gnu.org/`

1.1.1 SOFTWARE IDE

❖ A software IDE (<u>I</u>ntegrated <u>D</u>evelopment <u>E</u>nvironment), is an important modern day application, which includes

- ➢ The text editor, with colouring facilities
- ➢ The compiler and linker
- ➢ Further tools are commonly included, such as:
  - A file manager
  - Tools to browse different parts of your program
  - Debugging tools

❖ C++ IDEs include:

- ➢ Microsoft Visual Studio Express Editions
  - `http://www.microsoft.com/express/Downloads/`
- ➢ CodeBlocks
  - with the GCC (GNU C/C++ Compiler)
  - `www.codeblocks.org`
- ➢ Apple Mac: XCode `Apple App Store`

1.2 HELLO WORLD

❖ As we have stated, learning C++ is a fascinating, lifelong journey...

- ➢ Like "The Thousand Mile Journey," It must "Begin with a Single Step" as Lao Tzu stated

❖ Ever since Kernighan and Ritchie started their C Programming training by displaying "Hello World", the practice has become an integral part of software culture...why should we be any different?! :-)

❖ This project will display the message "Hello World" on a console output

❖ *Create a New Project called "HelloWorld"*

➢ *Create a new source file, naming it "HelloWorld.cpp"*

❖ *In this file, type the following code*

```cpp
#include <iostream>
using namespace std;

int main( )
{
    cout << "Hello World"<< endl;
    return 0;
}
```

Program Listing/Output 1.1: File HelloWorld.cpp The Hello World Program

❖ This program contains a Function, the **main** Function

➢ When running this program, this is the starting point

➢ This will be used extensively throughout this book

➢ Functions are the main building block of procedural programming

❖ The command cout <u>C</u>onsole <u>Out</u>, means pass the specified data to the Console Window

➢ The delimiters « are used as operators for splitting up this statement

❖ The command endl <u>End</u> <u>L</u>ine means End this line i.e. start a new line

❖ **C Note:** C programming uses the following:

	C	C++	MEANING
➢	printf		Displays string data to the console window
➢	sprintf	cout	Formats data to strings
➢	fprint		Formats data to a stream

❖ In C++, cout covers the abilities of all of these

➢ It also facilitates much neater, more concise syntax when displaying variables (section 2 onwards)

❖ *Build (i.e. Compile) this code*

➢ You should find No errors or warnings

➢ If there are any, it means there are compile time errors (discussed in section 1.5)

• *At this stage, check your typing*

❖ *Run this code*

➢ The console window will come up and display the words "Hello World"

➢ However, it will close immediately afterwards, so reading what was on the window is a near impossibility[2]

❖ So while this program does indeed run, it has a **runtime** problem we need to fix i.e. the skeleton program needs development

❖ *Modify the program so that it looks as follows:*

```
#include <iostream>
using namespace std;

int main( )
{
     cout << "Hello World"<< endl;
     while(true){}
     return 0;
}
```

Program Listing/Output 1.2: File HelloWorld_02.cpp The modified Hello World Program

❖ The line `while(true){}` ensures that the window stays open

➢ This is a while loop, covered in section 3.2.3...

• In this case it is saying do nothing indefinitely

• This ensures the window stays open

❖ While this program is very simple, the fact remains you have performed the four categories in building software

➢ Creating a Project

➢ Writing the Source Code

➢ Building the project, comprising

• Compiling each file

• Linking the files

➢ Running the Program

---

[2]This is the case with most IDEs, some, such as Codeblocks have the subsequent steps built in, so the command window does not disappear immediately

1.3 WRITING C++ SOURCE CODE

❖ It is essential to remember when you are a programmer, you have 2 audiences to please

➢ The computer

- This may comprise further hardware such as uploading the software to a mobile phone or tablet

➢ Your fellow humans

- This means primarily the people who will be maintaining and using your program
- This is frequently overlooked
- Just like when you
  ○ Write a report,
    ▪ You don't think "What do I want to write?"
    ▪ You think "What do my readers want to know?"
  ○ Give a presentation,
    ▪ You don't think "What do I want to say?"
    ▪ You think "What does my audience want to know?"
- When you are programming, you must also ask "How can I make my program as easy as possible to understand, maintain and prevent future problems"
- One of the humans who will be maintaining your programs is yourself! :-)

➢ Code which is clearly formatted is not only much more pleasurable to look at, it is far easier to maintain and debug (section 1.5)

1.3.1 COMMENTS

❖ Comments are the parts of source code which are for humans to read

➢ They are not looked at by the compiler

❖ They are of critical importance for any program

➢ All but the simplest programs are not self explanatory - people who maintain and develop your programs (which will include you!) need to understand them

➢ It is self contradictory to say "It is a good but poorly commented program"

❖ There are primarily two means of ensuring that text is a comment

➢ **Single line commenting:** The double forward slash // ensures that everything after it on the same line is a comment

```
// Everything on the rest of this line is a comment
```

➢ **Multiple line commenting:** A section can be
- Started with the forward slash followed by the asterisk /* and
- Ended with the reverse i.e. asterisk followed by the forward slash */

```
/*

        Everything in this multiple line
        section is a comment

*/
```

1.3.1.1 Style With Primitive Tools

❖ A major part of your development as a programmer is developing your commenting style

➢ In other words, this is an invaluable opportunity for you to add your personal touch to all your programs

❖ Since control characters and tags, which are used extensively in word processing and web design tools, need to be avoided, the following are not available:

➢ Fonts, font sizes, styles and colours
➢ Borders, tables

❖ What we have available are all the characters, along with the tab

❖ Over the years, programmers have come up with several ingenious ways of commenting code

➢ Common practices include:
- Starting each project with
  ○ An explanation of what it does
    ■ How the end human users are expected to interact with it
  ○ A list of who contributed what to the project and when

1.3.2 SOURCE CODE FILES

❖ There are two major categories of C++ source code file

➢ Source File ⠀⠀⠀`.cpp` extension

➢ Header File   .h or .hpp extension

❖ Further categories will depend on individual projects; a common one is

➢ Resource Files

  • These are further files which are presented to the audience when the program runs, such as

    ◦ Icons, .ico files
    ◦ A music file, such as .mp3 or .wav

❖ The details are covered in section 7, and followed for the rest of the course

1.4 SUBSEQUENT STAGES IN BUILDING A PROGRAM
1.4.1 COMPILATION

❖ The two most common tasks of the compiler can be categorised, using two P's as follows:

➢ **Parrot**: What you write in source code, it translates into a language which the runtime environment can understand

  • Traditionally, and for most modern day systems, this is/has been machine code for the microprocessors to execute
    ◦ Most C++ environments use this system

  • In some cases, this is another language, often called bytecode, for the runtime environment to interpret
    ◦ Java compilation produces bytecode which must be interpreted by the JVM Java Virtual Machine
    ◦ This is less common in C++

      ▪ In the Microsoft .NET environment, C++ compilation produces an MSIL (Microsoft Intermediate Language) file which is interpreted by the CLR (Common Language Runtime)

➢ **Pigeon**: When you define any entity in C++, such as variables or functions the compiler allocates (an) address(es) in runtime memory for it/them

1.4.1.1 Object Files

❖ In C++, the compilation process comprises creating an object file, which contains the code (mostly) for the hardware to execute

❖ The extensions for object files are:

➢ `.o`
➢ `.obj`

❖ The details will be covered in section 7

1.4.2 LINKING

❖ Linking is the practice of putting together i.e. linking the different object files to produce a working program

➢ This becomes an issue when the project comprises more than one file
  • i.e. All but the simplest of projects

❖ This will be covered in section 7

1.4.3 RUNNING

❖ The final stage is making our program do its job i.e. running it

❖ Errors at this stage are known as "run-time" errors

❖ Every time you launch an application

➢ On your computer, you are running an executable file
➢ On your another device, such as a smartphone, you are running a special executable file

1.5 DEBUGGING (1)

1.5.1 WHAT ARE BUGS/ERRORS?

❖ A bug/error is a fault in a program which prevents it from working properly

❖ Debugging is the practice of fixing bugs/error

❖ Following on from the 3 stages in building a program, there are 3 categories of error:

➢ Compile-Time Error
➢ Linker Error
➢ Runtime Error

❖ In order to illustrate the differences, as well as the similarities, between these, consider the following analogy:

➢ A group of people, who speak only English, need directions to get to their destination;

- They don't know what kind of place will meet all their requirements, so they each need to make a request which will be put together
- The only person who can help them speaks and understands only Swahili
- Fortunately, the following 2 people are present
  - Someone who can interpret from English to Swahili (the compiler)
  - Someone who can put together all the Swahili requests (the Linker)

➢ **Compile Time Error:** If there is an error in the English, such as a grammatical error or an unrecognised name, the interpreter will not be able to translate

➢ **Linker Error:** If all the individual translations are correct but they do not make sense when put together

➢ **Runtime Error:** If incorrect instructions/directions are given by the Swahili instructor i.e. enough was done to enable actions but these are wrong

1.5.2 COMPILE-TIME ERRORS

1.5.2.1 Compiler Errors

❖ Compiler errors are problems with the source code which prevent the compiler creating the object file

❖ Examples include

➢ Incorrect syntax
➢ Typos and misspellings of variable/function names
➢ Attempting to use a variable which has not yet been defined
➢ Attempting to call a function which has not yet been declared

1.5.2.2 Compiler Warning

❖ Warnings are issued when there is something which

➢ Can cause prospective trouble
➢ But will not prevent the program from running

❖ Common warnings include:

➢ Defining a variable then not using it
➢ Forgetting to initialise a variable

❖ Treat warnings with the same kind of compulsory need for elimination, as you do with compiler errors

1.5.3 RUNTIME ERRORS

❖ Runtime errors occur when the program is running e.g.

> ➢ Incorrect calculation results
> ➢ Failing to respond appropriately to user instructions
> > • Failing to open or close a program/feature requested
> > • Not acknowledging changes requested by the user
> > • Introducing changes, such as changes to settings, when updates are installed

2. DATA TYPES & OPERATORS

2.1 LEARNING BY EXAMPLE

2.1.1 PERFORMING CALCULATIONS

❖ Since the best, if not only, way to start learning is by example, let us look at a program which illustrates several fundamental concepts

> ➢ Two numbers will be added together
> ➢ The result will be displayed on the console

```
/*      ************************************************************
        ARITHMETIC
        ************************************************************
        -       Demonstrating the usage of Data and Operators in C++
*/
#include <iostream>
using namespace std;

int main( )
{
        //»»»»»»»»»»»»»»»»»»»»»»»»»»»»»»»»»»»»»»»»»»»»»»»»»»»»»»»»»»»»»»»»»
        //      Defining and Initialising the Variables
        //
        int myInt_A = 5;
        int myInt_B = 7;
        int myInt_C = 8;

        //»»»»»»»»»»»»»»»»»»»»»»»»»»»»»»»»»»»»»»»»»»»»»»»»»»»»»»»»»»»»»»»»»
        //      Carrying out the Calculations
        //
        int total_A_B = 0;
        total_A_B = myInt_A + myInt_B;

        //»»»»»»»»»»»»»»»»»»»»»»»»»»»»»»»»»»»»»»»»»»»»»»»»»»»»»»»»»»»»»»»»»
        //      Displaying the Results to the User
        //
```

```
        cout    << "The total of "<< myInt_A<< " and "<< myInt_B
                << " is "<< total_A_B
                << endl;

        //»»»»»»»»»»»»»»»»»»»»»»»»»»»»»»»»»»»»»»»»»»»»»»»»»»»»»»»»»»»»»»»»»»»»»
        cout    << endl<< endl<< "=============="
                << endl<< "End...Arithmetic, Getting Started"
                << endl;
        while(true){}
        return 0;

}//END...Function MAIN
```

```
The total of 5 and 7 is 12


==============
End...Arithmetic, Getting Started
```

Program Listing/Output 2.1: Getting Started: Defining and adding two variables

❖ This program, though simple, illustrates several fundamental concepts:

❖ The Line `int myInt_A = 5` actually performs 2 operations

➢ Defining the variable

- It names a variable " `myInt_A` " and asks the compiler to allocate a place in runtime memory for the variable
- It defines the variable `myInt_A` of type `int`
  ○ Amongst other things, this determines the amount of memory required, and what can and can't be done with this variable

Thus `myInt_A` is now an **instance** of type int

➢ Initialising the variable

- Note: as we will be discussing in section 4, it is usually advisable to perform one action per line
- However, this is the one notable exception, where it is advisable to perform 2 actions

2.1.2 INTERACTING WITH THE CONSOLE WINDOW

❖ Two of the most commonly used commands in C++ are

➢ `cout`   Used to display text in the console window

➢ `cin`   Used to acknowledge text, entered by the user, in the console window

- The variable placed after this will contain the entered value

❖ The following program shows how they can be used:

```
/*         *****************************************************************
           GREETING DEMO
           *****************************************************************
           -        Demonstrates the usage of the pervasive commands
                    - cout
                    - cin
                    for interacting with the Console Window
*/
#include <iostream>
#include <string>
using namespace std;

int main( )
{
       //»»»»»»»»»»»»»»»»»»»»»»»»»»»»»»»»»»»»»»»»»»»»»»»»»»»»»»»»»»»»»»»»»»»»»
       string name_User = " ";
       cout << "Please enter your name, then press return"<< endl;
       cin >> name_User;
       cout << "Hello "<< name_User<< ", hope you are well? "<< endl;


       //»»»»»»»»»»»»»»»»»»»»»»»»»»»»»»»»»»»»»»»»»»»»»»»»»»»»»»»»»»»»»»»»»»»»»
       cout << endl<< endl<< "==========="
            << endl
            << "END...C++ Demo: cout and cin Commands"
            << " "
            << endl;
       while(true){}
       return 0;

}//END...Function MAIN
```

```
Please enter your name, then press return
 John
 Hello John, hope you are well?


 ===========
 END...C++ Demo: cout and cin Commands
```

Program Listing/Output 2.1: GreetingDemo.cpp: using the cout and cin Commands

❖ Note that this program has used a variable of type "string"

> This holds "strings of text"

- i.e. Groups of characters ranging
  - from a single character
  - to a sentence, or even groups of sentences
- ➢ Since working with textual data is pervasive, this is one of the most commonly used variable types
- ➢ Note that this is not a primitive type (section 2.3 onwards), but one which requires the C++ Library inclusion

```
#include <string>
```

2.2 VARIABLES

❖ A variable has primarily 4 things associated with it

- ➢ A data type
  - This is of critical importance in C++, and can be
    - Primitive, covered in section 2.3
    - User defined, covered in the parts ADTs (section 17.2) and OOP (section 17 onwards)
- ➢ A name
  - You choose this
  - It is imperative to choose this wisely (section 2.2.1)
- ➢ An address
  - This will be a hexadecimal number indicating its location in runtime memory
  - The computer uses this address, often subtly, hiding it from the programmer
  - In memory management programming, the programmer (i.e. YOU! :-)) has/have access to this, using Pointers or References (section 10 onwards)
- ➢ A value
  - This is subject to change, being a "variable"

2.2.1 NAMING VARIABLES

2.2.1.1 Rules

❖ Variable names can comprise only the characters

- ➢ ASCII Characters
- ➢ Numerical characters

➢ the underscore _ character

❖ The following are forbidden:

➢ Punctuation marks such as commas, full stops,

➢ Mathematical operators such as brackets, + - * / symbols

➢ Spaces

❖ C++ has a number of keywords which are reserved - these cannot be duplicated as variable names - the full list is given in Appendix B

2.2.1.2 Guidelines

❖ Give your variables meaningful names

➢ Choose something descriptive e.g.

- If a variable is to represent the number of people visiting an exhibition, Use a name like `num_People`
- Avoid using a symbol or short name such as `a`, `b`, or `x`

❖ The underscore character _ is widely used by programmers

➢ It appears like a space to the human eye,

➢ Yet ensures the name remains one word

❖ Since spaces are not allowed, using a mixture of upper and lower case is highly recommended

➢ Having them in the same "word" is known as Camel Case

- `numberSpectators, findHighestFactor`

➢ Further separation using the underscore character is very common:

- `num_Employees, highest_PrimeFactor`

❖ Avoid starting any names with the underscore character, including a double underscore character

➢ e.g. `_num_People, __name_Person`

➢ Several preprocessor directives start with these (section 7 onwards)

- There is the possibility of a naming clash

2.2.2 INITIALISING A VARIABLE

❖ There are primarily 2 methods for initialising a variable

➢ Using Assignment Equals =

➢ Using Parentheses ()

2.2.2.1 Using Assignment Equals =

❖ The statement

```
a = b
```

sets the value of a equal to the value of b (see section 2.4.1.2.1)

2.2.2.2 Using Parentheses ()

❖ The following statement

```
a(b)
```

also sets the value of a equal to the value of b (see section 2.4.1.2.1)

2.2.2.3 Which is Preferable?

❖ This is one of several situations where it is down to personal preference, part of the art form of C++

❖ Amongst C++ programmers

➢ The assignment equals methods tends to be more popular since the statement is more readable

• It strongly resembles the equivalent mathematical statement

➢ A slight issue with using parentheses is that it can be confused for several other invocations in C++ including

• A function call (see section 5 onwards)
• A macro call (see section 5)
• Creation of an object with a constructor call (see section 18.5)

❖ There is one situation, regarding Smart Pointers, when the Parentheses must be used; the = operator is prohibited

➢ This is covered in Cybernetics in C++ - Expert

2.2.3 ENSURING READ ONLY STATUS USING CONST

❖ How can we ensure that once a "variable" has been initialised, it can no longer have a different value written to it?

➢ i.e. Ensure it has Read Only status?

❖ Put the keyword `const` in the definition

❖ There are a great many situations where this is essential

➢ When we are using mathematical/scientific constants such as

- PI π
- Planck's Constant

➤ When we want to prevent accidental modification

❖ Examples

```
const double PI = 3.142;

const double PLANCKS_CONSTANT = 4.136*10^-15;
```

❖ NB, It does not matter whether we put the `const` keyword before or after the data type, as long as it comes before the variable name:

➤ The statement

```
const double PI = 3.142;
```

or

```
double const PI = 3.142;
```

ensures that any subsequent attempt to modify it, such as

```
PI = 10.3;
```

will cause a compile-time error

❖ This is similar to writing data to ROM memory

➤ You can write to the memory the first time, the initialisation

➤ Thereafter it is Read-Only

❖ `const` is one of the most important qualifiers in C++

➤ Its usage is encouraged in C++ considerably more frequently than in C

➤ There are several situations where it is used, encountered in forthcoming chapters, then Rules of Thumb for its usage are covered in section 32.

2.3 PRIMITIVE DATA TYPES

❖ Primitive Data Types are the most fundamental data types which come built into the C++ environment

➤ The other category is User Defined Type

- ...when the programmer (i.e. you!:-)) creates a compound data type, such as a Struct or a Class, much of the time comprising at least some primitive types
- These will be covered from section 17.2 onwards

❖ The fundamental data types are

  ➢ char
  ➢ int
  ➢ double
  ➢ float
  ➢ bool

❖ The following qualifiers can be added to form a compound data type:

  ➢ signed, unsigned
  ➢ short, long

❖ Appendix A contains a listing of the primitive data types in C++

❖ Some programming languages, such as Perl, Python and MATLAB, hide data types from the programmer completely[3]

  ➢ On one hand, this creates a simpler programming environment
  ➢ The price paid is in efficiency
    • All data items effectively have to be of the most sophisticated type
      ○ E.g. Numerical data must be of type double, or equivalent
    • This means consuming more memory than required a significant proportion of the time

2.3.1 TYPE CASTING

❖ Type casting is the practice of changing the data type of one instance
❖ There are several sorts of cast

  ➢ This section will look at the simplest and most commonly used one, the Static Cast
  ➢ Further casts include the
    • Dynamic Cast
    • Cast in and out of read only status
  ➢ This is a common requirement, as the following example illustrates:

---

[3]One could argue there is a slight exception to this case when Object Oriented Programming is used in Perl and Python.

❖ Consider the following program

➢ (Q) Will it give the correct result?

➢ (A) No!

```
#include <iostream>
#include <string>
using namespace std;

int main( )
{
      int myInt_A = 11;
      int myInt_B = 2;
      int myInt_C = 0;

myInt_C = myInt_A / myInt_B;

}//END...Function MAIN
```

Program Listing/Output 2.2: This Program compiles fine but will it run correctly?

❖ Have you seen the problem?

➢ The variable `myInt_C` is of type int and so cannot take on a mixed number value

• The program will calculate its value to be 5, not 5.5

➢ It needs to be of type float or double

❖ The Static Cast command is used to convert from one primitive type to another

```
static_cast<newType>(variableName);
```

❖ Thus to convert the variable to type double, the following command is used:

```
static_cast<double>(pricePerTicket);
```

❖ The above program can be corrected by applying the following casts:

```
/*    ***********************************************************
      C++ STATIC CASTS
      ***********************************************************
      -       Demonstration: How Static Casts can be used to prevent
```

```
                common errors
*/
#include <iostream>
#include <string>
using namespace std;

int main( )
{
        //»»»»»»»»»»»»»»»»»»»»»»»»»»»»»»»»»»»»»»»»»»»»»»»»»»»»»»»»»»»»»
        int myInt_A = 11;
        int myInt_B = 2;

        double myDbl_C = 0;

        // static_cast<double>(myInt_C);
        myDbl_C = static_cast<double>(myInt_A) /
        static_cast<double>(myInt_B);

        cout << myInt_A<< " / "<< myInt_B<< " is "<< myDbl_C<< endl;


        //»»»»»»»»»»»»»»»»»»»»»»»»»»»»»»»»»»»»»»»»»»»»»»»»»»»»»»»»»»»»»
        cout << endl<< endl<< "==========="
             << endl
             << "END...C++ Demo: Static Casts"
             << " "
             << endl;
        while(true){}
        return 0;

}//END...Function MAIN
```

```
11 / 2 is 5.5


===========
END...C++ Demo: Static Casts
```

Program Listing/Output 2.3: Cast_Static_Test_Client.cpp

❖ As we see, this now gives the correct value of 5.5

❖ Note that all the data in the calculation had to be of type double

  ➢ All, in this case both, the Operands
  ➢ The Result

❖ If any one of these was of type int, the result would not have been correct[4]

2.3.1.4 C Style Casts

❖ C programming also uses casts heavily, but the style is different

```
(NewType) variableName;
```

➢ Example:

```
(double) myInt_B;
```

❖ Since C++ is fully compatible with C, this will still compile,

➢ However, these will work at runtime only with primitive variables

➢ There are several further casts in C++ (covered in Cybernetics in C++ - Expert) with which these will not work

❖ **Rule of Thumb:** In C++. avoid the C style casting whenever possible

2.4 OPERATORS

❖ This section looks at some of the most commonly used operators

❖ For a full list of C++ operators, and their explanations, see Appendix A.

2.4.1 Mathematical Operators

2.4.1.1 Arithmetic Operators

❖ All the standard arithmetic operators are used as standard in C++

➢	+	Addition
➢	-	Subtraction
➢	*	Multiplication
➢	/	Division
➢	%	Modulo Operator - the Remainder

2.4.1.2 Assignment & Equality Operators

❖ There are 2 operators which can be called "equals" in computer science

❖ They both resemble, but are different from the "equals" operator in mathematics

---

[4]Strictly speaking, different C++ environments can behave differently when data types are mixed; ensuring all the involved variables are of the appropriate type will ensure correct operation on all platforms.

2.4.1.2.1 Assignment Operations using =

❖ The assignment = operator is used to assign a value to a variable

```
numSpectators = 30000;
```

2.4.1.2.2 = Operator in Mathematics and Computer Science

❖ The assignment has been designed to resemble the mathematical = operator

> However, there are important differences
  • In mathematics, = means "is the same as"
  • In computer science, = is a movement operator

❖ Consider the following statement:

a	=	b
The ADDRESS		The VALUE
of variable a -		of variable b
an L-value		- an R-value

❖ The functionality of the Assignment Operator is to

> Take the VALUE of the Right Hand Side
> Put it into the ADDRESS of the Left Hand Side

❖ Thus the = operator has

> 2 inputs
  • The **L**-value   An Address, a   **L**eft hand side
                       Location, known at
                       Compile time
  • The **R**-value   Known at **R**untime   **R**ight hand side
> Functionality: Move the value into the address
> No Output(s)

❖ Consequently, the statement

```
c = c + 5
```

> is perfectly acceptable in programming
  • It is saying take the value of c + 5 and put it into the address of c
  • i.e. the value of c is increasing by 5
> Is not a mathematical statement

❖ In contrast, the statement

```
3 = c
```

➢ is perfectly acceptable in mathematics
➢ is not a computer science statement
  • It is saying take the value of c - OK so far...
  • Then place this in the address of 3...?"!
    ○ 3 is a VALUE, not a VARIABLE
      ■ It does not make sense to talk about its address

2.4.1.2.3 Equality Operations using ==

❖ The Equality Operator tests WHETHER or not two VALUES are EQUAL

➢ It returns a Boolean value, TRUE if they are equal, FALSE if they are not

❖ It is used as follows:

```
a == b
```

❖ Thus the == operator has

➢ 2 inputs
  • An r-**value**   Left hand side
  • An r-**value**   Right hand side
➢ Functionality: Compare whether they are equal
➢ Output: Boolean Variable
  • TRUE if the values are equal
  • FALSE if the values are unequal

2.4.1.2.4 Compound Assignment Operators

❖ Compound Assignment Operators are a shorthand means of replacing a variable with a modified value

➢ They comprise an operator being followed by the =

❖ Examples:

COMPOUND ASSIGNMENT	RENDERS SAME RESULT AS...
➢ a += 5	a = a + 5
➢ a -= 6	a = a - 6
➢ a *= b	a = a * b
➢ a /= 3	a = a / 3

2.4.1.2.5 Assignment vs Compound Assignment Operators

❖ While statements such as those above render the same results, the Compound Operator is sizeably more efficient

➢ The Intermediate Compiler Code is considerably different

❖ The statement a = a + 5 comprises the following actions:

➢ Look at the right hand side: go to the register containing the variable a
➢ Note the value stored here, store it in a temporary register
➢ Add 5 to the value in this register
➢ Look at the left hand side: go to the register containing the variable a
➢ Note its address
➢ Take the value at this address and add the value stored in the temporary register
➢ i.e. **2** visits to were required to the register containing the variable **a**

❖ The statement a += 5 comprises the following actions:

➢ Look at the left hand side: go to the register containing the variable a
➢ Take the value on the right hand side
➢ Add this to the value stored in a

❖ Since the latter is sizeably more efficient, statements of the nature of a = a + 5 are considered unacceptable in C++

❖ Note: Certain languages, such as Visual Basic and MATLAB, do not support compound assignments

➢ There, the inefficient form is the only choice
  • This is one of several reasons why these languages are unsuitable for high performance

2.4.1.3 Increment & Decrement Operators

❖ Adding and subtracting 1 from a variable are such common features that special operators have been devised for them

➢ Adding 1: Increment ++
➢ Subtracting 1: Decrement --

❖ The following sections look at the similarities and differences between the "pre-" and "post-" forms

2.4.1.3.1 Post Increment

❖ Post Increment means

> ➢ Use the old value of the variable
> ➢ Then add 1

❖ The syntax for the Post Increment is to place the ++ symbols **after** the variable name as follows:

```
variableName++
```

> ➢ Thus in order to perform the post-increment on the variable `num_Slots_CheckedSoFar`, the following will statement will be used:

```
num_Slots_CheckedSoFar++
```

2.4.1.3.2 Pre Increment

❖ Pre Increment means

> ➢ Add 1 to the variable
> ➢ Then use the new value

❖ The syntax for the Pre Increment is to place the ++ symbols **before** the variable name as follows:

```
++variableName
```

> ➢ Thus in order to perform the pre-increment on the variable `num_Slots_CheckedSoFar`, the following will statement will be used:

```
++num_Slots_CheckedSoFar
```

2.4.1.3.3 Similarities & Differences: Pre Increment & Post Increment

❖ When the only action which has to be performed in a statement is the increment of a variable, pre-increment and post-increment have the same effect

❖ However, when more than one action is to be performed, their differences become apparent

> ➢ Consider the following program:
>> • What values will the variables `p`, `q`, `r` and `s` have after execution of the following statements?
>>
>> ```
>> p = a++;
>> q = a;
>> s = ++b;
>> t = b;
>> ```

```
/*         *************************************************************
           C++ INCREMENT OPERATORS: Pre Increment and Post Increment
           *************************************************************
           -         This program demonstrates the usage of the C++
                     Increment operators
                     -       PreIncrement ++variableName
                     -       PostIncrement variableName++
                     showing their Similarities and Differences
           -         Both Add 1 to the Operand
                     -       PreIncrement
                             -       Uses the Old/Current Value
                             -       Then Increments
                     -       PostIncrement
                             -       Increments
                             -       Then Uses the New Value
*/
#include <iostream>
#include <string>
using namespace std;

int main( )
{
     //»»»»»»»»»»»»»»»»»»»»»»»»»»»»»»»»»»»»»»»»»»»»»»»»»»»»»»»»»»»»»»»»»»»
     int a = 3, b = 7,
         p = 0, q = 0, s = 0, t = 0;

     cout << "======================================="<< endl;
     cout << "Initial Values "<< endl;
     cout << "======================================="<< endl;
     cout << "a = "<< a<< endl;
     cout << "b = "<< b<< endl;
     cout << "p = "<< p<< endl;
     cout << "q = "<< q<< endl;
     cout << "s = "<< s<< endl;
     cout << "t = "<< t<< endl;
     cout << endl;

     //»»»»»»»»»»»»»»»»»»»»»»»»»»»»»»»»»»»»»»»»»»»»»»»»»»»»»»»»»»»»»»»»»»»
     //       Performing the Increments
     //       -       What values will p, q, s and t have after these
     //               operations
     //
     p = a++;
     q = a;
     s = ++b;
     t = b;

     //»»»»»»»»»»»»»»»»»»»»»»»»»»»»»»»»»»»»»»»»»»»»»»»»»»»»»»»»»»»»»»»»»»»
     cout << "======================================="<< endl;
     cout << "New Values "<< endl;
     cout << "---------------------------------------"<< endl;
     cout << "Performing the actions:"<< endl;
```

```
        cout << "p = a++; "<< endl;
        cout << "q = a; "<< endl;
        cout << "s = ++b; "<< endl;
        cout << "t = b; "<< endl;
        cout << "----------------------------------------"<< endl;
        cout << "a = "<< a<< endl;
        cout << "b = "<< b<< endl;
        cout << "p = "<< p<< endl;
        cout << "q = "<< q<< endl;
        cout << "s = "<< s<< endl;
        cout << "t = "<< t<< endl;

        //»»»»»»»»»»»»»»»»»»»»»»»»»»»»»»»»»»»»»»»»»»»»»»»»»»»»»»»»»»»»»»»»»
        cout    << endl<< endl<< "==========="
                << endl
                << "END...C++ Demo: Pre & Post Increments"
                << " "
                << endl;
        while(true){}
        return 0;

}//END...Function MAIN


/*
    *******************************************************************

    NB This has been given for demonstration purposes only
    -       Avoid writing code where unnecessary multiple actions take
            place per line

    *******************************************************************

*/


==========================================
Initial Values
==========================================
a = 3
b = 7
p = 0
q = 0
s = 0
t = 0


==========================================
New Values
------------------------------------------
```

```
Performing the actions:
p = a++;
q = a;
s = ++b;
t = b;
----------------------------------------
a = 4
b = 8
p = 3
q = 4
s = 8
t = 8


===========
END...C++ Demo: Pre & Post Increments
```

Program Listing/Output 2.4: Pre_Post_Incrment_Demo_Client.cpp

❖ This shows the difference

➢ The line

```
p = a++;
```

- means that
  - ○ p will take on the **old** value of a
  - ○ After this line, a will be incremented

➢ The line

```
s = ++b;
```

- means that
  - ○ s will take on the **new** incremented value of b

❖ In terms of efficiency, there is an essential difference:

➢ Since the post increment uses the old value, It has to be stored in memory

➢ Consequently, the post increment always requires (at least) one extra memory register on top of the pre increment

➢ **Rule of Thumb:** Prefer pre increment to post increment whenever possible

2.4.1.4 Bitwise Operators

❖ The **Bitwise** Operators are used to perform a logical operation on each BIT of a binary **number**

➤ Since all numbers have an underlying binary representation in digital technology, this can be applied to all numbers

❖ Examples:

➤ `874 | 484 = 1006`

➤ `874 & 484 = 352`

ORIGINALS	BASE-2/BINARY										RESULTS
BASE-10	**512**	**256**	**128**	**64**	**32**	**16**	**8**	**4**	**2**	**1**	**BASE-10**
874	1	1	0	1	1	0	1	0	1	0	
484		1	1	1	1	0	0	1	0	0	
874 | 484	1	1	1	1	1	0	1	1	1	0	1006
874 & 484	0	1	0	1	1	0	0	0	0	0	352

2.4.2 LOGICAL OPERATORS

❖ **Logical** Operators apply to **conditions**

❖ The most common ones are

➤ ‖   OR

➤ &&   AND

➤ !   NOT

❖ These are commonly used with conditional statements and loops (section 3 onwards)

❖ *Common Interview Question: What's the difference between & and && ?*

➤ *Answer: && is LOGICAL AND, & is BITWISE AND*

2.4.3 THE SIZEOF OPERATOR

❖ The sizeof operator is used to check the size of data, of both

➤ Data Instances

```
sizeof( dataInstanceName )
```

➤ Data Types

```
sizeof( dataType )
```

❖ The following program shows this in use:

```
/*      ************************************************************
        SIZEOF OPERATOR
        ************************************************************
        -       Using the sizeof Operator to find the size in Bytes of
                -       Data Instances
                -       Data Types
*/
#include <iostream>
#include <string>
using namespace std;

int main( )
{
        //»»»»»»»»»»»»»»»»»»»»»»»»»»»»»»»»»»»»»»»»»»»»»»»»»»»»»»»»»»»»»»»»»
        int myIntA = 5;
        float myFlt_B = 6.2;
        double myDbl_C = 7.5;

        signed int mySignedInt_D = -4.5;
        long int myLongInt_E = 234;

        cout << "======= SIZES (in Bytes) ======="<< endl;
        cout << "sizeof( myIntA ) = "<< sizeof( myIntA )<< endl;
        cout << "sizeof(myFlt_B) = "<< sizeof(myFlt_B)<< endl;
        cout << "sizeof(myDbl_C) = "<< sizeof(myDbl_C)<< endl;
        cout << "sizeof(mySignedInt_D) = "<< sizeof(mySignedInt_D)<<
                endl;
        cout << "sizeof(myLongInt_E) = "<< sizeof(myLongInt_E)<< endl;
        cout << "sizeof(int) = "<< sizeof(int)<< endl;
        cout << "sizeof(long int) = "<< sizeof(long int)<< endl;
        cout << "sizeof(long long int) = "<< sizeof(long long int)<<
                endl;


        //»»»»»»»»»»»»»»»»»»»»»»»»»»»»»»»»»»»»»»»»»»»»»»»»»»»»»»»»»»»»»»»»»
        cout << endl<< endl<< "============"
             << endl
             << "END...C++ Demo: SizeOf Operator"
             << " "
             << endl;
        while(true){}
        return 0;

}//END...Function MAIN
```

```
======= SIZES (in Bytes) =======
sizeof( myIntA ) = 4
sizeof(myFlt_B) = 4
```

```
sizeof(myDbl_C) = 8
sizeof(mySignedInt_D) = 4
sizeof(myLongInt_E) = 4
sizeof(int) = 4
sizeof(long int) = 4
sizeof(long long int) = 8


===========
END...C++ Demo: SizeOf Operator
```

Program Listing/Output 2.5: Sizeof_Operator_Client.cpp

3. CONTROL STATEMENTS

❖ Along with data and operators, control statements are of fundamental importance in programming

➢ Data and operators determine what the actions will be

➢ Control statements determine how they should be performed

❖ The two most common control statements, Conditional Statements and Loops, are often referred to as the "Blinking and Breathing" of programming

➢ Conditional Statements are used to control when and how an action should be performed under certain conditions

➢ Loops are used to control when/how an action should be performed more than once

3.1 CONDITIONAL STATEMENTS

3.1.1 IF STATEMENTS

❖ The simplest and very widely used conditional statement is the straight-forward IF statement:

❖ The "If" Clause is as follows:

```
IF( Condition )
    THEN Action if the Condition is TRUE
```

❖ Examples

➢ If I am running low on petrol

• I will fill up asap

➢ If I need to get to Knightsbridge, using London Underground

• I will make sure I am travelling on the Piccadilly Line

❖ In C++, the syntax is as follows:

```
if( /* CONDITION */ )
{
        // Action(s) if the CONDITION is TRUE;
}
```

❖ The following program shows how to display if a tank needs to be filled with petrol:

```
/*      ************************************************************
        IF STATEMENT
        ************************************************************
        -       This program shows how to implement and use
                the Simple (but essential!:-)) If Statement
*/
#include <iostream>
#include <string>
using namespace std;

int main( )
{
        //»»»»»»»»»»»»»»»»»»»»»»»»»»»»»»»»»»»»»»»»»»»»»»»»»»»»»»»»»»»
        double fuel_In_Tank_Current = 65.5;
        const double fuel_In_Tank_Alert = 20.0;

        //»»»»»»»»»»»»»»»»»»»»»»»»»»»»»»»»»»»»»»»»»»»»»»»»»»»»»»»»»»»
        if( fuel_In_Tank_Current <= fuel_In_Tank_Alert )
        {
                cout << "Fuel running Low: Fill up asap"<< endl;
        }

        //»»»»»»»»»»»»»»»»»»»»»»»»»»»»»»»»»»»»»»»»»»»»»»»»»»»»»»»»»»»
        cout    << endl<< endl<< "============"
                << endl
                << "END...C++ Demo: If statement"
                << " "
                << endl;
        while(true){}
        return 0;

}//END...Function MAIN
```

Program Listing/Output 3.1: If_Statement_Client.cpp

3.1.1.1 What is a condition?

❖ This is a simple but important question:

➢ A condition is something which returns a Boolean variable: TRUE or FALSE

➢ It makes sense to control actions according to conditions, only if the conditions are either true or false

➢ Thus conditional statements work with only Boolean variables

❖ Note that several modern compilers, including Microsoft and QT, perform implicit conversions when a non-Boolean type variable is placed as the condition

➢ They will assume
- Any non-zero value is TRUE
- Only zero is FALSE

➢ However, they are still working with Boolean variables

❖ (Q) What is the flaw in the following code fragment?

```
if( salary = 9800 )
{
        taxRate = 0.25;
}
```

(A) The equivalence equals == is required; this tests the **condition** whether two values are equal

```
if( salary == 9800 )
{
        taxRate = 0.25;
}
```

➢ The single = is assignment equals - this does not test a condition and is thus inappropriate in a conditional statement.

➢ It is fair to say all programmers independently rediscover this

❖ Most high performance languages, including C, Java and C#, use this system

❖ Confusion is caused because several languages, such as Visual Basic and MATLAB, still use the single = in conditional statements

3.1.2 THE IF, THEN, ELSE CLAUSE

❖ The If, Then, Else clause, which is used abundantly in computer science, IT, and all forms of logic, takes the following form:

```
IF( Condition )
```

> **THEN Action if the Condition is TRUE**
>
> **ELSE**
>
> **Action if the Condition is FALSE**

❖ Examples:

➤ If it is raining,
  - I will put up my umbrella
  - Else, I'll leave it in my bag

➤ If I will be travelling in London
  - I will use the London Underground
  - Else, I will be travelling by car

❖ In C++, the syntax is as follows:

```
if( /* CONDITION */ )
{
      // Action(s) if the CONDITION is TRUE;
}
 else
{
      // Action(s) if the CONDITION is FALSE;
}
```

❖ Note that the keyword "THEN," which was widely used in older languages, is not used in C++

➤ Most modern day languages, including C, Java and C# have also dropped it

➤ Some languages, such as Visual Basic, still use it

❖ The following program shows the If-Then-Else Clause in use

```
/*     ***********************************************************
       IF, THEN, ELSE CLAUSE
       ***********************************************************
       -      This program shows how to implement and use
              the If, Then, Else Clause
*/
#include <iostream>
#include <string>
using namespace std;

int main( )
{
      //»»»»»»»»»»»»»»»»»»»»»»»»»»»»»»»»»»»»»»»»»»»»»»»»»»»»»»»»»»»»
      int batch_A = 5;
      int batch_B = 7;
```

```
    int min_ContainerCapacity = 0;

    //»»»»»»»»»»»»»»»»»»»»»»»»»»»»»»»»»»»»»»»»»»»»»»»»»»»»»»»»»
    if( batch_A > batch_B )
    {
          min_ContainerCapacity = batch_A;
    }
    else
    {
          min_ContainerCapacity = batch_B;
    }

    cout << "The smallest container size required to fit batches"
         << " of size "<< batch_A<< " or "<< batch_B<< " is "
         << min_ContainerCapacity
         << endl;

    //»»»»»»»»»»»»»»»»»»»»»»»»»»»»»»»»»»»»»»»»»»»»»»»»»»»»»»»»»
    cout << endl<< endl<< "==========="
         << endl
         << "END...C++ Demo: If, Then, Else Clause"
         << " "
         << endl;
    while(true){}
    return 0;

}//END...Function MAIN
```

```
The smallest container size required to fit batches of size 5
or 7 is 7


===========
END...C++ Demo: If, Then, Else Clause
```

Program Listing/Output 3.2: If_Then_Else_Clause_Demo_Client.cpp

3.1.2.1 Alternative Form using the Ternary ? : Operator

❖ The same If, Then, Else clause can be implemented using the ternary operator ?:

❖ It now takes the following form:

```
                     ?    Action if the      :    Action if the
     Condition            Condition is            Condition is
                     •       TRUE          •          FALSE
```

❖ Note: This is the only ternary operator in C++ i.e. one which takes 3 inputs

❖ The following program shows this in action

```
/*      ***************************************************************
        TERNARY OPERATOR ?:
        ***************************************************************
        -       This program shows how to implement and use
                the If, Then, Else Clause using the ?: operator
        -       This is the only ternary operator in C++
                -       Takes 3 inputs
*/
#include <iostream>
#include <string>
using namespace std;

int main( )
{
        //»»»»»»»»»»»»»»»»»»»»»»»»»»»»»»»»»»»»»»»»»»»»»»»»»»»»»»»»»»»»»»»»»»
        int batch_A = 5;
        int batch_B = 7;
        int min_ContainerCapacity = 0;

        //»»»»»»»»»»»»»»»»»»»»»»»»»»»»»»»»»»»»»»»»»»»»»»»»»»»»»»»»»»»»»»»»»»
        min_ContainerCapacity
                = ( batch_A > batch_B ) ? ( batch_A ) : ( batch_B );

        cout << "The smallest container size required to fit batches"
                << " of size "<< batch_A<< " or "<< batch_B<< " is "
                << min_ContainerCapacity
                << endl;

        //»»»»»»»»»»»»»»»»»»»»»»»»»»»»»»»»»»»»»»»»»»»»»»»»»»»»»»»»»»»»»»»»»»
        cout << endl<< endl<< "==========="
                << endl
                << "END...C++ Demo: Ternary Operator ?: "
                << " "
                << endl;
        while(true){}
        return 0;

}//END...Function MAIN
```

```
The smallest container size required to fit batches of size 5
or 7 is 7


===========
END...C++ Demo: Cascaded/Nested If Statement
```

Program Listing/Output 3.3: TernaryOperator_Client.cpp

❖ After compilation, there is no difference in terms of memory consumption or speed

  ➢ i.e. which one is preferable is purely down to the programmer, and a matter of personal preference

❖ This latter form sometimes makes for a more compact statement, so can be useful when part of a longer statement e.g.

```
cout << "Larger of a and b: "<< ( a > b ) ? ( a ) : ( b )<< endl
```

3.1.3 CASCADED/NESTED IF STATEMENTS

❖ It is perfectly possible, in fact a common practice, to place a conditional statement inside another to produce a cascaded/nested statement (both terms are commonly used)

❖ This has many uses, one of which is to accommodate more than one condition

❖ The following program shows how further language greetings can be accounted for using nesting

```
/*     ************************************************************
       CASCADED IF STATEMENT
       ************************************************************
       -      This program shows how to implement and use
              a Cascaded/Nested If Statement
*/
#include <iostream>
#include <string>
using namespace std;

int main( )
{
       //»»»»»»»»»»»»»»»»»»»»»»»»»»»»»»»»»»»»»»»»»»»»»»»»»»»»»»»»»»»»»»
       //     Greeting
       //
       string name_User = " ";
       cout  << "Please enter your name"<< endl;
       cin >> name_User;

       cout  << "Hello "<< name_User<< endl;

       //»»»»»»»»»»»»»»»»»»»»»»»»»»»»»»»»»»»»»»»»»»»»»»»»»»»»»»»»»»»»»»
       //     Options - choosing a language
       //
       string option_User_Language = " ";
       cout     << "Please select a language"<< endl;
       cout     << "(E)\t  English"<< endl;
```

```
   cout    << "(B)\t  Bengali"<< endl;
   cout    << "(F)\t  French"<< endl;
   cout    << "(G)\t  German"<< endl;
   cout    << "(S)\t  Spanish"<< endl;

   cin >> option_User_Language;

   if( (option_User_Language == "E") )
   {
         cout << "Hello"<< endl;
   }
   else
   {
         if( option_User_Language == "B" )
         {
               cout << "Namaskar"<< endl;
   }
         else
         {
               if( option_User_Language == "F" )
               {
                     cout << "Bonjour"<< endl;
               }
               else
               {
                     if( option_User_Language == "G" )
                     {
                          cout << "Guten Tag"<< endl;
               }
                     else
                     {
                          if( option_User_Language == "S" )
                          {
                                cout   << "Hola"<< endl;
                          }
                          else
                          {
                                cout   << "Sorry, your input"<<
                                        " is not recognised"
                                        << endl;
                          }
                     }
               }
         }
   }
}

//»»»»»»»»»»»»»»»»»»»»»»»»»»»»»»»»»»»»»»»»»»»»»»»»»»»»»»»»»»»»»»»»»»»»»»»»»»
cout << endl<< endl<< "==========="
     << endl
     << "END...C++ Demo: Cascaded/Nested If Statement"
     << " "
     << endl;
while(true){}
```

```
  return 0;

}//END...Function MAIN
```

Program Listing/Output 3.4: Cascaded_If_Client.cpp

3.1.4 THE IF, ELSE IF, ELSE CLAUSE

❖ The Else If clause is added to the If Then Else clause in order to accommodate more than one condition

❖ Thus the If, Else If, Then, Else clause takes the following form:

IF(Condition 1)

   THEN Action if Condition 1 is TRUE

ELSE IF(Condition 2)

   THEN Action if Condition 2 is TRUE

ELSE IF(Condition 3)

   THEN Action if Condition 3 is TRUE

ELSE IF(Condition 4)

   THEN Action if Condition 4 is TRUE

   . . .

ELSE

   Action if ALL above Conditions are FALSE

❖ In C++, the syntax is as follows:

```
if( /* CONDITION 1 */ )
{
      // Action(s) if the CONDITION 1 is TRUE;
}
else if( /* CONDITION 2 */ )
{
      // Action(s) if the CONDITION 2 is TRUE;
}
else if( /* CONDITION 3 */ )
{
      // Action(s) if the CONDITION 3 is TRUE;
}
else if( /* CONDITION 4 */ )
{
      // Action(s) if the CONDITION 4 is TRUE;
}
```

```
     //...

else
{
     // Action(s) if the above CONDITIONS are FALSE;
}
```

❖ The program of section 3.1.3 can also be implemented using if, else if , else statements, as follows:

```
/*     ************************************************************
     IF, ELSE IF, ELSE STATEMENT
     ************************************************************
     -      This program shows how to implement and use
            the If, Else If, Else clause
*/
#include <iostream>
#include <string>
using namespace std;

int main( )
{
    //»»»»»»»»»»»»»»»»»»»»»»»»»»»»»»»»»»»»»»»»»»»»»»»»»»»»»»»»»»»»»»»»»
    //      Greeting
    //
    string name_User = " ";
    cout    << "Please enter your name"<< endl;
    cin >> name_User;

    cout    << "Hello "<< name_User<< endl;

    //»»»»»»»»»»»»»»»»»»»»»»»»»»»»»»»»»»»»»»»»»»»»»»»»»»»»»»»»»»»»»»»»»
    //      Options - choosing a language
    //
    string option_User_Language = " ";
    cout    << "Please select a language"<< endl;
    cout    << "(E)\t   English"<< endl;
    cout    << "(B)\t   Bengali"<< endl;
    cout    << "(F)\t   French"<< endl;
    cout    << "(G)\t   German"<< endl;
    cout    << "(S)\t   Spanish"<< endl;

    cin >> option_User_Language;

    if((option_User_Language == "E")||(option_User_Language == "e"))
    {
         cout << "Hello"<< endl;
    }
    else if( option_User_Language == "B" )
    {
```

```
                  cout << "Namaskar"<< endl;
      }
      else if( option_User_Language == "F" )
      {
            cout << "Bonjour"<< endl;
      }

      else if( option_User_Language == "G" )
      {
            cout << "Guten Tag"<< endl;
      }

      else if( option_User_Language == "S" )
      {
            cout << "Hola"<< endl;
      }

      else
      {
            cout << "Sorry, your input is not recognised"<< endl;
      }


      //»»»»»»»»»»»»»»»»»»»»»»»»»»»»»»»»»»»»»»»»»»»»»»»»»»»»»»»»»»»»»»»»»»»»»
      cout << endl<< endl<< "==========="
           << endl
           << "END...C++ Demo: If, Else If, Else Statement"
           << " "
           << endl;
      while(true){}
      return 0;

}//END...Function MAIN
```

Program Listing/Output 3.5: If_ElseIf_Else_Client.cpp

❖ On this occasion, it is fair to say that it has tidied things up considerably

❖ However, cascaded statements are often essential, such as when calculating taxes (see exercise)

3.1.5 SWITCH STATEMENT

❖ A switch statement is an alternative to the if, else if, else statements

❖ The structure of a switch statement is as follows

```
switch( variableName )
{
      case value_1:
            // Action(s) if the variableName == value_1;
      break;
```

```
    case value_2:
        // Action(s) if the variableName == value_2;
    break;

    case value_3:
        // Action(s) if the variableName == value_3;
    break;

    //...

    default:
    // Action(s) if variableName == None of the above;
    break;
}
```

❖ Thus an alternative means of implementing the greeting selector, written
in Program Listing/Output 3.5 using the if, else if, else clause, is the
following:

```
/*      ************************************************************
        SWITCH STATEMENT
        ************************************************************
        -       This program shows how to implement and use
                the Switch Statement
*/
#include <iostream>
#include <string>
using namespace std;

int main( )
{
    //»»»»»»»»»»»»»»»»»»»»»»»»»»»»»»»»»»»»»»»»»»»»»»»»»»»»»»»»»»»»»»»»»»
    string name_User = " ";
    cout    << "Please enter your name"<< endl;
    cin >> name_User;

    cout    << "Hello "<< name_User<< endl;

    //»»»»»»»»»»»»»»»»»»»»»»»»»»»»»»»»»»»»»»»»»»»»»»»»»»»»»»»»»»»»»»»»»»
    //      Options - choosing a language
    //
    char option_User_Language =  ' ';
    cout    << "Please select a language"<< endl;
    cout    << "(E)\t  English"<< endl;
    cout    << "(B)\t  Bengali"<< endl;
    cout    << "(F)\t  French"<< endl;
    cout    << "(G)\t  German"<< endl;
    cout    << "(S)\t  Spanish"<< endl;

    cin >> option_User_Language;
```

```
    switch( option_User_Language )
    {
         case 'E':
              cout << "Hello"<< endl;
              break;

         case 'B':
              cout << "Namaskar"<< endl;
              break;              .

         case 'F':
              cout << "Bonjour"<< endl;
              break;

         case 'G':
              cout << "Guten Tag"<< endl;
              break;

         case 'S':
              cout << "Hola"<< endl;
              break;

         default:
              cout << "Sorry, your input is not recognised"<< endl;
              break;

    }//End...switch - option_User_Language

    //»»»»»»»»»»»»»»»»»»»»»»»»»»»»»»»»»»»»»»»»»»»»»»»»»»»»»»»»»»»»
    cout     << endl<< endl<< "==========="
             << endl
             << "END...C++ Demo: Switch Statement"
             << " "
             << endl;
    while(true){}
    return 0;

}//END...Function MAIN
```

```
Please enter your name
John
Hello John
Please select a language
(E)    English
(B)    Bengali
(F)    French
(G)    German
(S)    Spanish
F
Bonjour
```

```
============
END...C++ Demo: Switch Statement
```

Program Listing/Output 3.6: SwitchStatement_Client.cpp

3.1.6 SWITCH STATEMENT VS IF, ELSE IF, ELSE CLAUSE

❖ Now that you have two ways of selecting options, the question is which should you use?

❖ Note that the `switch` statement works only for testing equality

➤ Thus if you want to test anything more sophisticated, such as if a salary falls between specified tax bands, you need to choose the if, else if, else clause

❖ When you do need to test for equality, it is a matter of personal preference

➤ Both are equally efficient as far as memory consumption and runtime speed are concerned

➤ The switch statement remains popular since several programmers find it more readable

• It is a popular choice for implementing user interfaces

❖ Since the if, else if, else clause works fine for equality, and all other tests, some programmers prefer to use only It, avoiding the switch statement altogether

EXERCISE

❖ *Write conditional statements which will determine which tax category a person falls in, using the following criteria:*

					TAX RATE
➤			Salary $<$	20000	30%
➤	20000	$<$	Salary $\leq$	30000	40%
➤	30000	$<$	Salary		50%

3.2 PROBLEM SOLVING WITH LOOPS

3.2.1 WHAT ARE LOOPS?

❖ Loops are used when an action needs to be performed **more than once**

❖ Each time the action is performed is called an Iteration

❖ Since this is used the bulk of the time in programming, loops are one of "Blinking and Breathing" of programming

❖ There are 3 loops which come under 2 categories

➢ When you **know** in advance/**can** evaluate the number of iterations

- **FOR** loop

➢ When you **don't know** in advance/**can't** evaluate the number of iterations - you test whether a **condition** has been met

- You test the condition **before** the first iteration
  ○ **WHILE** loop
  ○ The minimum number of iterations is 0
- You test the condition **after** the first iteration
  ○ **DO** loop...also called a **DO/WHILE** loop
  ○ The minimum number of iterations is 1

❖ In order to illustrate the differences, as well as the similarities, between these, consider the following analogy:

➢ I have asked you to run up and down the stairs (don't ask me why! :-))

- If you can deduce the *number* of times, if I have either
  ○ Given you a ballpark figure or
  ○ Said each run is to represent a conference room on the 1st floor this is a **for** loop
- If instead I say keep on running until it gets dark outside i.e. We have a *condition*, not a number...
  ○ If you check whether or not it is already dark before you start running, this is a **while** loop
  ○ If you run up and down once, then check whether or not it is dark, this is a **do** loop/ **do while** loop

3.2.2 FOR LOOP

❖ As discussed above, the for loop is used when we **know** in advance/**can** evaluate the number of iterations

❖ The syntax is as follows:

❖ In the following example, a for loop is being used to display a greeting more than once

for   (type
         variableName
         =   Initial   ;   Condition
         Value   for
                                       ending   ;   condition
                                                          for
                                                          iterating   )

{
                    Action(s) per iteration

}

```
/*     ****************************************************************
       FOR LOOP
       ****************************************************************
       -      This program shows how to implement and use
              For Loops
*/

#include <iostream>
using namespace std;

int main()
{
       //»»»»»»»»»»»»»»»»»»»»»»»»»»»»»»»»»»»»»»»»»»»»»»»»»»»»»»»»»»»»»»»»
       for( int i = 0; i < 10; ++i )
       {
              cout << "Hello for the "<< ( i + 1)<< "th time"<< endl;
       }


       //»»»»»»»»»»»»»»»»»»»»»»»»»»»»»»»»»»»»»»»»»»»»»»»»»»»»»»»»»»»»»»»»
       cout   << endl<< endl<< "==========="
              << endl
              << "END...C++ Demo: For Loop"
              << " "
              << endl;
     while(true){}
     return 0;

}//END...Main
```

```
Hello for the 1th time
Hello for the 2th time
Hello for the 3th time
Hello for the 4th time
Hello for the 5th time
Hello for the 6th time
Hello for the 7th time
Hello for the 8th time
Hello for the 9th time
Hello for the 10th time
```

```
============
END...C++ Demo: For Loop
```

❖ C Note: Being able to define and initialise the iteration variable is a new facility in C++

➢ In C, the definition has to be done outside the loop

  • Thus in the above example, the loop would look as follows:

```
int i = 0;
for( i = 0; i < 10; ++i )
{
        cout << "Hello for the "<< ( i + 1)<< "th time"
            << endl;
}
```

➢ A major advantage of defining and initialising within the loop is that scope of the variable becomes local only for the loop (see section 5.7 onwards for discussions on scope of a variable)

➢ This means
  • More efficient usage of runtime memory (section 14 onwards)
  • Better organisation for human programmers

❖ While the above program, Program Listing/Output 3.7, is working, it does have one display problem:

➢ It says 1th, 2th, 3th
➢ as opposed to 1st, 2nd, 3rd

❖ In order to fix this, a conditional statement is required inside the loop

EXERCISE

❖ *Add a conditional statement inside the loop so that it displays the appropriate suffix*

➢ *Note that you want to say 1st, 2nd 3rd*
➢ *but also 11th, 12th, 13th*
➢ *but then 21st, 22nd, 23rd...*
➢ *After 100, this pattern must repeat*
  • *101st, 102nd, 103rd*
  • *111th, 112th, 113th*
  • *121st, 122nd, 123rd*

3.2.3 WHILE LOOP

❖ As discussed above, the while loop is used when we **don't know** in advance/**can't** evaluate the number of iterations

➤ We test a condition
➤ before the first iteration

❖ The syntax is as follows:

```
while ( Condition for iterating )
{
        Action(s) per iteration, as long as
        the Condition is TRUE
}
```

❖ Consider the following situation: We are helping out in the design of a building, now investigating the dimensions of a wall:

➤ The architects have specified the following for aesthetic reasons
  • square
  • with dimensions being integers in cm,
➤ The civil engineers have specified the area needs to be
  • a minimum of 158600cm^2
  • but as close as possible to this
➤ We thus need a program which will find the lowest square number above 158600
  • Starting at 158600, we start counting up in integers
  • We test each one for whether or not it is square
  • When we encounter a square number, we stop iterating, then display it
➤ Since we
  • Don't know how many iterations we need in advance
  • Want to check the first number, in case it is a square, a **while** loop is required

❖ The following listing shows such a program:

```
/*      ***********************************************************
        WHILE LOOP
        ***********************************************************
        -       This program shows how to implement and use
                While Loops
*/
#include <iostream>
#include <math.h>
using namespace std;

int main( )
{
    //»»»»»»»»»»»»»»»»»»»»»»»»»»»»»»»»»»»»»»»»»»»»»»»»»»»»»»»»»»»»»»»»»»»»»
    cout << "This program finds the lowest square number "
        << "above 158600: "
        << endl;
    double currentNum = 158600.0;

    //»»»»»»»»»»»»»»»»»»»»»»»»»»»»»»»»»»»»»»»»»»»»»»»»»»»»»»»»»»»»»»»»»»»»»
    //   Test: Is the square root of the number an INTEGER
    //    -       Starting at 158600.0, keep iterating
    //    -       Calculate the square root
    //            -       Test if it is an integer by subtracting the
    //                    "floor" - the rounded down part of the number
    //                    floor(3.95) = 3, floor(5.2) = 5, floor(7.0) = 7
    //    -       Condition: Keep iterating till this condition is met
    //
    while(((sqrt(currentNum) ) - ( floor( sqrt( currentNum )))) != 0)
    {
            ++currentNum;
    }
    cout << currentNum<< " = "<< sqrt(currentNum)<< "^{}2"<< endl;

    //»»»»»»»»»»»»»»»»»»»»»»»»»»»»»»»»»»»»»»»»»»»»»»»»»»»»»»»»»»»»»»»»»»»»»
    cout   << endl<< endl<< "==========="
        << endl
        << "END...C++ Demo: While Loop"
        << " "
        << endl;
    while(true){}
    return 0;

}//END...Function MAIN
```

```
This program finds the lowest square number above 158600:
 159201 = 399^{}2


 ===========
```

```
END...C++ Demo: While Loop
```

Program Listing/Output 3.8: WhileLoop_Client.cpp

3.2.3.1 Empty While Loop

❖ We are now in a position to discuss the line we have been using to ensure the console window stays open:

```
while(true){}
```

➢ The "condition" between the parentheses "true" is always TRUE

➢ We have nothing between the braces { },

- i.e. we are saying "do nothing"

➢ Thus this statement is saying "Do nothing forever!"

- Consequently, the console stays open

EXERCISE

❖ *Using loops, write a program which will output all the 6th numbers between*

➢ *2000 and 10000*

➢ *Numbers entered by the user*

A 6th number is one which has an integer as its sixth root e.g. $64 = 2^6$, $729 = 3^6$

❖ *Write a program to output all the triangular numbers between*

➢ *2000 and 10000*

➢ *Numbers entered by the user*

3.2.4 DO LOOP/WHILE LOOP

❖ As discussed above, the do loop is used when we **don't know** in advance/**can't** evaluate the number of iterations

➢ We test a condition

➢ **after** the first iteration

❖ The syntax is as follows:

```
do

{
    Action(s) on the
    ❖    First Iteration
    ❖    Subsequent  iterations,  as  long  as
         the Condition is TRUE
}

while  (  Condition for iterating  )  ;
```

❖ Although the do loop is sizeably less common than the while loop, there are certain situations where it is essential

❖ Common usage: User Interface

➢ You want to present a list of options for your user, a "Home Menu"
  - (S)he selects one, the program goes off and does the activities with this option
  - When finished, you want the initial "Home" menu to be presented to the user again

➢ In designing the program, we can use elimination to decide which of the 3 loops we need:
  - We want to present the "Home" menu more than once - LOOP required
    ◦ We're presenting nothing else repeatedly, just this one menu, so ONE loop is needed
  - We cannot evaluate in advance how many times the user will use this menu
    ◦ NOT for loop
  - 0 iterations: Unacceptable...the menu must be displayed to the user at least ONCE, at the start
    ◦ NOT while Loop
  - Need: DO Loop

❖ The following program shows this:

> The menu is a list of languages for our user to choose from
  - The user has the option "Q" to quit
  - As long as this is not selected, the menu will be presented again
> The activity will be displaying Hello in this language

```
/*    ***********************************************************
      DO LOOP/ DO WHILE LOOP
      ***********************************************************
      -        This program shows how to implement and use
               Do Loops/Do...While Loops
               -      Displays a "Home" Menu, with language options
                      -        User chooses either a language or to Quit
                      -        If a language is chosen, Hello is displayed in
                               that language
                               -        The program then returns to displaying
                                        the Home Menu
                      -        If Quit is chosen
                               -        The program ends
*/
#include <iostream>
using namespace std;

int main( )
{
      char option_User_Language = ' ';

      do
      {
          // "Home" Menu
          cout    << "Please select a language"<< endl;
          cout    << "(E)\t   English"<< endl;
          cout    << "(B)\t   Bengali"<< endl;
          cout    << "(F)\t   French"<< endl;
          cout    << "(G)\t   German"<< endl;
          cout    << "(S)\t   Spanish"<< endl;
          cout    << "(Q)\t   Quit"<< endl;

          // Activity selected by user
          cin >> option_User_Language;

          switch( option_User_Language )
          {
                case 'E':
                        cout << "Hello"<< endl;
                        break;

                case 'B':
                        cout << "Namaskar"<< endl;
                        break;

                case 'F':
```

```
                        cout << "Bonjour"<< endl;
                        break;

                case 'G':
                        cout << "Guten Tag"<< endl;
                        break;

                case 'S':
                        cout << "Hola"<< endl;
                        break;

                case 'Q':
                        cout << "You have chosen to End this
                                    program..."<< endl;
                        break;

                default:
                        cout << "Sorry, your input is not recognised"
                << endl;
                        break;

        }// END: switch( option_User_Language )

  }
  while( option_User_Language != 'Q' );


  //==============================================================
  cout  << endl
        << "==============="
        << endl<< "END...C++ Demo Do Loops"<< endl;
  while(true){} // While loop:
  return 0;

}// END: MAIN Function
```

```
Please select a language
(E)     English
(B)     Bengali
(F)     French
(G)     German
(S)     Spanish
(Q)     Quit
E
Hello

Please select a language
(E)     English
(B)     Bengali
```

```
(F)       French
(G)       German
(S)       Spanish
(Q)       Quit
B
Namaskar

Please select a language
(E)       English
(B)       Bengali
(F)       French
(G)       German
(S)       Spanish
(Q)       Quit
F
Bonjour

Please select a language
(E)       English
(B)       Bengali
(F)       French
(G)       German
(S)       Spanish
(Q)       Quit
G
Guten Tag

Please select a language
(E)       English
(B)       Bengali
(F)       French
(G)       German
(S)       Spanish
(Q)       Quit
S
Hola

Please select a language
(E)       English
(B)       Bengali
(F)       French
(G)       German
(S)       Spanish
(Q)       Quit
Q
You have chosen to End this program...

===============
END...C++ Demo Do Loops
```

Program Listing/Output 3.9: WhileLoop_Client.cpp

EXERCISE: GENERATING PRIME NUMBERS

❖ *Using loops, write a program which will generate the*

  ➢ *First 100 prime numbers*
  ➢ *Number of prime numbers requested by the user*

❖ Tip: when designing your program, use elimination to decide what kind of loop you will need

3.3 C++ STATEMENTS & CONSTRUCTS

3.3.1 BLOCKS

❖ A BLOCK is an entity enclosed using curly braces { }

❖ We have so far come across the following examples:

  ➢ Conditional Statements
  ➢ Loops

❖ We will see the following in forthcoming chapters:

  ➢ Classes
  ➢ Functions
  ➢ ADTs

❖ Blocks are fundamentally important in programming

  ➢ They are used to structure a program
  ➢ They determine the scope of a variable, i.e. where it can be read/written to (section 5.7 onwards)

3.3.2 BREAK STATEMENT

❖ The `break` statement enables exiting from a block

❖ It is commonly used to exit from a loop during any one of its iterations

```
/*      ***************************************************************
        BREAK STATEMENT
        ***************************************************************
        -       Demonstrating the usage of the Break Statement in C++
                -       This forces an exit from the BLOCK in which it
                        is called
        -       In this case, it is called inside a for loop, on
                Iteration 9
```

```
                    -        The for loop stops iterating now
*/
#include <iostream>
using namespace std;

int main( )
{
        //»»»»»»»»»»»»»»»»»»»»»»»»»»»»»»»»»»»»»»»»»»»»»»»»»»»»»»»»»»»»»»»»»»»»»»»»»
        for( int i = 0; i < 12; ++i )
        {
                cout << "Starting iteration "<< i<< "\t";
                if( i == 9 )
                {
                        break;
                }
                cout << "Ending iteration "<< i<< endl;
        }

        //»»»»»»»»»»»»»»»»»»»»»»»»»»»»»»»»»»»»»»»»»»»»»»»»»»»»»»»»»»»»»»»»»»»»»»»»»
        cout    << endl<< endl<< "==========="
                << endl
                << endl<< "END...C++ Demo Break statement"<< endl
                << " "
                << endl;
        while(true){}
        return 0;

}//END...Function MAIN
```

```
Starting iteration 0     Ending iteration 0
Starting iteration 1     Ending iteration 1
Starting iteration 2     Ending iteration 2
Starting iteration 3     Ending iteration 3
Starting iteration 4     Ending iteration 4
Starting iteration 5     Ending iteration 5
Starting iteration 6     Ending iteration 6
Starting iteration 7     Ending iteration 7
Starting iteration 8     Ending iteration 8
Starting iteration 9

===========

END...C++ Demo Break statement
```

Program Listing/Output 3.10: Break_Statement_Client.cpp Demonstration of the C++ break statement

3.3.3 CONTINUE STATEMENT

❖ The continue statement in a loop forces the program to

➤ Start the next iteration

➤ thus abandoning any remaining instructions in the current iteration

❖ In this sense it is the opposite of `break`

```cpp
#include <iostream>
using namespace std;

int main( )
{
        //===========================================================
        for( int i = 0; i < 12; ++i )
        {
                cout << "Starting iteration "<< i<< endl;
                continue;

                cout << "Finishing iteration "<< i<< endl;
                                                       // This section
                                                       // is never
                                                       // reached

        }

        //===========================================================
        cout      << endl
                  << "==============="
                  << endl<< "END...C++ Demo Continue statement"<< endl;
        while(true){}
        return 0;
}
```

```
Starting iteration 0
Starting iteration 1
Starting iteration 2
Starting iteration 3
Starting iteration 4
Starting iteration 5
Starting iteration 6
Starting iteration 7
Starting iteration 8
Starting iteration 9
Starting iteration 10
Starting iteration 11

===============
END...C++ Demo Continue statement
```

Program Listing/Output 3.11: Continue_statement.cpp Demonstration of the C++ continue statement

❖ The continue statement is frequently used inside a conditional statement, to ensure certain instructions are bypassed under certain conditions

3.3.4 THE GOTO STATEMENT

❖ *This section is primarily for reference - avoid using the* goto *statement in your C++ code*

❖ Since one of the ambitions of C++ was and still is to make it a fully featured programming language, it supports a jump statement goto

➢ It was widely used in languages from the past such as

  • Assembler
  • Basic

➢ In C++, it is useful to be able to recognise this since you may well be developing a program

  • Which has legacy C code
  • Which in turn may have legacy Assembler code which still has this statement

❖ The goto statement works with a label

➢ A label is any name followed by a colon
➢ The syntax is as follows:

```
goto labelName;
```

  • This will get the program to jump to the label, missing out any intermediate statements

❖ The following program shows how it can be used for

➢ One action

  • This is a hypothetical example, a conditional statement will be far better

➢ Multiple actions

  • This is again a hypothetical example, a for loop will be far better

```
/*      ****************************************************************
        GOTO Statement - Avoid in C and C++
        ****************************************************************
*/
#include <iostream>
using namespace std;
```

```
int main( )
{
    //»»»»»»»»»»»»»»»»»»»»»»»»»»»»»»»»»»»»»»»»»»»»»»»»»»»»»»»»»»»»»»
    cout << "======== Goto Statement: Single Action ========"
        << endl;

    goto myGreeting;
    cout << "Sorry, this greeting is never reached"<< endl;
    myGreeting:
        cout << "Hello, my Greeting has been reached"
            << endl<< endl;

    //»»»»»»»»»»»»»»»»»»»»»»»»»»»»»»»»»»»»»»»»»»»»»»»»»»»»»»»»»»»»»»
    cout << "======== Goto Statement: Multiple Actions ========"
        << endl;

    int i = 1;
    repeat:
        cout << i<< ")\t Hello" << endl;
        ++i;
        if( i < 12 )
        {
            goto repeat;
        }

    //»»»»»»»»»»»»»»»»»»»»»»»»»»»»»»»»»»»»»»»»»»»»»»»»»»»»»»»»»»»»»»
    cout   << endl<< endl<< "==========="
        << endl
        << "END...C++ Demo: goto Statement - Avoid in C and C++"
        << " "
        << endl;
    while(true){}
    return 0;

}//END...Function MAIN
```

```
======== Goto Statement: Single Action ========
 Hello, my Greeting has been reached

 ======== Goto Statement: Multiple Actions ========
 1)      Hello
 2)      Hello
 3)      Hello
 4)      Hello
 5)      Hello
 6)      Hello
 7)      Hello
 8)      Hello
 9)      Hello
 10)     Hello
```

```
11)     Hello

============
END...C++ Demo: goto Statement - Avoid in C and C++
```

Program Listing/Output 3.12: GotoStatement_Client.cpp

3.3.4.1 Prospective Evils of the goto statement

❖ The `goto` statement is more a leftover from previous languages than a core C/C++ concept; it is fundamentally flawed for use in C++

❖ It is likely to destroy the structure of any program

> The importance of good organisation cannot be overstated

> C++ provides several mechanisms:
  • Functions in procedural programming
  • Classes and Objects in object oriented programming
  • Templates in generic programming

> The `goto` statement is likely to come into conflict with these

❖ It is weak

> It can be used only within a function (section 5 onwards), not between functions

❖ **Rule of Thumb:** Avoid the goto statement whenever possible

3.3.5 DROPPING THE BRACES {} FOR 1 ACTION

❖ In C and C++, it is possible to eliminate the braces when the block contains only one statement

❖ For example:

```
if( salary < 9800 )
    cout << "Below minimum wage"<< endl;
```

will have the same meaning as:

```
if( salary < 9800 )
{
    cout << "Below minimum wage"<< endl;
}
```

❖ Some programmers argue that an advantage is it lowers the number of lines in the program

➢ One of the few methods which can be used to assess how elegantly a program has been written is the number of lines

❖ However, this should be done with caution if at all:

3.3.5.1 Potential Hazards in dropping the Braces

❖ This method only ever works if we have ONE statement

❖ When we need to add more, we mustn't forget to put in the braces, otherwise the code will take on a different meaning

3.3.5.1.1 Breaking at the Wrong Place

❖ Consider the following fragment of code

```
if( salary < 25000

        tax_Rate = 0.25;

cout << "The Tax Rate is 25%"<< endl;
```

➢ At first glance, it can be interpreted to mean:

```
if( salary < 25000 )

{

        tax_Rate = 0.25;

        cout << "The Tax Rate is 25%"<< endl;

}
```

➢ However, it actually means the following:

```
if( salary < 25000 )

{

        tax_Rate = 0.25;

}

cout << "The Tax Rate is 25%"<< endl;
```

• i.e. The display will happen regardless, and will often be wrong

❖ In 1990, a network collapse took place at AT&T

➢ The cause turned out to be as follows:

➢ A switch statement contained several if statements

➢ A break statement

○ was expected to end an inner if statement

○ instead, it ended the outer switch statement

➢ See http://users.csc.calpoly.edu/~jdalbey/SWE/Papers/att_collapse.html for details

3.3.5.1.2 Dangling Else Clause Issue

❖ A major problem is the Dangling Else Clause where in a multiple level If, Then, Else clause, there can be ambiguity over which If statement an Else case belongs to

❖ Consider the following statement:

```
IF a THEN IF b THEN c ELSE d
```

➢ Either of the following interpretations fit this specification:

• Interpretation 1: The "d" goes with the "b" If statement

```
if(a)
{
        if(b)
        {
                c;
        }
        else
        {
                d;
        }
}
```

• Interpretation 2: The "d" goes with the "a" If statement

```
if(a)
{
        if(b)
        {
                c;
        }
        }
        else
        {
                d;
        }
}
```

➤ With the design of C++ compilers, the following (appallingly written! :-)) fragment

```
if(a) if(b) c; else d;
```

will actually bind to Interpretation 2

❖ This reemphasises the importance of formatting code and being meticulous with every kind of bracket

➤ In this case, using the braces properly performs the disambiguation

❖ The Fish & Chips Problem: This issue is analogous to an ambiguity in real life:

➤ Imagine you are a waiter in a restaurant; a customer says to you "Fish and chips twice"
➤ There is ambiguity over whether this means
  • 2 helpings of fish and 2 helpings of chips?
    ○ i.e. (Fish and chips) twice
  • 1 helping of fish and 2 helpings of chips?
    ○ i.e. Fish and (chips twice)
➤ As above, the usage of brackets removes the ambiguities

3.3.5.2 Guideline with Braces

❖ This reiterates the importance of making your code as clear as possible for maintenance purposes

❖ **Rule of Thumb:** Always put in the braces

➤ Not only does this ensure far more safety when developing the code
➤ It makes even the simplest code more readable and thus easier to maintain and debug

3.4 CREATING ALIASES IN C & C++

3.4.1 DEFINING CONSTANTS USING #DEFINE - AVOID IN C++

❖ Using the `#define` command to create a "macro" alias is supported in C++

➤ This is mostly to ensure backward compatibility with C where it is very popular
➤ Its usage is strongly discouraged in C++ (see section 6.2)

❖ The syntax is as follows

```
#define ALIAS_NAME alias
```

➢ Thus it is used as follows:

```
#define STADIUM_CAPACITY 35000
```

- After this, whenever the compiler see the term STADIUM_CAPACITY, it will replace it with 35000

3.4.1.1 Issues with Macro Aliases

❖ There are several problems with macro aliases which are major issues in C++

➢ We have no sense of data type

- In the above example, STADIUM_CAPACITY has no type whatsoever i.e. it is not of type int, double, signed double or anything else
- In C programming, this is not ideal but not a large problem
- In C++, where type flexibility is supported through more than one kind of Polymorphism (section 6.2.3 and section 23 onwards), this is widely considered unacceptable

➢ Once compiled, the runtime program will contain only the alias; the alias name is lost

- i.e. In the above example, only the raw number 35000 will be present at runtime, not the name STADIUM_CAPACITY
- This is particularly unsuitable in debugging

3.4.2 TYPEDEF

❖ "Typedef" is used to create an alias for a data type

➢ It is far stronger than the #define macro
➢ It is used considerably in C++

❖ The syntax is as follows

```
typedef aliasTarget aliasName;
```

➢ Note that this time the terminator is required

3.4.2.1 Cannot modify Typedef

❖ Note that typedef does more than just a simple text replacement

❖ Once something has been set in the typedef statement, it cannot be modified in subsequent statements

❖ This is not the case for macro text replacements with #define

> These are just raw text replacements

❖ The following example shows what is and is not allowed

```
#include <iostream>
using namespace std;

int main()
{
        //»»»»»»»»»»»»»»»»»»»»»»»»»»»»»»»»»»»»»»»»»»»»»»»»»»»»»»»»»»»»»»»»»»»»»»»»»»»»»
        typedef int wholenumber ;
        // signed wholenumber num_Spectators;    // ILLEGAL: cannot
                                                  // subsequently
                                                  // modify a
                                                  // typedef alias

        // Reminder: Avoid #define in C++ wherever possible
        #define WHOLENUM int
        unsigned WHOLENUM num_Seats;              // OK: can
                                                  // subsequently
                                                  // modify a
                                                  // #define alias

        //...

}//END...Function MAIN
```

Program Listing/Output 3.13: Lack of Changeability of typedef. Changeability of #define

4. PROGRAMMING PRACTICE

4.1 GOOD VS BAD PRACTICES

❖ As we have been discussing, it is important to remember BOTH your audiences when programming

> Your Computer
> Your fellow humans
  • This is commonly forgotten

❖ Consider the following two programs, Program Listing/Output 4.1 and Program Listing/Output 4.2:

> They are identical as far as the runtime environment of the computer is concerned[5]

---

[5]There is a difference in the text that is displayed on the console window but this is still considered to be the same operation.

➢ As a human, decide which one you would rather be maintaining and
debugging! :-)

```cpp
#include <iostream>
#include <string>
using namespace std;
int main( ){
double x1,x2,y1,y2;
cout<<"x1="; cin>>x1;
cout<<"y1="; cin>>y1;
x2=-y1;
y2=-x1;
cout << "x2="<<x2<<", y2="<<y2;
while(true){} return 0;}
```

```
x1=6
y1=7
x2=-7, y2=-6
```

Program Listing/Output 4.1: Reflect_LineYequalsMinusX_Bad_Client.cpp

```cpp
/*      ************************************************************
        Program:        Coordinate Geometry Transformation
                        Reflection in the line y = -x
        ************************************************************
        -       User Interaction:
                -       (S)he will be prompted to enter the initial x, y
                        coordinates
                -       The coordinates of the original point, reflected in
                        the line y = -x will be displayed
        -       Algorithm: Transformation in the line y = -x
                - Comprises Multiplication by the matrix

                                |0  -1|
                                |-1   0|

                - Thus the following is carried out:

                        |0 -1| |x|  = |  (0)x + (-1)y |  = | -y|
                        |-1 0| |y|    |  (-1)x + (0)y |    | -x|
*/
#include <iostream>
#include <string>
using namespace std;
```

```
int main( )
{

      //»»»»»»»»»»»»»»»»»»»»»»»»»»»»»»»»»»»»»»»»»»»»»»»»»»»»»»»»»»»»»»»»»»»»»»»
      cout   << "==========="<< endl
             << "Program, Coordinate Geometry Transformation: "
             << "Reflection in the Line y = -x"<< endl
             << "==========="<< endl;

      //»»»»»»»»»»»»»»»»»»»»»»»»»»»»»»»»»»»»»»»»»»»»»»»»»»»»»»»»»»»»»»»»»»»»»»»
      // Setting up the Initial Coordinates
      //
      double x_original = 0.0, x_transformed = 0.0;
      double y_original = 0.0, y_transformed = 0.0;

      cout   << "Please enter the initial x-coordinate, "
             << "then press Return"<< endl;
      cin    >> x_original;

      cout   << "Please enter the initial y-coordinate, "
             << "then press Return"<< endl;
      cin    >> y_original;

      //»»»»»»»»»»»»»»»»»»»»»»»»»»»»»»»»»»»»»»»»»»»»»»»»»»»»»»»»»»»»»»»»»»»»»»»
      //     Transforming the Coordinates
      //
      x_transformed = y_original * (-1);
      y_transformed = x_original * (-1);

      //»»»»»»»»»»»»»»»»»»»»»»»»»»»»»»»»»»»»»»»»»»»»»»»»»»»»»»»»»»»»»»»»»»»»»»»
      //     Displaying the Results
      //
      cout   << "-----------"<< endl
             << "Point ("<< x_original<< ", "<<  y_original<< ")"
             << " reflected in the line y = -x is "
             << "("<< x_transformed<< ", "<<  y_transformed<< ")"
             << endl;

      //»»»»»»»»»»»»»»»»»»»»»»»»»»»»»»»»»»»»»»»»»»»»»»»»»»»»»»»»»»»»»»»»»»»»»»»
      cout   << endl<< endl<< "==========="<< endl
             << "END...Program, Coordinate Geometry Transformation: "
             << "Reflection in the Line y = -x"<< endl;
      while(true){}
      return 0;
}//END...Function MAIN
```

```
===========
Program, Coordinate Geometry Transformation: Reflection in the Line y
    = -x


===========
Please enter the initial x-coordinate, then press Return
6
Please enter the initial y-coordinate, then press Return
7


-----------
Point (6, 7) reflected in the line y = -x is (-7, -6)


===========
END...Program, Coordinate Geometry Transformation: Reflection in the
    Line y = -x
```

Program Listing/Output 4.2: Reflect_LineYequalsMinusX_Good_Client.cpp

4.2 GUIDELINES

❖ These have been placed here, at the end of the first part, because it is important we stick to good programming practices from an early stage

❖ Several of the entities mentioned here will be discussed in detail in forthcoming chapters

4.2.1 ORGANISATION

❖ One of the crucial building blocks of software success is good organisation

❖ The well structured program will be far easier to maintain, develop and essentially debug

❖ C++ provides several means of organising programs including

➢ File Management
➢ Classes & Objects
➢ Templates

4.2.2 NAMING OF VARIABLES

❖ The following practices are widely recommended and followed in the software industry:

| | START WITH | | CASE:REST OF NAME | EXAMPLES |
	Case	Word		

❖ Variables
  ➢ Local
  ➢ Global
  ➢ Data
    Members
	Lower	Noun	Mixture	`num_People`
				`x_Transformed`
				`square_NumbersFoundSoFar`

❖ Functions
  ➢ Global
  ➢ Member
    Functions
	Lower	Verb	Mixture	`calc_Temperature`
				`negate_Current`
				`check_IsHallLargeEnough`

❖ Class
❖ ADT
	Upper	Noun	Mixture	`Employee`
				`Vehicle`
				`DialogueBox`

❖ Constants	Upper	Noun	Upper	`PI`
				`PLANCKS_CONSTANT`
				`STADIUM_CAPACITY`

4.2.3 FORMATTING OF SOURCE CODE

❖ Ensure you do not attempt too many actions in one statement

❖ Make your program readable by ensuring no one line is too long

  ➢ When a statement needs to contain a lot of text, make it run over multiple lines using the backslash \ character

4.3 DEBUGGING

❖ Most graphical IDEs come with sets of debugging tools

  ➢ This section covers the tools in MS Visual Studio

  ➢ The Arithmetic example will be used

❖ Debugging tools enable stopping a program running at a desired part, and then analysing bits of it thoroughly

4.3.1 BREAKPOINTS

❖ Breakpoints are tools used to stop a running program i.e. to break the running

❖ In the Arithmetic example, placing a breakpoint at Line 21 will stop the program after the 3 variables `myInt_A`, `myInt_B` and `myInt_C` have been initialised:

```
Arithmetic_Client.cpp  ×
(Unknown Scope)
   1  ⊟/*    ************************************************
   2  │        ARITHMETIC
   3  │      ************************************************
   4  │      -   Demonstrating the usage of Data and Operators in C++
   5  │  */
   6  │  #include <iostream>
   7  │  using namespace std;
   8  │
   9  ⊟int main( )
  10  │  {
  11  ⊟      //»»»»»»»»»»»»»»»»»»»»»»»»»»»»»»»»»»»»»»»»»»»»»»»»»»
  12  │      //  Defining and Initialising the Variables
  13  │      //
  14  │      int myInt_A = 5;
  15  │      int myInt_B = 7;
  16  │      int myInt_C = 8;
  17  │
  18  ⊟      //»»»»»»»»»»»»»»»»»»»»»»»»»»»»»»»»»»»»»»»»»»»»»»»»»»
  19  │      //  Carrying out the Calculations
  20  │      //
● 21  │      int total_A_B = 0;
  22  │      total_A_B = myInt_A + myInt_B;
  23  │
```

Program Listing/Output 4.3   Placing Breakpoints in programs

4.3.2 VIEWING VARIABLES

❖ When the program is set running, it stops at Line 21

❖ The variables, as they are in runtime memory, can be viewed, at this stage:

➢ In MS Visual Studio, the "Locals" window shows the variables which are used in main()

➢ Section 5.7 onwards will discuss the meaning of "Local" variables, along with other levels of scope

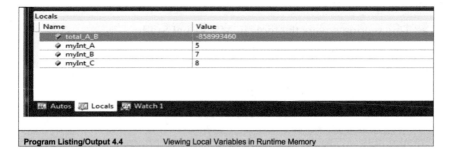

Locals	
Name	Value
total_A_B	-858993460
myInt_A	5
myInt_B	7
myInt_C	8

Autos   Locals   Watch 1

Program Listing/Output 4.4   Viewing Local Variables in Runtime Memory

❖ Note that the variable total_A_B has a garbage value, since we have broken the program before it has been initialised

❖ This will be put right shortly

4.3.3 SINGLE STEPPING

❖ Single Stepping is a means of making a program run one line at a time, moving forwards only when the instructed by the programmer (i.e. you!:-))

❖ There are 2 categories

➢ **Step Over:** This comprises carrying out the action(s) of the line then moving on to the next

➢ **Step Into:** This comprises stepping into any entity in that line, such as a function, and moving to that part of the program (section 5.8 onwards)

❖ By stepping over each line in this program (Debug Menu, Step Over or F10),

➢ The program will be executed line-by-line

➢ The changing values for the variables in runtime memory can be viewed

PROCEDURAL PROGRAMMING

❖ Procedural programming is not only an essential and widely used paradigm in its own right, but one that underpins several further paradigms, including

➤ Object Oriented Programming
➤ Generic Programming

❖ It first came to the masses when Fortran was introduced in 1957; it revolutionised not only but the programming world but the world in general

➤ Firstly, it enabled "Functional thinking"
  • In several walks of life, including mathematics,
    ○ We do not want to focus on individual variables/entities
    ○ We want to think of "procedurally" i.e. we have some inputs, we perform an operation, we get a result
➤ Secondly, it enabled writing mathematical operations precisely the way people like to think about them e.g. a = b + c
  • Fortran is a portmanteau for <u>fo</u>rmula <u>trans</u>lation

❖ Pascal took this line of thinking one step further

❖ The C programming language was introduced 1972, and again revolutionised the world

➤ Its original objective was for implementing UNIX operating systems
➤ Its immense power quickly made it applicable in several further fields
➤ To this day, it remains the leading language in embedded systems

❖ C++ strongly supports the Procedural Programming Paradigm

➤ The bulk of procedural programming in C++ is C programming
➤ However, as we are about to find out, there are further features which C++ supports which C does not

❖ There are several beauties of Procedural Programming including

➢ Organisation of Programs
➢ Bringing one level of abstraction - "Procedural Thinking," a method that is used a lot outside the programming world
➢ Reusability
➢ Introducing the idea of "Scope of Variables"

5. FUNCTIONS

5.1 PROCEDURAL THINKING

❖ Procedural programming brought a new line of thinking "Procedural Thinking" into programming

❖ It primarily comprises looking at systems which have

➢ (An) Input(s)
➢ A Process
➢ An Output

❖ It is very powerful and widely used outside the Programming World

5.1.1 PROCEDURAL THINKING OUTSIDE THE PROGRAMMING WORLD

❖ Microwave Oven

➢ Input: food and/or drinks
➢ You get it to perform a functionality - heat the contents
➢ The result is hotter food/drink

❖ Printer

➢ You send a file from a computer
➢ The printer carries out its process
➢ The output is a printed document

❖ Nearly all forms of Electronic Hardware:

➢ Signals are input to a circuit
➢ A process is carried out
➢ Some form of response, such as an amplified signal, is output

❖ Industry/Assembly Line

➢ At each stage different materials are put in

➢ A process is carried out

➢ Some form of hardware is output

❖ Mathematics

➢ Arithmetic: focuses on individual numbers

➢ Classical Algebra

• Uses
  ○ Variables
  ○ Functions, which
    ■ Take inputs,
    ■ Perform a functionality,
    ■ Have a result

• The focus is on the function, not the individual values which are input or output

EXERCISE

❖ *This list can be extended almost indefinitely. See if you can come up with 3 further examples*

5.1.2 PROCEDURAL THINKING INSIDE THE PROGRAMMING WORLD

❖ The entity which is used to implement procedural programming in C and C++ is the **Function**

❖ Other languages use both this, and different names for similar concepts:

➢ Pascal uses the term "Procedure"

➢ MATLAB also uses the term "Function"

➢ Visual Basic and Fortran (Fortran II onwards) use the terms "Subroutine and Function"

5.2 WHAT ARE FUNCTIONS?

❖ Functions are

➢ Self contained blocks of code implemented to perform a special purpose

➢ Required to divide up and organise all but the most trivial of programs

➢ The fundamental building block of Procedural Programming

❖ Functions can have

➢ 0 or more input(s)

• There is no universal limit to the number of inputs; practical limitations will depend on individual systems

➢ 0 or 1 output

❖ This chapter will illustrate implementing procedural programs in the following sections, with functions which have

➢ No Inputs, No Output

➢ Some Inputs, No Output

➢ Some Inputs, An Output

❖ Functions must have a name

❖ They are recognised by the compiler using the parentheses () operator

➢ Note that this is one of several uses of the parentheses operator (see section 31.4)

❖ Before we look at these, we need to look at the 3 entities required by functions:

➢ Definition

➢ Declaration

➢ Invocation/Call

5.2.1 FUNCTION NAMING

❖ The high importance of meaningful names is more important than ever with functions

❖ General advice for function names includes:

❖ Start them with A VERB

➢ since Functions are all about doing things

➢ Some programmers like to start function names with a run-in like "fn_" to remind everyone who looks at them that it is a function

➤ This often works in C

➤ In C++, there are several further entities which perform functionalities such as

- Member Functions (section 17 onwards)
- Member Functions of inner classes (section 20.10 onwards)
- Functors (Cybernetics in C++ - Expert),
  so to have a separate run-in for each might be more confusing than helpful

❖ Examples:

Action of the function	Suggested Function Name
➤ Calculates required input voltage	`calc_Voltage_In_Required`
➤ Calculates recommended output voltage	`calc_Voltage_Out_Recommended`
➤ Tells us whether the project is over budget	`check_IsProject_OverBudget`

EXERCISE: NAMING FUNCTIONS

❖ *Suggest names for functions which perform the following purposes:*

➤ *Tells you whether a year is a leap year*

➤ *Calculates the VAT payable*

➤ *Calculates the Net Salary*

➤ *Tells you whether the stadium capacity is sufficient*

➤ *Tells you whether a meeting can go ahead or not*

➤ *Displays how much battery power is left*

5.2.2 FUNCTION DEFINITIONS

❖ Just like with variables, the definition of a function

➤ Is where

- The characteristics are set by the programmer
- Runtime Memory is allocated for it

➤ Should happen only once

❖ The anatomy of a function definition is as follows

```
Output  functionName        (  InputTypes  and  Input  )
Type                           Names  separated  by
                               commas
```

{

        // Actions performed by the function

}

❖ It is the parentheses operator ( ) which tells the compiler that the entity is a function

5.2.3 FUNCTION DECLARATIONS

❖ The declaration is a means of telling the compiler that a function, which has been defined, is going to be used

❖ The anatomy of a function declaration is as follows

➢ Note that it needs to end with the terminator

```
Output   functionName    (  InputTypes  separated  )  ;
Type                         by commas
```

❖ Functions almost always need to be declared and defined separately

➢ They are usually written in different files (discussed in detail section 7)

➢ There is a rare situation whereby if a definition appears before any reference to it, it can act as the declaration as well

• This works only for the most trivial programs. where a function is never referred to in any other file - so is best avoided

❖ The declaration is in fact made up of the "Signature" of the function

5.2.3.1 Signature of a Function (1)

❖ The signature of a function is a means of identifying a function

➢ Just like the signature of a human is an entity which determines individuality, due to its uniqueness

❖ From above, it is apparent that the signature of a function comprises:

➢ The Function Name

➢ The Output TYPE

➢ The Input TYPES

❖ The names of the input ARE NOT part of the signature

❖ The signature of a function is even more important in C++ than in C, since C++ supports

➢ Function Overloading

➢ Object Oriented Programming

5.2.4 FUNCTION INVOCATION

❖ The Function Invocation, also known as a Function Call, is the usage of the function

❖ It requires

➢ In all cases:

  • The function name
  • The parentheses operators()

➢ When the function has inputs, **values** for each of these, of the appropriate type

➢ When the function has an output, to set a variable, of the appropriate type. where the output **value** will be placed

5.2.4.1 Caution: Circular Reference

❖ A potential problem is the circular reference when functions just call each other indefinitely

➢ This will lead to overflow of the runtime stack (see section 14.2)

❖ There are 3 major categories:

➢ A function calling itself indefinitely

  • Note that there is the practice "Recursion" where a function calls itself, but it is important to ensure base cases are appropriately defined (section 8)

➢ Two functions calling each other indefinitely

➢ More than two functions forming a "circle"

5.2.5 DECLARATIONS VS DEFINITIONS

❖ Declarations and Definitions are commonly confused

➤ They are different but overlapping concepts

❖ Common Interview Question

➤ (Q) What is the difference between a Declaration and a Definition?
...After all, A Definition is always a Declaration but a Declaration is not always (the bulk of the time is not) a Definition

➤ (A) A Function Declaration is like a **Customs Declaration**

- It is not the luggage itself, but a description of the luggage
- It can, and often does, appear multiple times
- It consumes no space, does not weigh anything

➤ The definition/luggage

- Occurs once
- Consumes space/memory

➤ There is a special case when the Luggage/Definitions can act as "All the Declarations"

- In particularly small airports,
  - ○ Luggage is collected straight from the aeroplane,
  - ○ Customs is little more than a checkout desk,
  - ○ Customs officers seeing your luggage don't need any further declaration
- In only the most trivial of programs, comprising only one file
  - ○ If the definitions of the functions come before all their invocations...
  - ○ ...there is no need for further declarations
- This special case is rare and occurs only for the smallest airports/projects
  - ○ Assume separate declarations and definitions will be needed

5.3 FUNCTIONS: NO INPUTS, NO OUTPUT

❖ The simplest, but still widely used kind of function is one which only performs a functionality

❖ The definition will take the following form

```
void functionName()
{
      // Actions performed by the function
}
```

> ➤ The keyword `void` specifies that there is no output
> ➤ The fact that there is nothing between the parentheses specifies that there are no inputs
>   • It is optional to also put the keyword `void` in here
>   • However, it is redundant and so not a common practice

❖ The declaration will take the following form:

```
void functionName();
```

❖ In order to call/invoke functions of this nature, the following are required:

> ➤ The function name
> ➤ The parentheses operator

❖ The call will thus take the following form:

```
functionName();
```

❖ The following program shows the implementation of functions with no inputs or output, along with their declarations and invocations:

```
/*     ************************************************************
       C++ Functions, No Inputs, No Output
       ************************************************************
       -        This program shows how to
                -      Define
                -      Declare
                -      Call
                functions with No Inputs and No Output
 */
#include <iostream>
using namespace std;

// Function DECLARATIONS
void greet_English();
void greet_Bengali();
void greet_French();


int main()
{
      // Function Invocations
      greet_Bengali();
      greet_English();
```

```
        greet_Bengali();
        greet_French();
        greet_English();

        //»»»»»»»»»»»»»»»»»»»»»»»»»»»»»»»»»»»»»»»»»»»»»»»»»»»»»»»»»»»»»»»»»»»
        cout  << endl<< endl<< "==========="
              << endl
              << "END...C++ Demo: Functions, No Inputs, No Output"
              << " "
              << endl;
        while(true){}
        return 0;

}//END...Function MAIN


// Function DEFINITIONS
void greet_English()
{
        cout  << "Hello"<< endl;

}//END...Function greet_English

void greet_Bengali()
{
        cout  << "Namaskar"<< endl;

}//END...Function greet_Bengali

void greet_French()
{
        cout  << "Bonjour"<< endl;

}//END...Function greet_French
```

```
Namaskar
Hello
Namaskar
Bonjour
Hello
===========
END...C++ Demo: Functions, No Inputs, No Output
```

Program Listing/Output 5.1: Functions_NoInputs_NoOutput_Client.cpp

EXERCISE: GREETING YOUR USER

❖ *Write a function which*

➢ *Asks your user to select a language (1) English, (2) French, (3) German*

➤ *Asks your user for his/her name*
➤ *Greets them by displaying*
  • *"Hello" if English*
  • *"Bonjour" if French*
  • *"Guten Tag" if German*
  • *followed by the name*

❖ *Ensure you appropriately*

➤ *Choose the function name*
➤ *Format and comment your code*

5.4 FUNCTIONS: SOME INPUTS, NO OUTPUT

❖ Extending the last section, what do we have to do when we want to have inputs to our functions?

❖ Function Definition: we need inputs, also known as parameters, with

➤ Names
➤ Types

It will thus take the following form:

```
void functionName( input1_type input1_Name, input2_type input2_Name
                  ... )
{
        // Actions performed by the function
}
```

❖ The declaration will take the following form (discussed below, section 5.4.1):

```
void functionName(input1_type, input2_type... );
```

❖ In order to call/invoke functions of this nature, the following are required, along with the function name

➤ The input VALUES
  • between the parentheses operator

❖ All the other rules we have covered in the previous section, regarding functions with no inputs, apply here

❖ The following program shows the implementation of functions with inputs but no output, along with their declarations and invocation:

```cpp
/*      ************************************************************
        C++ Functions, Inputs, No Output
        ************************************************************
        -       This program shows how to
                -       Define
                -       Declare
                -       Call
                functions with
                -       Inputs
                -       No Output
*/
#include <iostream>
#include <string>
using namespace std;



// »»»»»»»»»»»»»»»»»»»»»»»»»»»»»»»»»»»»»»»»»»»»»»»»»»»»»»»»»»»»»»»»»»»»»»»
// Function DECLARATIONS i.e. Function SIGNATURES
//
void greet_English_Name( string );
void greet_Bengali_Name( string name_IN );
void greet_Bengali_Name_Multiple( string name_IN, int num_Times );


int main( )
{
        //»»»»»»»»»»»»»»»»»»»»»»»»»»»»»»»»»»»»»»»»»»»»»»»»»»»»»»»»»»»»»»
        string name_person_01 = "John";
        string name_person_02 = "Mary";
        string name_person_03 = "Rachel";
        int myInt_A = 5;

        //»»»»»»»»»»»»»»»»»»»»»»»»»»»»»»»»»»»»»»»»»»»»»»»»»»»»»»»»»»»»»»
        //      Function INVOCATIONS with input VALUES
        //
        greet_English_Name( name_person_01 );
        greet_English_Name( "Jagdice" );

        greet_Bengali_Name( name_person_02 );
        greet_Bengali_Name( "Pedro" );

        greet_Bengali_Name_Multiple( name_person_03, myInt_A );
        greet_Bengali_Name_Multiple( "Miranda", 7 );


        //»»»»»»»»»»»»»»»»»»»»»»»»»»»»»»»»»»»»»»»»»»»»»»»»»»»»»»»»»»»»»»
        cout << endl<< endl<< "==========="
             << endl
             << "END...C++ Demo: Functions, Inputs, No Output"
             << " "
             << endl;
        while(true){} // While loop:
```

```
        return 0;
}

//»»»»»»»»»»»»»»»»»»»»»»»»»»»»»»»»»»»»»»»»»»»»»»»»»»»»»»»»»»»»»»»»»»»»»»»»»»»»»»»»
// Function DEFINITIONS

//»»»»»»»»»»»»»»»»»»»»»»»»»»»»»»»»»»»»»»»»»»»»»»»»»»»»»»»»»»»»»»»»»»»»»»»»»»»»»»»»
void greet_English_Name( string name_IN )
{
        cout << endl<< "------------"<< endl;
        cout << "Hello "<< name_IN<< ", how are you?"<< endl;

}//END...Function greet_English_Name

//»»»»»»»»»»»»»»»»»»»»»»»»»»»»»»»»»»»»»»»»»»»»»»»»»»»»»»»»»»»»»»»»»»»»»»»»»»»»»»»»
void greet_Bengali_Name( string name_IN )
{
        cout << endl<< "------------"<< endl;
        cout << "Namaskar "<< name_IN<< ", kamon acho?"<< endl;

}//END...Function greet_Bengali_Name

//»»»»»»»»»»»»»»»»»»»»»»»»»»»»»»»»»»»»»»»»»»»»»»»»»»»»»»»»»»»»»»»»»»»»»»»»»»»»»»»»
void greet_Bengali_Name_Multiple( string name_IN, int num_Times )
{
        cout << endl<< "------------"<< endl;
        for( int i = 1; i < ( num_Times + 1 ); ++i )
        {
                cout << "For the "<< i<< "th time:"
                     << " Namaskar "<< name_IN
                     << ", kamon acho?"
                     << endl;
        }

}//END...Function greet_Bengali_Name_Multiple
```

```
------------
Hello John, how are you?

------------
Hello Jagdice, how are you?

------------
Namaskar Mary, kamon acho?

------------
Namaskar Pedro, kamon acho?

------------
For the 1th time: Namaskar Rachel, kamon acho?
For the 2th time: Namaskar Rachel, kamon acho?
For the 3th time: Namaskar Rachel, kamon acho?
```

```
For the 4th time: Namaskar Rachel, kamon acho?
For the 5th time: Namaskar Rachel, kamon acho?


------------
For the 1th time: Namaskar Miranda, kamon acho?
For the 2th time: Namaskar Miranda, kamon acho?
For the 3th time: Namaskar Miranda, kamon acho?
For the 4th time: Namaskar Miranda, kamon acho?
For the 5th time: Namaskar Miranda, kamon acho?
For the 6th time: Namaskar Miranda, kamon acho?
For the 7th time: Namaskar Miranda, kamon acho?



===========
END...C++ Demo: Functions, Inputs, No Output
```

Program Listing/Output 5.2: Functions_Inputs_NoOutput_Client.cpp

5.4.1 SIGNATURE OF A FUNCTION (2)

❖ Revision: The signature of a function comprises the 3 components

  ➢ The Function Name
  ➢ The Output Type (section 5.5 onwards)
  ➢ The Input Types

❖ Note that one thing it **does not include** is the input names

  ➢ It is a common misconception that the input names are required
  ➢ Anything that is put immediately after each type name (but before the comma/closing parenthesis) is redundant
    • It ignored by the compiler
    • Thus it does no harm to have the input names but it is not a requirement

❖ In procedural programming

  ➢ It is a common practice to
    • Copy the first line, the prototype of the definition
    • Paste it where you want to have the declaration
    • Add a terminator on the end
  ➢ This has the added advantage that it reminds all humans who will be maintaining and debugging the program what the names of the inputs are

5.4.2 FORMAL PARAMETERS VS. ACTUAL PARAMETERS

❖ As we are seeing, when we say "Function Input," there is ambiguity, since this can mean both the

  ➢ Inputs to the definition
  ➢ Inputs to the invocation

❖ For this reason, the following terminology is often used

  ➢ **Formal** Parameters: Inputs to the Function **Definition**
  ➢ **Actual** Parameters: Inputs to the Function **Invocation**

5.4.3 PASSING A COPY OF THE VALUE

❖ As discussed in section 2.2.2.1, when we state the name of a variable, it can mean two possible things, depending on context:

  ➢ The Value
  ➢ The Address

❖ When a variable is passed as an **actual** parameter, it is passed by **value**

  ➢ Thus in the above example
    • When the variable `name_person_01` is passed to the function `greet_English_Name`
      ○ Only the value "`John`" is passed, not the address of the variable
    • It would be identical to the following:
      ```
      greet_English_Name( "John" );
      ```

5.4.4 NO PREDETERMINED ORDER OF EVALUATION OF ACTUAL PARAMETERS

❖ It is a common misconception, to give into the first glance assumption that actual parameters are evaluated left to right

❖ This is not necessarily the case

  ➢ Individual compilers set their own rules, and even these are not guaranteed to be consistent

❖ Common Interview Question:

  ➢ (Q) What is potentially wrong with the following?
    • You have a function, with the following prototype:

```
void myFunc(int myInt_IN_A, int myInt_IN_B);
```

- You want to call it, with the 2nd parameter having a value 1 more than the 1st
- The following code is suggested

```
int a = 45;
myFunc( a, ++a );
```

➢ (A) There is no guarantee that the leftmost parameter will be evaluated first

- If the 2nd is evaluated first, the values passed will be as follows:

```
myFunc( 46, 46 )
```

  ○ The 2nd parameter is evaluated as 1 more than 45
    ■ This is the new value of the variable a
  ○ This is input as the 1st parameter

➢ In this situation, the following code would serve the intended purpose:

```
int a = 45;
int b = ++a;
myFunc( a, b );
```

5.4.5 MODIFYING FORMAL PARAMETERS

5.4.5.1 Default Values

❖ New to C++, it is possible to assign default values to the input parameters

❖ These need to be specified, using the assignment operator =, in the declaration

❖ When calling the function:

  ➢ Leaving the actual parameter blank will mean the default will be used
  ➢ If the actual parameter is specified, this will be used

```
/*      ************************************************************
        FUNCTIONS WITH DEFAULT INPUT PARAMETERS
        ************************************************************
*/
#include <iostream>
#include <string>
using namespace std;

void greet_English_French(      int num_EnglishGreetings = 3,
                                int num_FrenchGreetings = 2
                    );
```

```cpp
int main( )
{
    //»»»»»»»»»»»»»»»»»»»»»»»»»»»»»»»»»»»»»»»»»»»»»»»»»»»»»»»»»»»»»»»»»»»»
    cout << "==========================================" << endl
         << "Default Values: 3 for English, 2 for French" << endl
         << "greet_English_French();"                     << endl
         << "==========================================" << endl;
    greet_English_French();

    //»»»»»»»»»»»»»»»»»»»»»»»»»»»»»»»»»»»»»»»»»»»»»»»»»»»»»»»»»»»»»»»»»»»»
    cout << "==========================================" << endl
         << "Chosen values"                              << endl
         << "==========================================" << endl;

    cout << "-----------------------------------"        << endl
         << "7 for English,Default 2 for French "        << endl
         << "greet_English_French(7);"                   << endl
         << "-----------------------------------"        << endl;
    greet_English_French(7);
    cout << endl;

    cout << "-----------------------------------"        << endl
         << "7 for English, 5 for French"                << endl
         << "greet_English_French( 7, 5  ) ;"            << endl
         << "-----------------------------------"        << endl;
    greet_English_French( 7, 5 );
    cout << endl;


    //»»»»»»»»»»»»»»»»»»»»»»»»»»»»»»»»»»»»»»»»»»»»»»»»»»»»»»»»»»»»»»»»»»»»
    cout << endl<< endl<< "==========="
         << endl
         << "END...C++ Demo: Functions with Default Input Values"
         << " "
         << endl;
    while(true){}
    return 0;

}//END...Function MAIN



void greet_English_French(      int num_EnglishGreetings,
                                int num_FrenchGreetings
                        )
{
    for( int i = 0;  i < num_EnglishGreetings; ++i )
    {
            cout << ( i + 1 ) << ")\t Hello"  << endl;
    }

    for( int i = 0;  i < num_FrenchGreetings; ++i )
```

```
        {
                cout << ( i + 1 ) << ")\t Bonjour" << endl;
        }

}//END...Function greet_English_French
```

```
=============================================
Default Values: 3 for English, 2 for French
greet_English_French();
=============================================
1)      Hello
2)      Hello
3)      Hello
1)      Bonjour
2)      Bonjour
=============================================
Chosen values
=============================================
-------------------------------------
7 for English,Default 2 for French
greet_English_French(7);
-------------------------------------
1)      Hello
2)      Hello
3)      Hello
4)      Hello
5)      Hello
6)      Hello
7)      Hello
1)      Bonjour
2)      Bonjour


-----------------------------------
7 for English, 5 for French
greet_English_French( 7, 5  ) ;
-----------------------------------
1)      Hello
2)      Hello
3)      Hello
4)      Hello
5)      Hello
6)      Hello
7)      Hello
1)      Bonjour
2)      Bonjour
3)      Bonjour
4)      Bonjour
5)      Bonjour
============
END...C++ Demo: Functions with Default Input Values
```

Program Listing/Output 5.3: Functions_Inputs_DefaultValue_Client.cpp

❖ Note: Only the "last" parameters can have defaults assigned to them

➤ In the above example, it is not possible to

- Use the default for `num_EnglishGreetings`, the first parameter
- Specify a value for `num_FrenchGreetings`, the second parameter

```
greet_English_French( , 7 )
```

will cause a compile-time error

5.4.5.2 Const Formal Parameters

❖ The input value of a function can be assigned read-only status by preceding it with the `const` keyword

➤ The function is still invoked with actual parameters, as usual
➤ The difference is that within the body of the function

- Only the value of the actual parameter can be used
- i.e. The function cannot modify the value

```cpp
/*          *****************************************************
            FUNCTIONS WITH CONST INPUT PARAMETERS
            *****************************************************
*/
#include <iostream>
#include <string>
using namespace std;

void greet_English_German( const int, int );


int main( )
{
        //»»»»»»»»»»»»»»»»»»»»»»»»»»»»»»»»»»»»»»»»»»»»»»»»»»»»»»»»»»»»
        greet_English_German( 7, 8 );


        //»»»»»»»»»»»»»»»»»»»»»»»»»»»»»»»»»»»»»»»»»»»»»»»»»»»»»»»»»»»»
        cout  << endl<< endl<< "============"
              << endl
              << "END...C++ Demo: Functions with CONST Inputs"
              << " "
              << endl;
        while(true){}
        return 0;

}//END...Function MAIN
```

```
void greet_English_German(     const int num_EnglishGreetings,
                               int num_GermanGreetings
                         )
{
      // num_EnglishGreetings = 5;  // ILLEGAL num_EnglishGreetings
                                    // is a READ-ONLY input

      num_GermanGreetings = 3;     // OK

      for( int i = 0; i < num_EnglishGreetings; ++i )
      {
             cout << ( i + 1 ) << ")\t Hello"<< endl;
      }

      for( int i = 0; i < num_GermanGreetings; ++i )
      {
             cout << ( i + 1 ) << ")\t Guten Tag"<< endl;
      }

}//END...Function greet_English_French
```

```
1)      Hello
2)      Hello
3)      Hello
4)      Hello
5)      Hello
6)      Hello
7)      Hello
1)      Guten Tag
2)      Guten Tag
3)      Guten Tag


===========
END...C++ Demo: Functions with CONST Inputs
```

Program Listing/Output 5.4: Functions_Inputs_Const_Client.cpp

5.5 FUNCTIONS: SOME INPUTS, AN OUTPUT

❖ To have a fully functional function, this section looks at how an output can be **returned** from a function

❖ Function Definition: we need, as before inputs, also known as parameters, with

  ➢ The function name
  ➢ Inputs, with names and types

➢ The type of the output
It will thus take the following form:

```
outputType functionName( input1_type input1_Name,
                         input2_type input2_Name  ... )
{
        // Action(s) performed by the function

        return output;
}
```

➢ C++ Can't return more than one value
  • Most other languages also can't inc C functions, Java methods, C#
    methods, Python functions and methods
  • Swift and MATLAB can

❖ The declaration will take the following form (discussed above,
section 5.4.1):

```
outputType functionName( input1_type, input2_type... );
```

❖ The following program shows the implementation of a function which
returns an output, along with its declaration and invocation:

```
/*     ************************************************************
       C++ Functions, Inputs, Output
       ************************************************************
       -          This program shows how to
                  -       Define
                  -       Declare
                  -       Call
                  functions with
                  -       Inputs
                  -       An Output
*/
#include <iostream>
#include <string>
using namespace std;

//»»»»»»»»»»»»»»»»»»»»»»»»»»»»»»»»»»»»»»»»»»»»»»»»»»»»»»»»»»»»»»»»»»»
// Function DECLARATION i.e. Function SIGNATURE
//
int calc_TotalCost( int, int );

//»»»»»»»»»»»»»»»»»»»»»»»»»»»»»»»»»»»»»»»»»»»»»»»»»»»»»»»»»»»»»»»»»»»
int main()
{
      int cost_Ticket_LondonOlympic_Cycling = 100;
      int size_SmithTours_Group = 34;
```

```cpp
        int cost_Cycling_Event = 0;
        // Function CALL
        cost_Cycling_Event
               = calc_TotalCost( size_SmithTours_Group,
        cost_Ticket_LondonOlympic_Cycling );

        // Displaying the result
        cout    << "Proposed London Olympic Tour, cycling event: "
                << "for "<< size_SmithTours_Group<< " people, "
                << "each ticket costing GBP "
                << cost_Ticket_LondonOlympic_Cycling
                << ", the total cost will be GBP "
                << cost_Cycling_Event
                << endl;

    //»»»»»»»»»»»»»»»»»»»»»»»»»»»»»»»»»»»»»»»»»»»»»»»»»»»»»»»»»»»»»»»»»»
     cout << endl<< endl<< "==========="
            << endl
            << "END...C++ Demo: Functions, Inputs, Output"
            << " "
            << endl;
     while(true){}
     return 0;
}

//»»»»»»»»»»»»»»»»»»»»»»»»»»»»»»»»»»»»»»»»»»»»»»»»»»»»»»»»»»»»»»»»»»»»
// Function DEFINITION

//»»»»»»»»»»»»»»»»»»»»»»»»»»»»»»»»»»»»»»»»»»»»»»»»»»»»»»»»»»»»»»»»»»»»
int calc_TotalCost( int num_Tickets_In, int cost_PerTicket_In )
{
     int totalCost = 0;
     totalCost = num_Tickets_In * cost_PerTicket_In;

     return (totalCost);

}//END...Function calc_TotalCost
```

```
Proposed London Olympic Tour, cycling event: for 34 people, each
ticket costing GBP 100, the total cost will be GBP 3400

===========
END...C++ Demo: Functions, Inputs, Output
```

Program Listing/Output 5.5: Functions_Inputs_Output_Client.cpp

5.5.1 RETURNING A VALUE

❖ It is important to note that what is returned is the VALUE of the output

❖ Thus in Program Listing/Output 5.5, what is returned is

> ➤ The Value 3400
> ➤ NOT the address of the variable `totalCost`

❖ Summarising, this does indeed mean that a function

> ➤ Takes **values** as inputs (section 5.4.3)
> ➤ Returns a **value** as the output

❖ If either is to pass the address instead, a pointer or reference is required (section 10 onwards)

5.6 THE MAIN FUNCTION

❖ It is now apparent that the "main" section which has been used in our programs so far is in fact a function

> ➤ It is a special one which is called automatically when the program begins

❖ In C programming, it takes on the following form:

```
void main()
{
      // Actions in main
      // ...

      // No return statement
}
```

❖ In C++, as we have been seeing, the form is slightly different

```
int main()
{
      // Actions in main
      // ...

      // Return statement required
      return 0;
}
```

> ➤ The reason is that since C++ supports object oriented programming, the internal engineering has to be slightly different

5.7 SCOPE OF A VARIABLE

❖ The scope of a variable is the region of the program in which it is accessible

❖ In procedural programming there are two possibilities

> ➤ Global Variables
> ➤ Local Variables

❖ The following program shows both being used:

```cpp
/*      ************************************************************
        SCOPE OF A VARIABLE
        in Procedural Programming: Local vs Global
        ************************************************************
 */
#include <iostream>
using namespace std;

//++++++++++++++++++++++++++++++++++++++++++++++++++++++++++++++++++++
// Function DECLARATIONS
//      - Signatures are different
//
int calc_Tax_Tickets( int, int );
double calc_Tax_Service( double, int );

//»»»»»»»»»»»»»»»»»»»»»»»»»»»»»»»»»»»»»»»»»»»»»»»»»»»»»»»»»»»»»V»»
// Global Variable - Avoid wherever possible
//
double tax_Rate = 0.3;

//++++++++++++++++++++++++++++++++++++++++++++++++++++++++++++++++++++
int main()
{
        //»»»»»»»»»»»»»»»»»»»»»»»»»»»»»»»»»»»»»»»»»»»»»»»»»»»»»»»»»»»»
        // Function Invocations

        //============================================================
        double charge_IT_perHour = 250;
        int num_Hours_IT = 12;

        double tax_IT = calc_Tax_Service(charge_IT_perHour,
        num_Hours_IT);

        cout    << "IT Services: The tax payable, for "
                << num_Hours_IT<< " hours, at a charge of "
                << charge_IT_perHour<< " per hour will be "
                << tax_IT
                << endl;
        //============================================================
        int cost_Ticket_rugbyMatch = 55;
```

```cpp
        int num_People_rugbyMatch = 22;

        int tax_rugbyMatch = calc_Tax_Tickets( cost_Ticket_rugbyMatch,
                                               num_People_rugbyMatch);

        cout << "Rugby Match: The tax payable, to take "
             << num_People_rugbyMatch<< "people, each ticket costing"
             << cost_Ticket_rugbyMatch<< " will be "
             << tax_rugbyMatch
             << endl;

        //»»»»»»»»»»»»»»»»»»»»»»»»»»»»»»»»»»»»»»»»»»»»»»»»»»»»»»»»»»»»»»»»»»
        cout << endl<< endl<< "==========="
             << endl
             << "END...C++ Demo: Local and Global Variables"
             << " "
             << endl;
        while(true){}
        return 0;

}//END...Function MAIN


//+++++++++++++++++++++++++++++++++++++++++++++++++++++++++++++++++++
// Function DEFINITIONS
//
int calc_Tax_Tickets( int ticketPrice, int numPeople )
{
        int total_Price = ticketPrice * numPeople;
        int tax_Amount = tax_Rate * total_Price;

        tax_Rate = 0.35;
                // tax_Rate is a GLOBAL variable - can be changed here

        // total_Income = 750.0;
                   // ILLEGAL - total_Income is LOCAL to the function
                   // calc_Tax_Service - cannot be accessed here

        return ( tax_Amount );

}//END Function int calc_Tax( int, int )

double calc_Tax_Service( double chargePerHour, int num_Hours )
{
        double total_Income = chargePerHour *

        static_cast<double>(num_Hours);
        double tax_Amount = tax_Rate * total_Income;

        tax_Rate = 0.35;
              // tax_Rate is a GLOBAL variable - can be changed here

        // total_Price = 750.0;
```

```
                        // ILLEGAL - total_Price is LOCAL to the function
                        // calc_Tax_Tickets - cannot be accessed here

        return ( tax_Amount );

}//END Function double calc_Tax( double, int )
```

```
IT Services: The tax payable, for 12 hours, at a charge of
250 per hour will be 900
Rugby Match: The tax payable, to take 22 people, each ticket
costing 55 will be 423


===========
END...C++ Demo: Local and Global Variables
```

Program Listing/Output 5.6: Scope_Local_Global_Variables_Client.cpp

❖ Note: Scope of a variable/entity is one of the most important concepts in any programming language

➢ It plays a crucial role in object oriented programming (section 17 onwards)

5.7.1 GLOBAL VARIABLES

❖ Global variables are accessible by any function from anywhere in the program[1]

❖ In Program Listing 2.1, the variable tax_Rate is global so can be, and in this case has been, modified by both functions

➢ The value of tax_Rate used in the function call calc_Tax_Service is the original value 0.3

➢ The value of tax_Rate used in the function call calc_Tax_Tickets is the value 0.35, set in the function calc_Tax_Service

5.7.1.1 Prospective Evils of Global Variables

❖ Global Variables are invariably messy

➢ ...as programmers have been finding out to their peril over decades, however organised one is with their program

❖ When an error is caused by a global variable having an incorrect value

---

[1]A small restriction can be applied using the qualifier static (the same keyword, one of 5 uses in C++), covered in detail in section 7.3.2

> Any function, or simple operation could have changed their value

- Modern day industrial programs typically comprise upwards of several hundred functions
- The fact that any function can call any combination of other functions any time, mean there is a potentially unlimited number of processes that could be the culprit
- Consequently debugging, finding individual problems, can become like finding a needle in a haystack

❖ **Rule of Thumb**: AVOID global variables whenever possible

❖ A far better way of implementing this problem is using object oriented programming (section 17 onwards)

5.7.2 LOCAL VARIABLES

❖ A Local Variable is a variable which is defined inside a function and can be accessed i.e. read and written to, only in there

❖ In Program Listing/Output 5.6,

> The local variables of the function `calc_Tax_Tickets` are

- `total_Price`
- `tax_Amount`

> The local variables of the function `calc_Tax_Service` are

- `total_Income`
- `tax_Amount`

> They can be accessed only within the functions themselves

- Thus the duplicate name `tax_Amount` is not an issue
  - There is no place in the program where both can be accessed

5.7.2.1 Automatic Local Variables

❖ An automatic local variable is one which is "reset" every time a the program exits from a function

❖ The bulk of variables are like this, so this is the default

> It is still possible to precede the variable with the qualifier `auto`

> However, this is redundant since this is the default, so is a very rare practice

❖ In Program Listing/Output 5.6, the local variables of both the functions `calc_Tax_Tickets` and `calc_Tax_Service` are automatic

> We do not want previous values of

- `total_Price` and `tax_Amount`
- `total_Income` and `tax_Amount`

to affect current calculations

5.7.2.2 *Static Local Variables*

❖ A `static` local variable is one which retains its value between function calls

❖ This means that it stays in scope throughout the duration of the program

❖ The following program demonstrates one of its most common uses: counting the number of times the function has been called

```
/*      *******************************************************
        STATIC Local Variables
        *******************************************************
        -       This program shows how to implement and use
                STATIC Local Variables
                -       i.e. Variables which retain their value between
                        function calls
*/
#include <iostream>
using namespace std;

//+++++++++++++++++++++++++++++++++++++++++++++++++++++++++++++++++++++
// Function DECLARATION
//
void greet_English();


//+++++++++++++++++++++++++++++++++++++++++++++++++++++++++++++++++++++
int main()
{
        for( int i = 0; i < 10; ++i )
        {
                greet_English();
        }

        //»»»»»»»»»»»»»»»»»»»»»»»»»»»»»»»»»»»»»»»»»»»»»»»»»»»»»»»
        cout << endl<< endl<< "==========="
                << endl
                << "END...C++ Demo: Functions, Static Local Variable"
                << " "
                << endl;
```

```
        while(true){}
        return 0;

}//END...Function MAIN


//+++++++++++++++++++++++++++++++++++++++++++++++++++++++++++++++++++++
// Function DEFINITIONS
//
void greet_English()
{
        // STATIC Local Variable
        // - Set to 0 only on the FIRST call
        // - Retains its value between calls thereafter
        //
        static int num_TimesCalled = 0;

        cout << "Hello, how are you?"<< endl;
        cout << "This function has been called "
             << num_TimesCalled<< " time(s)."
             << endl<< endl;

             // Increments every time the function is called
             ++num_TimesCalled;

}//END...Function greet_English
```

```
Hello, how are you?
This function has been called 1 time(s).

Hello, how are you?
This function has been called 2 time(s).

Hello, how are you?
This function has been called 3 time(s).

Hello, how are you?
This function has been called 4 time(s).

Hello, how are you?
This function has been called 5 time(s).

Hello, how are you?
This function has been called 6 time(s).

Hello, how are you?
This function has been called 7 time(s).

Hello, how are you?
This function has been called 8 time(s).

Hello, how are you?
```

```
This function has been called 9 time(s).

Hello, how are you?
This function has been called 10 time(s).

===========
END...C++ Demo: Functions, Static Local Variable
```

Program Listing/Output 5.7: Static_Local_Variable_Client.cpp

5.7.2.2.1 Initialising Static Local Variables

❖ Note that initialising static local variables is every bit as important as initialising automatic local variables

❖ In Program Listing/Output 5.7, the line

```
static int num_TimesCalled = 0;
```

ensures that it is initialised to zero

➢ This happens only in the FIRST function call

➢ The `static` qualifier ensures that in subsequent calls, the =0 is ignored, and the variable retains its value from the previous call

5.7.2.2.2 Caution with Static Local Variables

❖ Since they stay in scope throughout the duration of a program, it is important to use static local variables with caution

❖ There is high variance in the duration of programs

➢ A smart phone app might be used for a few minutes and then closed

➢ A personal computer operating system will be rebooted every day or every few days

➢ Software inside a microprocessor which is buried under a sea bed, used to signal to submarines, will be rebooted only every few decades

❖ If the static variable is of type int, so consumes ~4Bytes in most systems, this is not much of an issue

❖ If it is a large variable e.g. a matrix which represents a high resolution picture, having too many can be an issue

5.8 DEBUGGING: SINGLE STEP IN

❖ We are now in a position to look at what is meant by "Step-In" when single stepping, (following section 4.3)

❖ Consider again the program from section 5.5

➢ A breakpoint can be placed at Line 25, just as the variable definitions begin

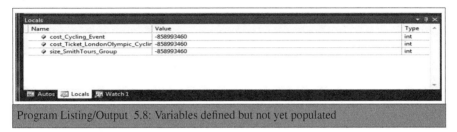

Program Listing/Output 5.8: Variables defined but not yet populated

❖ After stepping over the next two lines, these variables are populated as required:

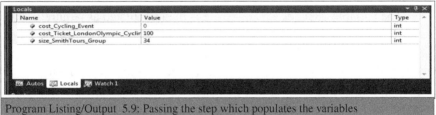

Program Listing/Output 5.9: Passing the step which populates the variables

❖ When Stepping Into (as opposed to Stepping Over) Line 29, the program jumps to the function definition in Lines 59-66

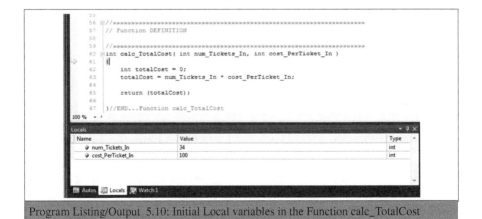

Program Listing/Output 5.10: Initial Local variables in the Function calc_TotalCost

❖ Now the function can be single stepped, just like the rest of the program

➤ When you step over the next line, the local variable `totalCost` is added

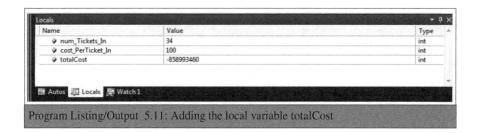

Program Listing/Output 5.11: Adding the local variable totalCost

❖ After single stepping the rest of this function, the program jumps back to the part of main() where this was called

➤ The variable `cost_Cycling_Event` is then populated as required

5.8.1 THE DEBUG/CALL STACK

❖ The Call Stack shows the sequence of functions being stepped into

➤ In this case, there is only 1 function
➤ When the function calls further functions, these will be added

❖ A LIFO Last In First Out Mechanism is manifest here

➤ The Last Function stepped into will be the first function stepped out of
➤ Overall, Functions will be Stepped out of in the reverse order to which they are stepped into
➤ This is known as a Stack

❖ The Stack metaphor is commonly encountered in computer science

➤ The Runtime Stack is one of 2 categories of runtime memory (section 14 onwards)
➤ The Stack Data Structure is a commonly used special purpose data structure (section 28 and Cybernetics in C++ - Expert)

5.9 OVERLOADING FUNCTIONS

❖ Overloading is the practice of giving the same name or symbol(s) to different entities

➢ This is widely used in C++, in several situations (discussed in various forthcoming sections)

❖ One facility, not available in C, newly introduced into C++, is the ability to give the same name to different functions

❖ The compiler identifies them as different as long as there is at least one difference in the signatures i.e.

➢ Output

- A different output type

➢ Inputs:

- The number of inputs
- The input types and their order

❖ The following program shows this in practice:

➢ There are 2 functions with the name `calc_Tax`

```
int calc_Tax( int ticketPrice, int numPeople );
double calc_Tax( double chargePerHour, int num_Hours )
```

➢ The fact that there are differences between

- The types of the inputs
- The types of the output

mean that this program will run fine

```
/*       ************************************************************
         C++ Function OVERLOADING
         ************************************************************
         -        This program shows how to Overload functions
         -        i.e. Implement functions with the SAME NAME
*/
#include <iostream>
using namespace std;

//++++++++++++++++++++++++++++++++++++++++++++++++++++++++++++++++++++
// Function DECLARATIONS
//       -        Signatures are different
//
int calc_Tax( int, int );
double calc_Tax( double, int );

//++++++++++++++++++++++++++++++++++++++++++++++++++++++++++++++++++++
int main()
{
      //»»»»»»»»»»»»»»»»»»»»»»»»»»»»»»»»»»»»»»»»»»»»»»»»»»»»»»»»»»»»»»»»»
```

```cpp
    // Function Invocations

    //=============================================================
    double charge_IT_perHour = 250;
    int num_Hours_IT = 12;

    // Compiler knows which one to call since there is ONE function
    // -    called calc_Tax which
    //          -     takes in a double and an int in that order
    //          -     returns a double
    //
    double tax_IT = calc_Tax( charge_IT_perHour, num_Hours_IT );

    //=============================================================
    int cost_Ticket_rugbyMatch = 55;
    int num_People_rugbyMatch = 22;

    //     Compiler knows which one to call since there is ONE
    function
    //     -     called calc_Tax which
    //          -     takes in two ints
    //          -     returns an int
    //
    int tax_rugbyMatch = calc_Tax( cost_Ticket_rugbyMatch,
                                   num_People_rugbyMatch
                                 );

    //»»»»»»»»»»»»»»»»»»»»»»»»»»»»»»»»»»»»»»»»»»»»»»»»»»»»»»»»»»»»»»»»
    cout   << endl<< endl<< "============"
           << endl
           << "END...C++ Demo: Overloading Functions"
           << " "
           << endl;
    while(true){}
    return 0;

}//END...Function MAIN


//++++++++++++++++++++++++++++++++++++++++++++++++++++++++++++++++++
// Function DEFINITIONS
//
int calc_Tax( int ticketPrice, int numPeople )
{
       int total_Price = ticketPrice * numPeople;
       int tax_Rate = 0.3;
       int tax_Amount = tax_Rate * total_Price;
       return ( tax_Amount );

}//END Function int calc_Tax( int, int )

double calc_Tax( double chargePerHour, int num_Hours )
{
```

```
        double total_Income = chargePerHour *
         static_cast<double>(num_Hours);
        double tax_Rate = 0.25;
        double tax_Amount = tax_Rate *  total_Income;
        return ( tax_Amount );

}//END Function double calc_Tax( double, int )
```

Program Listing/Output 5.12: Overloading_Functions_Client.cpp

5.9.1 PROSPECTIVE EVILS OF FUNCTION OVERLOADING

❖ Function Overloading undeniably introduces some programmer unfriendliness

  ➢ Having the same name for different functions can cause confusion

❖ Function Overloading was implemented in C++ for Object Oriented Programming

  ➢ Here, a major feature is overloading constructors (section 18.9.4)

❖ **Rule of Thumb:** In Procedural Programming, AVOID Function Overloading whenever possible

6. INLINE FUNCTIONS AND MACROS

❖ A general rule of life is that you do not want to pay more for the overhead than for the object/action itself

  ➢ You will not purchase a £200 laptop, then carry it around in a £500 case!

❖ This idea is very important in the programming world as well

6.1 FUNCTION INVOCATION OVERHEAD

❖ Whenever a function is invoked, at least 2 fetch instructions are involved at runtime

  ➢ One of these is to get to the function definition
  ➢ The other is to return from it

❖ For the best part, if the function is substantial, this is a minor issue

❖ If the function is very short, e.g. it performs just a single calculation, the operation becomes less than the total overhead

❖ However, being able to manage and run our program using procedural systems remains of the highest importance

➢ We want to be able to reuse our short functions, with just as much versatility as our longer ones

➢ We do not want to compromise our ability to design and manage programs using procedural methods

❖ The question now is thus "How can we avoid this problem of the overhead being too high, yet maintain the procedural organisation?"

➢ C programmers use macros

➢ C++ programmers recommend using inline functions

6.2 C MACROS - FOR THE BEST PART AVOID IN C++

❖ Macros are used in C programming, to ensure text replacement

➢ During compilation, the precompiler goes through the code and replaces every piece of text in the macro with what has been specified

❖ The `#define` command is used

➢ This is a preprocessor command

❖ Since C++ is backward compatible with C, not least because a significant proportion of C++ programs started their lives as C programs, macros do compile and run in C++

❖ C++ programmers advise avoiding them wherever possible

6.2.1 CREATING A MACRO

❖ There are primarily 2 kinds of macro

➢ Macros for creating a simple alias (section 3.4.1)

```
#define ALIAS_NAME alias
```

e.g.

```
#define STADIUM_CAPACITY 35000
```

➢ Macros for creating an alias for a functionality:

```
#define MACRO_NAME(inputName) Functionality
```

e.g.

```
#define CALC_DOUBLE(x) x+x
```

❖ **When the macro call is written, using the command of the form:**

```
CALC_DOUBLE(my_Int_A)
```

...where `my_Int_A` .is a previously defined variable.

➤ It will be replaced by the text:

```
my_Int_A + my_Int_A
```

❖ **The following program shows how this can be used**

```
/*      ************************************************************
        C STYLE MACROS
        ************************************************************
        -       This program shows how to implement and use
                C style Macros
        -       Avoid wherever possible in C++
*/
#include <iostream>
using namespace std;

//++++++++++++++++++++++++++++++++++++++++++++++++++++++++++++++++++
// MACRO
//
#define CALC_DOUBLE(x) x+x


//++++++++++++++++++++++++++++++++++++++++++++++++++++++++++++++++++
int main()
{
        int my_Int_A = 7;
        double my_Dbl_A = 5.84;

        cout << my_Int_A<< " doubled is "<< CALC_DOUBLE(my_Int_A)
             << endl;

        cout << my_Dbl_A<< " doubled is "<< CALC_DOUBLE(my_Dbl_A)
             << endl;


        //»»»»»»»»»»»»»»»»»»»»»»»»»»»»»»»»»»»»»»»»»»»»»»»»»»»»»»»»»»»»»»
        cout << endl<< endl<< "============"
             << endl
             << "END...Demo: C Macros - avoid wherever possible in C++"
             << " "
             << endl;
        while(true){}
        return 0;
```

```
}//END...Function MAIN
```

```
7 doubled is 14
5.84 doubled is 11.68


===========
END...Demo: C Macros - avoid wherever possible in C++
```

Program Listing/Output 6.1: Macro_Demo_Client.cpp

6.2.2 CONDITIONAL PREPROCESSOR DIRECTIVES

❖ C, and therefore C++, provide a mechanism for creating #define aliases under specified conditions

6.2.2.1 The #ifdef, #ifndef #endif Clause

❖ The #ifdef, #ifndef #endif clause is very similar to the if, then, else clause

❖ Along with the #define, the following preprocessor directives have been included:

➢ #ifdef   This says that if something has been macro-defined, perform the actions in the subsequent lines in the clause

➢ #ifndef   Conversely, this says that if something has not yet been macro-defined, perform the actions in the subsequent lines in the clause

➢ #endif   Ends the "clause"

❖ Consequently, the following is possible

```
/*        ************************************************************
          C STYLE MACRO CONDITIONAL DEFINITIONS
          ************************************************************
          -    This program shows how to implement and use conditional
               macro definitions
          -    Avoid whenever possible in C++
*/
#include <iostream>
using namespace std;

#define NUM_PEOPLE 50
#define PRICE_PER_TICKET 5

//»»»»»»»»»»»» The #ifdef, #ifndef, #endif Clauses »»»»»»»»»»»»»»»»»»»»»»»»
#ifndef PRICE_PER_TICKET
     #define PRICE_PER_TICKET 10
     #define HALL_CAPACITY 20000
```

```
#endif

#ifndef        HALL_CAPACITY
       #define NUM_SEATS 400
#endif

#ifndef NUM_COFFEE_ORDERS
       #define NUM_BEVERAGE_ORDERS 500
#endif

#ifdef PRICE_PER_TICKET
       #define PRICE_TICKET_CARNET 90
#endif
//»»»»»»»»»»»»»»»»»»»»»»»»»»»»»»»»»»»»»»»»»»»»»»»»»»»»»»»»»»»»»»»»»»»»»»

int main( )
{

    cout << "NUM_PEOPLE is "<< NUM_PEOPLE<< endl;
    cout << "PRICE_PER_TICKET is "<< PRICE_PER_TICKET<< endl;

    // Error: HALL_CAPACITY has not been macro-defined
    // since PRICE_PER_TICKET has been defined
    // cout << "HALL_CAPACITY is "<< HALL_CAPACITY<< endl;

    // NUM_SEATS gets macro-defined since HALL_CAPACITY has not been
    // macro-defined
    cout << "NUM_SEATS is "<< NUM_SEATS<< endl;

    // NUM_SEATS gets macro-defined since HALL_CAPACITY has not been
    // macro-defined
    cout << "NUM_SEATS is "<< NUM_SEATS<< endl;

    // PRICE_TICKET_CARNET gets macro-defined since PRICE_PER_TICKET
    // has been macro-defined
    cout << "PRICE_TICKET_CARNET is "<< PRICE_TICKET_CARNET<< endl;


    //»»»»»»»»»»»»»»»»»»»»»»»»»»»»»»»»»»»»»»»»»»»»»»»»»»»»»»»»»»»»»»»»
    cout << endl<< endl<< "============"<< endl
        << "END...Demo #ifdef, #ifndef, #endif Clauses"<< endl;
    while(true){}
    return 0;

}//END...Function MAIN
```

```
NUM_PEOPLE is 50
PRICE_PER_TICKET is 5
NUM_SEATS is 400
NUM_SEATS is 400
PRICE_TICKET_CARNET is 90
```

```
===========
END...Demo #ifdef, #ifndef, #endif Clauses
```

Program Listing/Output 6.2: Demonstrating the #ifdef, #ifndef, endif clause

6.2.3 PROSPECTIVE EVILS OF MACROS

❖ Macros were, and still are, widely used in C programming

❖ C++ programmers are, almost without exception, strongly opposed to macros

➢ We have invented the word "Macrophobia!"

Macros are obnoxious, smelly, sheet-hogging bedfellows for several reasons, most of which are related to the fact that they are a glorified text substitution facility whose effects are applied during preprocessing, before any C++ syntax and semantic rules can even begin to apply.

Herb Sutter,
Lead Designer for C++/CX and C++ AMP

The first rule about macros is: Don't use them unless you have to. Almost every macro demonstrates a flaw in the programming language, in the program, or in the programmer.

Bjarne Stroustrup,
Creator of C++

❖ There are several problems associated with macros in C++...

6.2.3.1 Losing track of types

❖ The main problem here is that we have completely lost track of types

➢ In the above example, we can no longer check the types of the input
➢ The output has no type whatsoever

❖ In C++, the concept of **Polymorphism** is pervasive, where **flexibility with types** is facilitated (covered in sections 23 and 27)

➢ Consequently, not keeping track of types will almost certainly lead to irreparable bugs

❖ The workings of the macro comprise the **pre-**compiler replacing the macro with the alias

➤ i.e. even the rest of the compiler, let alone all parts of the environment after this stage, will not be able to see the name of the macro

❖ Consequently, for all but the simplest macros, having nothing but the alias to work with, debugging can become a considerable obstacle

❖ The fact that macros are nothing but a raw text replacement often causes errors when using even simple operators

➤ In the following examples,
  • The macro used above is being used here:
    ```
    #define CALC_DOUBLE(x) x+x
    ```
  • Assume `myInt_A` to be a variable with the current value of 5

❖ **BODMAS/PEMDAS:** If we enter

```
CALC_DOUBLE( myInt_A ) * 3
```

➤ We expect the answer 30
➤ Instead we will get the answer 20, since the text replacement will be
```
5 + 5 * 3
```

➤ While parentheses will prevent this problem, they will not prevent the following:

❖ If someone enters `CALC_DOUBLE( myInt_A++ )`, this will be replaced as
```
5++ + 5++
```

which is undefined

❖ If someone tries to enters `CALC_DOUBLE( myInt_A & 1 )`

➤ this will be replaced as
```
myInt_A & 1 + myInt_A & 1
```

➤ this will be
```
myInt_A & ( 1 + myInt_A ) & 1
```

➤ not
```
( myInt_A & 1 ) + ( myInt_A & 1 )
```

6.2.4 RULES OF THUMB WITH MACROS

❖ There are a few rare problems when macros are one of the few mechanisms which provide a solution

➢ There is one situation, when managing programs over multiple files, when macros provide a possible solution

• This is covered in section 7

❖ Otherwise, for the best part, avoid macros whenever possible

6.3 INLINE FUNCTIONS

❖ The inline function is a special function which also performs text replacement, but in a more sophisticated means than macros

❖ It was introduced to C++ in order to facilitate the dual problem of

➢ A function overhead being greater than the operation

➢ Macros losing track of types, and their further problems

6.3.1 DEFINING & DECLARING INLINE FUNCTIONS

❖ Defining and declaring inline functions is virtually identical to defining and declaring "normal/non-inline" functions - just precede each with the keyword `inline`

➢ The syntax is as follows

```
inline returnType inlineFunctionName( inputTypes  and inputNames )
{
        // Functionalities;

}
```

❖ Thus if we want to improve the program in Program Listing/Output 6.1 using an inline function, the definition will be as follows:

```
inline int calc_double( int int_IN_A )
{
        return ( int_IN_A + int_IN_A );

}//END...Inline Function calc_double
```

❖ The declaration will be as follows:

```
inline int calc_double( int );
```

6.3.2 USING INLINE FUNCTIONS

❖ Calling an inline function is identical to calling a "normal/non-inline" function

❖ Thus, in order to call the inline function described above, we use a command like:

```
calc_double(my_Int_A)
```

...where `my_Int_A` .is a previously defined variable.

➢ Provided inline status has been granted (section 6.3.3), then this will be replaced with the following code:

```
my_Int_A + my_Int_A
```

❖ The following program shows how to define, declare and call an inline function:

```
/*        ************************************************************
          C++ INLINE Functions
          ************************************************************
          -        This program shows how to implement and use
                   Inline Functions
*/
#include <iostream>
using namespace std;

//++++++++++++++++++++++++++++++++++++++++++++++++++++++++++++++++++++
// Inline Function DECLARATION
//
inline int calc_double( int );

//++++++++++++++++++++++++++++++++++++++++++++++++++++++++++++++++++++
int main()
{
      int my_Int_A = 7;

      cout   << "my_Int_A = "<< my_Int_A<< endl;

      cout   << "...doubled using calc_double(my_Int_A) = "
             << calc_double(my_Int_A)
             << endl;

      // Correct answer when combined with other mathematical operations
      // - problem with macros has been eliminated
      //
      cout   << "...doubled then tripled using "
             << " calc_double(my_Int_A) * 3 = "
             << calc_double(my_Int_A) * 3
```

```
            << endl;

    //»»»»»»»»»»»»»»»»»»»»»»»»»»»»»»»»»»»»»»»»»»»»»»»»»»»»»»»»»»»»»»»»»»
    cout    << endl<< endl<< "==========="
            << endl
            << "END...C++ Demo: Inline Functions"
            << " "
            << endl;
        while(true){}
        return 0;

}//END...Function MAIN

//++++++++++++++++++++++++++++++++++++++++++++++++++++++++++++++++++++
// Inline Function DEFINITION
//
inline int calc_double( int int_IN_A )
{
        return ( int_IN_A + int_IN_A );

}//END...Inline Function calc_double
```

```
my_Int_A = 7
...doubled using calc_double(my_Int_A) = 14
...doubled then tripled using  calc_double(my_Int_A) * 3 = 42

===========
END...C++ Demo: Inline Functions
```

Program Listing/Output 6.3: InlineFunction_Demo_Client.cpp

6.3.3 THE INLINE KEYWORD: A REQUEST, NOT AN INSTRUCTION

❖ The inline keyword is unusual in that it does not guarantee that the function will be inline

❖ The compiler will decide whether to make it an inline or "normal" function

❖ Note that requests are sizeably less common than instructions

➢ When you have even a statement like `int myInt_A`, you are **instructing** the compiler to place `myInt_A` in memory, as an integer

6.3.4 MACRO PROBLEMS NOT MANIFEST IN INLINE FUNCTIONS

❖ A crucial aspect of inline functions is that the type remains specified, just like "normal/non-inline" functions

➤ This fits in well with Polymorphism in C++, where type flexibility is pervasive

❖ In addition, as the above example shows, when combining an inline function call with other mathematical operations, the operator precedence problem, manifest with macros, does not occur

7. PROGRAM MANAGEMENT OVER MULTIPLE FILES

7.1 SOFTWARE PROJECTS

❖ We have looked at splitting a program into functions

❖ Keeping all our code in one file quickly becomes cumbersome

❖ What we need to do is create a project

➤ This will comprise having our code in different files

➤ Until fairly recently, we would have to create a file known as a **makefile**, which would list the files and their paths

• This would be presented to the compiler and linker, so they would know which files to look at

➤ In modern times, graphical IDEs generally hide this from the programmer

• Instead, all we have to do is click on the File Management Window and click on which file we want to add: Compiler, Linker

❖ The various stages in developing a C++ project, along with the type of file at each stage, have been shown in Figure 7.1

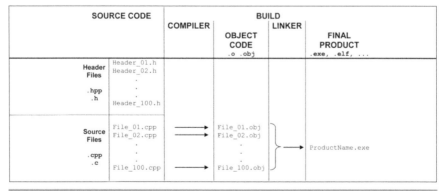

Figure 7.1: Files in a C++ Project and the different stages in Building

7.1.1 SOURCE CODE

❖ Source code is the C++ code itself, written by humans

❖ It consists primarily of 2 categories of file:

> ➢ Header files
>   • These are generally used for declarations
>   • They have the extension `.h`
>   ○ A few C++ systems use `.hpp` for C++ header files, and `.h` for C header files
> ➢ Source files
>   • These are used generally used for definitions
>   • They have the extension
>   ○ `.cpp` for C++ files
>   ○ `.c` for C files

❖ The following program shows how a project can be managed over multiple files:

```cpp
// TIP: Demo only, avoid using \#define - prefer using const

#ifndef RATES_TAX_INTEREST_H
#define RATES_TAX_INTEREST_H

#define TAX_THESHOLD_LOWER                  20000
#define TAX_THESHOLD_UPPER                  30000

#define TAX_RATE_LOWER                      0.3
#define TAX_RATE_INTERMEDIATE               0.4
#define TAX_RATE_UPPER                      0.5

#define INTEREST_THESHOLD_LOWER             3000
#define INTEREST_THESHOLD_UPPER             10000

#define RATE_INTEREST_LOWER                 0.001
#define RATE_INTEREST_INTERMEDIATE          0.0015
#define RATE_INTEREST_UPPER                 0.002


struct Client
{
        double salary_Gross;
        double tax_Rate;
        double salary_Net;

        double bank_Balance_Gross;
        double interest_Rate;
```

```
        double bank_Balance_Net;
};

#endif
```

```
#include "Rates_Tax_Interest.h"

double calc_TaxRate( Client client_IN_A );
double calc_Salary_Net( Client client_IN_A );
```

```
#include "TaxCalculations.h"


double calc_TaxRate( Client client_IN_A )
{
        double taxRate = 0.0;
        if( client_IN_A.salary_Gross < TAX_THESHOLD_LOWER )
        {
                taxRate = TAX_RATE_LOWER;
        }
        else if( client_IN_A.salary_Gross < TAX_RATE_INTERMEDIATE )
        {
                taxRate = TAX_RATE_INTERMEDIATE;
        }
        else
        {
        taxRate = TAX_RATE_UPPER;
        }
        return taxRate;

}// END...Function calc_TaxRate

double calc_Salary_Net( Client client_IN_A )
{
        double taxAmount = client_IN_A.salary_Gross *
                                client_IN_A.tax_Rate;
        double salary_Net = client_IN_A.salary_Gross - taxAmount;
        return salary_Net;


}// END...Function calc_TaxRate
```

```
#include <iostream>
#include <string>
using namespace std;

#include "Rates_Tax_Interest.h"
#include "TaxCalculations.h"

int main()
{
    //»»»»»»»»»»»»»»»»»»»»»»»»»»»»»»»»»»»»»»»»»»»»»»»»»»»»»»»»»»»»»»»»»»»»
    Client johnSmith;

    johnSmith.salary_Gross = 22000.0;
    johnSmith.tax_Rate = calc_TaxRate( johnSmith );
    johnSmith.salary_Net = calc_Salary_Net( johnSmith );

    cout    << endl
            << "===================="<< endl
            < "Hello, I am John Smith "<< endl
            << "my Gross Salary is "<< johnSmith.salary_Gross<< endl
            << "my tax Rate is "<< johnSmith.tax_Rate<< "%"<< endl
            << "thus my Net Salary is "<< johnSmith.salary_Net<< endl
            << "===================="<< endl;

    //»»»»»»»»»»»»»»»»»»»»»»»»»»»»»»»»»»»»»»»»»»»»»»»»»»»»»»»»»»»»»»»»»»»»
    cout    << endl<< endl<< "========================"
            << endl
            << "END...Demo, Multiple File Project"
            << endl;
    while(true){}
    return 0;

}//END...Function MAIN
```

```
====================
Hello, I am John Smith
my Gross Salary is 22000
my tax Rate is 0.5%
thus my Net Salary is 11000
====================


========================
END...Demo, Multiple File Project
```

Program Listing/Output 7.4: MultipleFileProject_Client.cpp

7.2 BUILDING C++ PROJECTS

❖ There are primarily 2 stages in Building a project:

➢ Compilation

➢ Linking

❖ The command "Build" means the project gets compiled and linked from scratch

❖ There is also the command "Make" which is used progressively less in modern times

➢ Until recently, carrying out a "Build" on most projects from scratch was prohibitively expensive

➢ Instead, a "Make" would be carried out

• The idea was that only the files which have been modified since the last make would be compiled and added to the project

○ i.e. A Make is a partial build

• Nowadays, Make tends to be used on particularly large projects

7.2.1 COMPILATION

❖ The role of the Compiler is to

➢ Take the **Source** files

• Something humans can understand

➢ Translate them into **Object** files

• Something computers can understand

• These usually have `.o` or `.obj` extensions

• Part of this comprises stating the memory addresses of the data items

❖ Only the source files i.e. `.cpp` and `.c` if leftover from C, get compiled

❖ Header files do not get compiled

➢ The data in them is "read" and used by the source files

❖ **Compile-Time errors** occur when the compiler is unable to translate the source code into object code

❖ The compiler is actually a series of programs

➢ The initial part is called the Pre-compiler

➢ Preprocessor Directives are issued to the pre-compiler

- The idea is that certain commands/ideas can be established as early as possible
- The most common examples are:
  - #define
  - #pragma

7.2.2 LINKING

❖ The role of the Linker is to

➤ Take the Object files

➤ Put them together i.e. link them into an executable product

  - The most common are for computers with a `.exe` extension
  - Embedded systems often use `.elf`

❖ **Linker Errors** thus occur when there is a problem of communicating data between object files

❖ A common reason is the Visibility of variables...

7.3 CONTROLLING VISIBILITY

7.3.1 CREATING EXTERNAL VISIBILITY USING EXTERN

❖ A frequent need will be to use a variable defined in another file

❖ The keyword `extern` signals to the

➤ Compiler that the variable is not in this file, so it should continue creating the object file

➤ Linker that the variable has been defined elsewhere in memory

❖ The following program shows this in action

```cpp
#include <iostream>
#include <string>
using namespace std;

// Global Variables - AVOID whenever possible
//                  - Demo only

string joeBloggs_OfficeLocation = "London";
```
Program Listing/Output 7.5: ContactDetails.cpp

```
#include <iostream>
#include <string>
using namespace std;

// OK - joeBloggs_OfficeLocation has been defined
//      in the file ContactDetails.cpp
extern string joeBloggs_OfficeLocation;

//    LINKER ERROR
//    amandaTaylor_OfficeLocation has not been defined in any
//    other file in this project
//
extern string amandaTaylor_OfficeLocation;

int main()
{
    //»»»»»»»»»»»»»»»»»»»»»»»»»»»»»»»»»»»»»»»»»»»»»»»»»»»»»»»»»»»»»»»»»»
    cout << "Joe Bloggs works in "<< joeBloggs_OfficeLocation
        << endl;


    //»»»»»»»»»»»»»»»»»»»»»»»»»»»»»»»»»»»»»»»»»»»»»»»»»»»»»»»»»»»»»»»»»»
    cout << endl<< endl<< "========================"
        << endl
        << "END...Demo, Static Global Variable"
        << endl;
    while(true){}
    return 0;

}//END...Function MAIN
```

```
Joe Bloggs works in London


========================
END...Demo, Static Global Variable
```

Program Listing/Output 7.6: Extern_Demo_Client.cpp

7.3.1.1 Linker Error caused by a Lack of Definition

❖ A Linker Error will result if an attempt is made to include a variable, using extern, which has not been defined in the project

❖ The following program shows this:

```
#include <iostream>
#include <string>
using namespace std;

// Global Variables - AVOID whenever possible
```

```
//      - Demo only

string joeBloggs_OfficeLocation = "London";
```

Program Listing/Output 7.7: ContactDetails.cpp

```cpp
#include <iostream>
#include <string>
using namespace std;

// OK - joeBloggs_OfficeLocation has been defined
//          in the file ContactDetails.cpp
extern string joeBloggs_OfficeLocation;

//      *** LINKER ERROR ***
//      amandaTaylor_OfficeLocation has not been defined in any
//      other file in this project
//
extern string amandaTaylor_OfficeLocation;

int main()
{
        cout << "Joe Bloggs works in "<< joeBloggs_OfficeLocation
            < endl;

        // *** LINKER ERROR ***
        // Attempting to use a variable not defined anywhere in the
        // project
        cout << "Amanda Taylor works in "<<
                amandaTaylor_OfficeLocation
            << endl;


        //»»»»»»»»»»»»»»»»»»»»»»»»»»»»»»»»»»»»»»»»»»»»»»»»»»»»»»»»»»»»
        cout << endl<< endl<< "========================"
            << endl
            << "END...Demo, Extern - NO External Global Variable"
            << endl;
        while(true){}
        return 0;

}//END...Function MAIN
```

Program Listing/Output 7.8: Extern_NoVar_Demo_Client.cpp

7.3.2 HIDING EXTERNAL VISIBILITY USING STATIC

❖ The visibility of any global variable or global function can be restricted to the file it is in by prefixing it with the keyword `static`

> ➤ This is commonly used in C programming in order to reduce the exposure of global variables

- In C++, this is still advised against
  - ○ Object Oriented Programming (section 17 onwards) provides a far better solution

> ➤ It is less commonly used with functions, since allowing many different parts of the program to use them usually does not "mess them up"

❖ If any attempt is made to access from another file them using extern, a linker error results

❖ The following program shows this in action:

```cpp
#include <iostream>
#include <string>
using namespace std;

// Global Variables - AVOID whenever possible
//                  - Demo only

string joeBloggs_OfficeLocation = "London";

static string joeBloggs_HomeLocation = "Ilford";
```

Program Listing/Output 7.9: ContactDetails.cpp

```cpp
#include <iostream>
#include <string>
using namespace std;

//    OK
extern string joeBloggs_OfficeLocation;

//    LINKER ERROR
//    joeBloggs_HomeLocation is STATIC in the file ContactDetails.cpp
//    -       Thus it cannot be accessed here, or anywhere else outside
//            that file
extern string joeBloggs_HomeLocation;

int main()
{
    //»»»»»»»»»»»»»»»»»»»»»»»»»»»»»»»»»»»»»»»»»»»»»»»»»»»»»»»»»»»»»»»»»»»»»»»»
    //      OK - joeBloggs_OfficeLocation is in another file in this
    //      project and is not static
    cout    << "Joe Bloggs works in "<< joeBloggs_OfficeLocation
            << endl;

    //      LINKER ERROR
    //      Attempting to use a variable defined as static in another
```

```
//      file
cout    << "Joe Bloggs works in "<< joeBloggs_HomeLocation
        << endl;

//»»»»»»»»»»»»»»»»»»»»»»»»»»»»»»»»»»»»»»»»»»»»»»»»»»»»»»»»»»»»
cout    << endl<< endl<< "========================"
        << endl
        << "END...Demo, Extern"
        << endl;
while(true){}
return 0;

}//END...Function MAIN
```

Program Listing/Output 7.10: StaticGlobalVariable_Demo_Client.cpp

7.3.2.1 Duplication of the keyword static

❖ The same keyword static has no less than 5 different meanings in C++ (section 30.1)

❖ We have now encountered 3 of them:
  ➤ static local variable   Retains its value between function calls
  ➤ static global variable   Visible only in the file where it was defined
  ➤ static global function   Visible only in the file where it was defined

7.4 NAMESPACES

❖ A namespace is a part of a program which encapsulates its contents from the rest of the project

❖ They are useful when duplicate names are used

  ➤ As we know, two variables in the same project cannot have the same name

  ➤ However, if each one is defined inside a separate namespace, then it is all right

  ➤ They are especially useful when different teams, often from different companies, are contributing to the same project

    ● If there is a common/obvious name for a variable which is likely to be duplicated by someone else, it is advisable to define it in a namespace

❖ A namespace is defined as follows

```
namespace NameOfNamespace
```

```
{
        // Data types and names
}
```

❖ They can be used in two ways

➤ **Using the keyword "using"**

- The first way is to say, at the beginning of the file that the namespace will be used, by writing:

```
using namespace NameOfNamespace;
```

- After that, the data can be used like any defined outside a namespace

➤ **Using the Scope Resolution Operator**

- The second way is to declare all the namespace name and data item when using the data, by stating:
  - The Name of the namespace
  - The Scope Resolution Operator ::
  - The name of the data item

```
NameOfNamespace::dataItem
```

❖ The following program shows this in practice:

```
namespace Data_Dog
{

      int num_Legs = 4;
      int num_Tails = 1;
      int num_Eyes = 2;
      int num_GillSlits = 0;
      double body_Temperature = 38.0;

}//END...namespace Data_Dog


namespace Data_Shark
{
      int num_Legs = 0;
      int num_Tails = 1;
      int num_Eyes = 2;
      int num_GillSlits = 7;
      double body_Temperature = 22.5;

}//END...namespace Data_Dog
```

Program Listing/Output 7.11: Animal_Data.h

```
#include <iostream>
#include <string>
using namespace std;

#include "Animal_Data.h"

using namespace Data_Dog;

int main()
{
    cout << "Lassie has "<< num_Legs<< " legs."<< endl;

    cout << "Jaws has a body temperature of "
        << Data_Shark::body_Temperature
        << " deg C"
        << endl;

        //»»»»»»»»»»»»»»»»»»»»»»»»»»»»»»»»»»»»»»»»»»»»»»»»»»»»»»»»»»»»»
        cout << endl<< endl<< "========================"
        << endl
        << "END...Demo, NameSpaces in C++"
        << endl;
    while(true){}
    return 0;

}//END...Function MAIN
```

```
Lassie has 4 legs.
Jaws has a body temperature of 22.5 deg C


========================
END...Demo, NameSpaces in C++
```

Program Listing/Output 7.12: NameSpace_Demo_Client.cpp

❖ While namespaces can seem to provide a means of organising data, in C++, you are advised to use the Object Oriented Programming entity, the Class, for this (section 17 onwards)

7.5 POTENTIAL PROBLEM: MULTIPLE INCLUSIONS

❖ An issue with including files is that (a) file(s) can end up getting included more than once

➢ This leads to the same problem of multiple definitions
➢ Most modern programming environments now give linker errors

❖ Consider the project which has the following files:

➢ `Rates_Tax_Interest.h`

➢ `TaxCalculations.h`

➢ `TaxCalculations.cpp`

➢ `InterestCalculations.h`

➢ `InterestCalculations.cpp`

➢ `Office.cpp`

❖ The following problem arises:

➢ `TaxCalculations.cpp` needs to access `Rates_Tax_Interest.h`

➢ `InterestCalculations.cpp` also needs to access `Rates_Tax_Interest.h`

➢ As soon as `Office.cpp` includes both

   • `TaxCalculations.h`
   • `InterestCalculations.h`

there will be 2 inclusions of `Rates_Tax_Interest.h`

❖ This potential problem will be manifest in all but the simplest projects

7.6 PREVENTING MULTIPLE INCLUSIONS

❖ There are two common solutions which ensure that once an inclusion has taken place, the same action will not be performed subsequently

➢ Using Macros

➢ Using preprocessor `#pragma` directives

7.6.1 ENSURING SINGLE INCLUSIONS USING MACROS

❖ This problem can be solved using macros

➢ For the best part, C++ programmers suffer from "macrophobia" and advise avoiding macros wherever possible (discussed in section 6.2.3)

➢ However, this is the one occasion where "macros" can be used

7.6.1.1 Using #ifndef to prevent Multiple Inclusions

❖ The standard technique used throughout C and C++ programming is to place all the declarations in a header file inside a #ifndef #endif clause

➢ A common practice for naming the macro is the filename, followed by "_H"

- - e.g. if the name of the file is `InterestCalculations.h`, a macro name could be `INTERESTCALCULATIONS_H`
- ➢ Consequently, even though a header file might be included more than once
  - - As soon as the precompiler sees that the macro has already been defined, the declarations will not be declared again
- ❖ Reengineering our project as follows will ensure the project compiles and links without any multiple inclusion problems:

```
// TIP: Demo only, avoid using #define for consts - prefer using
      const

#ifndef RATES_TAX_INTEREST_H
#define RATES_TAX_INTEREST_H

#define TAX_THESHOLD_LOWER             20000
#define TAX_THESHOLD_UPPER             30000

#define TAX_RATE_LOWER                 0.3
#define TAX_RATE_INTERMEDIATE          0.4
#define TAX_RATE_UPPER                 0.5

#define INTEREST_THESHOLD_LOWER        3000
#define INTEREST_THESHOLD_UPPER        10000

#define RATE_INTEREST_LOWER            0.001
#define RATE_INTEREST_INTERMEDIATE     0.0015
#define RATE_INTEREST_UPPER            0.002


struct Client
{
      double salary_Gross;
      double tax_Rate;
      double salary_Net;

      double bank_Balance_Gross;
      double interest_Rate;
      double bank_Balance_Net;
};

#endif
```

Program Listing/Output 7.13: Rates_Tax_Interest.h

```
#include "Rates_Tax_Interest.h"

#ifndef TAXCALCULATIONS_H
#define TAXCALCULATIONS_H


double calc_TaxRate( Client client_IN_A );
double calc_Salary_Net( Client client_IN_A );

#endif
```

Program Listing/Output 7.14: TaxCalculations.h

```
#include "Rates_Tax_Interest.h"

#ifndef INTERESTCALCULATIONS_H
#define INTERESTCALCULATIONS_H

double calc_InterestRate( Client client_IN_A );
double calc_Salary_Net( Client client_IN_A );

#endif
```

Program Listing/Output 7.15: InterestCalculations.h

```
#include "TaxCalculations.h"


double calc_TaxRate( Client client_IN_A )
{
      double taxRate = 0.0;
      if( client_IN_A.salary_Gross < TAX_THESHOLD_LOWER )
      {
            taxRate = TAX_RATE_LOWER;
      }
      else if( client_IN_A.salary_Gross < TAX_RATE_INTERMEDIATE )
      {
            taxRate = TAX_RATE_INTERMEDIATE;
      }
      else
      {
      taxRate = TAX_RATE_UPPER;
      }
      return taxRate;

}// END...Function calc_TaxRate

double calc_Salary_Net( Client client_IN_A )
{
      double taxAmount = client_IN_A.salary_Gross *
```

```
client_IN_A.tax_Rate;
      double salary_Net = client_IN_A.salary_Gross - taxAmount;
      return salary_Net;


}// END...Function calc_TaxRate
```

Program Listing/Output 7.16: TaxCalculations.cpp

```
#include "InterestCalculations.h"


double calc_InterestRate( Client client_IN_A )
{
      double interestRate = 0.0;
      if( client_IN_A.salary_Gross < TAX_THESHOLD_LOWER )
      {
            interestRate = TAX_RATE_LOWER;
      }
      else if( client_IN_A.salary_Gross < TAX_RATE_INTERMEDIATE )
      {
            interestRate = TAX_RATE_INTERMEDIATE;
      }
      else
      {
            interestRate = TAX_RATE_UPPER;
      }
      return interestRate;

}// END...Function calc_InterestRate

double calc_BankBalance_Net( Client client_IN_A )
{
      double interestAmount = client_IN_A.bank_Balance_Gross *
                              client_IN_A.interest_Rate;
      double bank_Balance_Net = client_IN_A.bank_Balance_Gross +
                                interestAmount;
      return bank_Balance_Net;

}// END...Function calc_BankBalance_Net
```

Program Listing/Output 7.17: InterestCalculations.cpp

```
#include <iostream>
using namespace std;

#include "InterestCalculations.h"
#include "TaxCalculations.h"

int main( )
{
      Client maryPearce;
```

```
        // Code using the Interest and Tax Calculation functions

        //»»»»»»»»»»»»»»»»»»»»»»»»»»»»»»»»»»»»»»»»»»»»»»»»»»»»»»»»»»»»»»
        cout << endl<< endl<< "==========="<< endl
             << "END...Demo Procedural Project over Multiple Files"
             << "...corrected using Macros"
             << endl;
        while(true){}
        return 0;

}//END...Function MAIN
```

Program Listing/Output 7.18: MultipleFilesCorrected_Macro_Client.cpp

7.6.2 Ensuring Single Inclusions using Preprocessor Directives

❖ An alternative is to use the preprocessor directive `#pragma`

➢ This is a means of providing extra information to the compiler, on top of the language specifications

❖ The command

```
#pragma once
```

ensures that the file it is in gets included only once in the project

❖ Consequently, this can be used, as an alternative to the #define macros, as a means of preventing multiple file inclusions:

```
// TIP: Demo only, avoid using #define for constants - prefer using
//     const

#pragma once

#define TAX_THESHOLD_LOWER              20000
#define TAX_THESHOLD_UPPER              30000

#define TAX_RATE_LOWER                  0.3
#define TAX_RATE_INTERMEDIATE           0.4
#define TAX_RATE_UPPER                  0.5

#define INTEREST_THESHOLD_LOWER         3000
#define INTEREST_THESHOLD_UPPER         10000

#define RATE_INTEREST_LOWER             0.001
#define RATE_INTEREST_INTERMEDIATE      0.0015
#define RATE_INTEREST_UPPER             0.002


struct Client
```

```
{
      double salary_Gross;
      double tax_Rate;
      double salary_Net;

      double bank_Balance_Gross;
      double interest_Rate;
      double bank_Balance_Net;
};
```

```
#pragma once

#include "Rates_Tax_Interest.h"

double calc_TaxRate( Client client_IN_A );
double calc_Salary_Net( Client client_IN_A );
```

```
#pragma once

#include "Rates_Tax_Interest.h"

double calc_InterestRate( Client client_IN_A );
double calc_Salary_Net( Client client_IN_A );
```

```
#include "TaxCalculations.h"


double calc_TaxRate( Client client_IN_A )
{
      double taxRate = 0.0;
      if( client_IN_A.salary_Gross < TAX_THESHOLD_LOWER )
      {
            taxRate = TAX_RATE_LOWER;
      }
      else if( client_IN_A.salary_Gross < TAX_RATE_INTERMEDIATE )
      {
            taxRate = TAX_RATE_INTERMEDIATE;
      }
      else
      {
            taxRate = TAX_RATE_UPPER;
      }
      return taxRate;

}// END...Function calc_TaxRate
```

```
double calc_Salary_Net( Client client_IN_A )
{
      double taxAmount = client_IN_A.salary_Gross *
                                      client_IN_A.tax_Rate;
      double salary_Net = client_IN_A.salary_Gross - taxAmount;
      return salary_Net;


}// END...Function calc_TaxRate
```

Program Listing/Output 7.22: TaxCalculations.cpp

```
#include "InterestCalculations.h"


double calc_InterestRate( Client client_IN_A )
{
      double interestRate = 0.0;
      if( client_IN_A.salary_Gross < TAX_THESHOLD_LOWER )
      {
            interestRate = TAX_RATE_LOWER;
      }
      else if( client_IN_A.salary_Gross < TAX_RATE_INTERMEDIATE )
      {
            interestRate = TAX_RATE_INTERMEDIATE;
      }
      else
      {
            interestRate = TAX_RATE_UPPER;
      }
      return interestRate;

}// END...Function calc_InterestRate

double calc_BankBalance_Net( Client client_IN_A )
{
        double interestAmount = client_IN_A.bank_Balance_Gross *
                                client_IN_A.interest_Rate;
        double bank_Balance_Net =  client_IN_A.bank_Balance_Gross +
                                interestAmount;
   return bank_Balance_Net;


}// END...Function calc_BankBalance_Net
```

Program Listing/Output 7.23: InterestCalculations.cpp

```
#include <iostream>
using namespace std;

#include "InterestCalculations.h"
#include "TaxCalculations.h"
```

```
int main( )
{
        Client maryPearce;

        // Code using the Interest and Tax Calculation functions

        //»»»»»»»»»»»»»»»»»»»»»»»»»»»»»»»»»»»»»»»»»»»»»»»»»»»»»»»»»
        cout << endl<< endl<< "============"<< endl
             << "END...Demo Procedural Project over Multiple Files"
             << "...corrected using Pragma"
             << endl;
        while(true){}
        return 0;

}//END...Function MAIN
```

Program Listing/Output 7.24: MultipleFilesCorrectedPragma_Client.cpp

8. RECURSION

8.1 WHAT IS RECURSION?

❖ Recursion is a special line of procedural thinking, where a procedure call involves the procedure itself on a different scale

➢ An essential component is the "Base Case" which determines when the sequence of procedure calls should stop

❖ It is also known by the name "Divide and Conquer" Algorithms

➢ You Divide a problem into smaller sub-problems

• Often called "Optimal Sub-Structure"

➢ Then Conquer each part

❖ This is a powerful means of thinking which is used considerably outside the programming world, as well as inside it

8.1.1 RECURSION OUTSIDE THE PROGRAMMING WORLD

❖ Looking up words in a dictionary

➢ You do not start at the beginning and browse through until you find the word

➢ You perform an alphabetical search on the first letter

➢ Once you have found this, you perform an alphabetical search on the second letter

- i.e. this is the same procedure as for the first letter but on a reduced scale
- ➢ You perform alphabetical searches on subsequent letters until you reach the
- ➢ Base Case: In this case: **2** possibilities
  - Either you find the word
  - You acknowledge that the word is not in the dictionary

❖ Recursive Abbreviations:

  ➢ **GNU**: <u>G</u>NU's <u>N</u>ot <u>U</u>nix
  ➢ **PHP**: <u>P</u>HP <u>H</u>ypertext <u>P</u>rocessor

❖ Mathematics

  ➢ In classical algebra, the practice of Induction comprises
  - Looking at an equation for one variable
  - Then replacing the variable with the original function
  - Then replacing the function with the newly formed function...and so on ...
  ➢ Chain Rule: Function of a Function...
  - Computer Science: Functions call functions
  - Using Functors (Cybernetics in C++ - Expert), functionality can be the input of another functionality
  ➢ Iterative Functions:
  - These comprise writing an iterative equation, of a variable in terms of its previous value(s)
  - When you want to use iteration to find the solution of a problem

❖ Business/Entrepreneurship

  ➢ You have to spend money in order to make money

❖ Business/Enterprises:

  ➢ Price comparison of price comparison services

❖ Chemical Engineering

  ➢ Contact Process: Sulphuric Acid is produced using Sulphuric Acid
  ➢ Haber Process: Ammonia is produced using Ammonia

❖ Celebrity Status: Being famous for being famous

❖ Recursive Tasks

> ➤ Recursive Tasks involve applying themselves
> ➤ There is often an extra awkward challenge involved
> ➤ Examples include:
>> • Writing a report on how to write reports
>> • Giving a presentation on how to give presentations
>> • Thinking about thinking
>> • Using a web browser to download a web browser
>> • Building a software IDE i.e. writing software for writing software

❖ The prefix "meta" is often used to denote a recursive entity

> ➤ Meta-analysis: An analysis of an Analysis(es)
> ➤ Metadata: Data about Data
> ➤ Metastatistics: Statistics about statistics
>> • Central Limit Theorem: The distribution of sample means will always be Normal/Gaussian

❖ Time Management in software and IT is a recursive problem:

Project work will take longer than you think, even after you have taken into account that it will take longer than you think! :-)

Chris Brown

❖ Science & Cosmology

The most incomphensible thing abut the Universe is that it is comprehensible

Albert Einstein

> ➤ Is the difference between Theory and Practice greater in Practice than in Theory?
> ➤ Does the rule "Every Rule has some Exceptions" have some Exceptions?
> ➤ Russell's Paradox focuses on Sets of Sets
>> • Let R = the set of all sets that are not members of themselves.
>> • If R is not a member of itself,
>>> ○ its definition dictates that it must contain itself,
>> • If it contains itself
>>> ○ It contradicts its own definition as the set of all sets that are not members of themselves

➢ Is the idea "Theories do not have to be correct in order to be useful," a theory which might not be correct?

➢ Do the Laws of Evolution themselves evolve?

➢ Is the statistic "62.8% of statistics are made up" made up?

➢ Since the study of consciousness (like quantum science) is mostly counter-intuitive

- Does the Study of Common Sense require you to abandon Common Sense?

➢ Perfectionism

- Since Perfect is by definition an unreachable standard, is a perfect objective one that does not aim for perfection?

➢ Political/War Statements

- "Freedom is not Free"
- "Peace cannot be attained using only peaceful means"

➢ Communications: Is saying "Words cannot describe how I feel" using words to describe a feeling, albeit a strong one

❖ Government & Politics

➢ Policies of policy making

➢ 2015 UK General Election polls showed that polls can be highly unreliable

❖ Anecdotes/Comical Ironies

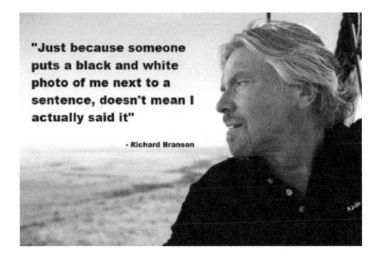

EXERCISE

❖ *This list can be extended almost indefinitely. See if you can come up with 3 further examples of Recursion.*

8.1.2 RECURSION INSIDE THE PROGRAMMING WORLD

❖ In procedural programming, recursion comprises

➢ Writing a function, the body of which calls itself

❖ It is useful for algorithms on certain data structures (covered in detail in Cybernetics in C++ - Expert)

➢ Tree data structures are made up of "sub-trees", which in turn are made up of "sub-sub trees"...

• The same algorithm, such as search or deletion, is applied at one scale, then another

➢ Sorting algorithms, such as quicksort and mergesort, use recursion

• They divide a data structure into component parts, then sort these out
• These sorted parts are then assembled into further sorted parts, continuing until the entire data structure is sorted

❖ Digital Signal Processing

➢ Digital Filters use this principal; they are very widely used, examples include:

• Cancelling out noise in communication systems
• Deciphering between foetal and maternal heartbeats

❖ **Metaprgramming**

➢ The practice of developing a program which creates or manipulates another program

• The bulk of C compilers, are themselves, at least in part, C programs

8.2 CAUTION WITH RECURSION

❖ While recursion is sometimes essential, it can be risky

❖ Note that the base case is essential

❖ If the function enters a case when a base case has not been defined

➢ It will carry on calling itself indefinitely

➢ This will lead to runtime stack overflow (section 14.2)

  • i.e. the program will crash

❖ **Rule of Thumb:** Use recursion only when essential, avoid it whenever possible

8.3 COMPUTING FACTORIALS

❖ In mathematics, the factorial of a positive integer is found by multiplying it by all integers between itself and 1

➢ 1!		1
➢ 2!	2 x 1	2
➢ 3!	3 x 2 x 1	6
➢ 5!	5 x 4 x 3 x 2 x 1	24
➢ 10!	10 x 9 x 8 x 7 x 6 x 5 x 4 x 3 x 2 x 1	3,628,800

❖ In order for this method to be valid, the following (counter intuitive! :-)) base case is defined:

0! = 1

➢ Without this, it could be argued that all factorials would end in multiplication by 0, rendering the entire concept invalid

8.3.1 ITERATIVE ALGORTHM: USING LOOPS

❖ Since this is a repeated action, where the number of iterations is known, a for loop can be used:

```
/*      *************************************************************
        FACTORIAL - using Loops
        *************************************************************
*/
#include <iostream>
using namespace std;


int calc_Factorial( int );


int main()
{
        //»»»»»»»»»»»»»»»»»»»»»»»»»»»»»»»»»»»»»»»»»»»»»»»»»»»»»»»»»»»»»»»»»
        cout    << "0! = "<< calc_Factorial(0)<< endl;
        cout    << "1! = "<< calc_Factorial(1)<< endl;
        cout    << "2! = "<< calc_Factorial(2)<< endl;
        cout    << "3! = "<< calc_Factorial(3)<< endl;
```

```
        cout    << "5! = "<< calc_Factorial(5)<< endl;
        cout    << "10! = "<< calc_Factorial(10)<< endl;

        //»»»»»»»»»»»»»»»»»»»»»»»»»»»»»»»»»»»»»»»»»»»»»»»»»»»»»»»»»»»
        cout    << endl<< endl<< "==========="
                << endl
                << "END...Demo, Computing Factorials using Loops"
                << " "
                << endl;
        while(true){}
        return 0;

}//END...Function MAIN


int calc_Factorial( int int_IN_A )
{
        int int_IN_A_factorial = 1;

        if( int_IN_A != 0 ) // Base Case 0! = 1
        {
                // All other cases
                for( int i = int_IN_A; i > 0; --i )
                {
                        //cout << i<< endl;
                        int_IN_A_factorial *= i;
                }
        }
        return (int_IN_A_factorial);


}//END...Function calc_Factorial
```

```
0!  = 1
1!  = 1
2!  = 2
3!  = 6
5!  = 120
10! = 3628800


===========
END...Demo, Computing Factorials using Loops
```

Program Listing/Output 8.1: Factorial_Loop_Client.cpp - Computing a factorial using a for loop

8.3.2 USING RECURSION

❖ There is an alternative way to think about Factorials, in terms of Recursion

❖ Looking at the example of 10! :

> ➢ 10! = 10 x 9 x 8 x 7 x 6 x 5 x 4 x 3 x 2 x 1
> ➢ However 9 x 8 x 7 x 6 x 5 x 4 x 3 x 2 x 1 = 9!
> ➢ Thus 10! = 10 x 9!
> ➢ But what is 9! ?....
>> • 9! = 9 x 8 x 7 x 6 x 5 x 4 x 3 x 2 x 1
>> • i.e. 9 x 8!
> ➢ Summarising:
>> • 10!   = 10 x 9!
>>   = 10 x 9 x 8!
>>   = 10 x 9 x 8 x 7!
>>   ...
>>   = 10 x 9 x 8 x 7 x 6 x 5 x 4 x 3 x 2 x 1!
>> • **Base Case**: The procedure must stop when we get down to 1
> ➢ Since this is true for any positive integer, the following recursive algorithm can be assembled for computing factorials:
>> • n! = n x (n-1)!
>>   = n x (n-1) x (n-2)!
>>   = n x (n-1) x (n-2) x (n-3)!
>>   ...
>>   = n x (n-1) x (n-2) x (n-3) x (n-4) x... x 2 x 1!
>>
>> • **Base Case**: In the general case as well, the procedure must stop when we get down to 1

❖ In order to implement this, we require a Recursive Function which:

> ➢ Checks for the base case at each iteration
> ➢ If not, it
>> • Calls itself with the input being 1 less than the last input
>> • Multiplies the result of this with the current product:

```
int calc_Factorial( int int_IN_A )
{
        if( int_IN_A == 0 ) // Base Case
        {
```

```
                 return (1);
        }
        else
        {
                 return( int_IN_A * calc_Factorial( int_IN_A - 1 ) );
        }
}//END...Function calc_Factorial
```

Program Listing/Output 8.2: calc_Factorial: Function to Calculate Factorials
RECURSIVELY

❖ The following program shows this in action:

```
/*     **************************************************************
       FACTORIAL - using Recursion
       **************************************************************
*/
#include <iostream>
using namespace std;

int calc_Factorial( int );


int main()
{
       //»»»»»»»»»»»»»»»»»»»»»»»»»»»»»»»»»»»»»»»»»»»»»»»»»»»»»»»»»»
       cout   << "0! = "<< calc_Factorial(0)<< endl;
       cout   << "1! = "<< calc_Factorial(1)<< endl;
       cout   << "2! = "<< calc_Factorial(2)<< endl;
       cout   << "3! = "<< calc_Factorial(3)<< endl;
       cout   << "5! = "<< calc_Factorial(5)<< endl;
       cout   << "10! = "<< calc_Factorial(10)<< endl;


       //»»»»»»»»»»»»»»»»»»»»»»»»»»»»»»»»»»»»»»»»»»»»»»»»»»»»»»»»»»
       cout   << endl<< endl<< "==========="
              << endl
              << "END...Demo, Computing Factorials using Recursion"
              << " "
              << endl;
       while(true){}
       return 0;

}//END...Function MAIN


int calc_Factorial( int int_IN_A )
{
       if( int_IN_A == 0 )
       {
              return (1);
       }
```

```
        else
        {
                return( int_IN_A * calc_Factorial( int_IN_A - 1 ) );
        }
}//END...Function calc_Factorial
```

```
0!  = 1
1!  = 1
2!  = 2
3!  = 6
5!  = 120
10! = 3628800


===========
END...Demo, Computing Factorials using Recursion
```

Program Listing/Output 8.3: Factorial_Recursion_Client.cpp - Computing a factorial using Recursion

8.4 FIBONACCI SEQUENCE

❖ The Fibonacci Sequence is a famous series which comprises adding the number before, starting with two 1's:

```
1 1 2 3 5 8 13 21 34 55 89
```

➢ This was first put together by Italian mathematician Fibonacci to analyse the growth of a rabbit population,

❖ Recursion can be used here since

➢ Each term is the sum of the previous two terms
➢ Base Cases:
   • The first term is 1
   • The second term is also 1

❖ The algorithm is as follows

➢ Fibonacci(n) = Fibonacci(n-1) + Fibonacci(n-2)
➢ Fibonacci(0) = 1
➢ Fibonacci(1) = 1

❖ This can be implemented as follows:

```
/*       ************************************************************
         FIBONACCI SERIES using RECURSION
         ************************************************************
         -      This program implements the famous Fibonacci Sequence:
                Each term is the sum of the previous two
*/
#include <iostream>
#include <string>
using namespace std;

long int generate_Fibonacci_Number( int );

int main( )
{
        //»»»»»»»»»»»»»»»»»»»»»»»»»»»»»»»»»»»»»»»»»»»»»»»»»»»»»»»»»»»»»»»»»»»»
        cout << "=============================================="<< endl
             << "Generating terms of the Fibonacci Sequence..."<< endl
             << "=============================================="<< endl
             << endl;
        //»»»»»»»»»»»»»»»»»»»»»»»»»»»»»»»»»»»»»»»»»»»»»»»»»»»»»»»»»»»»»»»»»»»»
        //         Displaying the First 11 Fibonacci terms
        //
        for( int i = 0; i < 11; ++i )
        {
                cout   << "fibonacci( "<< i<< " ) = "
                       << generate_Fibonacci_Number (i)<< endl;
        }

        //»»»»»»»»»»»»»»»»»»»»»»»»»»»»»»»»»»»»»»»»»»»»»»»»»»»»»»»»»»»»»»»»»»»»
        //      Displaying higher Fibonacci terms
        //
        cout  << "fibonacci( 15 ) = "<< generate_Fibonacci_Number( 15 )
              << endl;
        cout  << "fibonacci( 20 ) = "<< generate_Fibonacci_Number( 20 )
              << endl;
        cout  << "fibonacci( 30 ) = "<< generate_Fibonacci_Number( 30 )
              << endl;
        cout  << "fibonacci( 35 ) = "<< generate_Fibonacci_Number( 35 )
              << endl;

        //»»»»»»»»»»»»»»»»»»»»»»»»»»»»»»»»»»»»»»»»»»»»»»»»»»»»»»»»»»»»»»»»»»»»
        cout  << endl<< endl<< "============"
              << endl
              << "END...C++ Demo: Fibonacci Series using Recursion"
              << " "
              << endl;
        while(true){}
        return 0;

}//END...Function MAIN


/*++++++++++++++++++++++++++++++++++++++++++++++++++++++++++++++++++++
       FUNCTION: generate_Fibonacci_Number
       -       Returns the term of the Fibonacci Sequence for that
               particular index
```

```
|       -       Input: Index
*/
long int generate_Fibonacci_Number( int myINT_IN )
{
        //»»»»»»»»»»»»»»»»»»»»»»»»»»»»»»»»»»»»»»»»»»»»»»»»»»»»»»»»»»»»»»»»»»
        if( ( myINT_IN == 0 ) || ( myINT_IN == 1 ) )
        {
                return myINT_IN;
        }
        //»»»»»»»»»»»»»»»»»»»»»»»»»»»»»»»»»»»»»»»»»»»»»»»»»»»»»»»»»»»»»»»»»»
        else
        {
                return(       generate_Fibonacci_Number( myINT_IN - 1 ) +
                              generate_Fibonacci_Number( myINT_IN - 2 )
                        );
        }
}//END...generate_Fibonacci_Number
```

```
================================================
Generating terms of the Fibonacci Sequence...
================================================

fibonacci( 0 )  = 0
fibonacci( 1 )  = 1
fibonacci( 2 )  = 1
fibonacci( 3 )  = 2
fibonacci( 4 )  = 3
fibonacci( 5 )  = 5
fibonacci( 6 )  = 8
fibonacci( 7 )  = 13
fibonacci( 8 )  = 21
fibonacci( 9 )  = 34
fibonacci( 10 ) = 55
fibonacci( 15 ) = 610
fibonacci( 20 ) = 6765
fibonacci( 30 ) = 832040
fibonacci( 35 ) = 9227465


===========
END...C++ Demo: Fibonacci Series using Recursion
```

Program Listing/Output 8.4: Fibonacci_Recursion_Client.cpp

❖ Note that the execution time computational complexity rapidly rises as the input number goes up

➢ Even high specification computers will require several minutes to compute `Fibonacci(45)`

• Something like `Fibonacci(50)` might cause the program to crash

8.4.1 APPLICATIONS OF THE FIBONACCI SEQUENCE

❖ It has found a wide range of applications since its discovery

❖ It has been used for analysing populations of most breeding animals, including humans

❖ The ratio between subsequent terms converges to the Golden Ratio:

➢ Golden Ratio $\varphi = \frac{1+\sqrt{5}}{2}$ = 1.61803398874989...(irrational)

➢ It is believed to be aesthetically pleasing

➢ Architects have been using it as the ratio between the width and height of buildings for centuries

   • The Parthenon in Athens, along with a lot of other Greek buildings, is known to use this

➢ In music, it is hypothesised (and disputed) that certain great pieces use the Golden Ratio, for things like the gap between sequences of keys and boundaries

   • Works by Le Mer and Chopin have been suggested to use this

8.5 GENERATING PERMUTATIONS

❖ Permutations are ways of arranging items where order **does** matter

❖ Combinations are ways of arranging items where order **does not** matter

❖ Consider the set of characters

```
S A T R
```

➢ The following

```
RATS STAR TARS ARTS
```

are considered to be
   • The same combination of letters
   • Different permutations of letters

❖ For N distinct items, there are N! possible permutations

➢ For every repetition, N! must be divided by the factorial of the size each of each repeated set

❖ In order to generate the possible permutaions

➢ The Iterative Approach is to use Brute Force i.e. list every possible combination

➢ An alternative is the Recursive approach

8.5.1 RECURSIVE

❖ The Recursive Approach, to find all Permutations, for a word of size N, is as follows

➢ Find the all Permutations of the rightmost N-1 characters
➢ Rotate all N characters
- This means
  ○ Shift all characters, except the leftmost, one position left
  ○ Move the leftmost character to the end
    ▪ i.e. It "rotates" to the right
➢ Repeat this procedure N times

❖ The question now: How do find all Permutations of the rightmost N-1 characters?

➢ Carry out this procedure for the N-1 characters
➢ This will comprise finding all permutations of the rightmost N-2 characters
- This in turn will comprise finding all permutations of the rightmost N-3 characters
➢ Thus Recursion is being employed
➢ The Base case: Set of characters, size 1
- The character itself is "All Permutations of this string"

8.5.2 THE NEXT PERMUTATION ALGORITHM

❖ The following program shows this in action
❖ Note that this uses

➢ A functionality called `next_permutation`
- This is from the C++ Standard Library (covered in Cybernetics in C++ - Expert)
➢ Arrays, which will be covered in section 9 onwards

```
/*    ************************************************************
      GENERATING PERMUTATIONS
      ************************************************************
      -      This program generates the 4! (=24) permutations of
             C A T S
      -      Uses the next_permutation functionality
             -      Found in the Standard Library
*/
```

```cpp
#include <iostream>
#include <string>
#include <algorithm>
using namespace std;

int main( )
{
    //»»»»»»»»»»»»»»»»»»»»»»»»»»»»»»»»»»»»»»»»»»»»»»»»»»»»»»»»»»»»»»»»»»
    cout    << "===================="<< endl
            << "STRING PERMUTATIONS" << endl
            << "===================="<< endl
            << endl;

    char myStr_A[] = {'c', 'a', 't', 's'};
    int length_myStr_A = sizeof(myStr_A) / sizeof(char);
    cout << "length_myStr_A = "<< length_myStr_A<< endl;

    cout << "The 4! possible permutations of C A T S:"<< endl ;
    do
    {
    cout << myStr_A[0]<< ' '
         << myStr_A[1]<< ' '
         << myStr_A[2]<< ' '
         << myStr_A[3]<< ' '
         << endl;
    }
    while ( next_permutation( myStr_A, myStr_A + length_myStr_A));

    //»»»»»»»»»»»»»»»»»»»»»»»»»»»»»»»»»»»»»»»»»»»»»»»»»»»»»»»»»»»»»»»»»»
    cout << endl<< endl<< "==========="
         << endl
         << "END...C++ Demo: Next Permutation Algorithm"
         << " "
         << endl;
    while(true){}
    return 0;

}//END...Function MAIN
```

```
====================
STRING PERMUTATIONS
====================

length_myStr_A = 4
The 4! possible permutations of C A T S:
c a t s
c s a t
c s t a
c t a s
c t s a
s a c t
s a t c
```

```
s  c  a  t
s  c  t  a
s  t  a  c
s  t  c  a
t  a  c  s
t  a  s  c
t  c  a  s
t  c  s  a
t  s  a  c
t  s  c  a


===========
END...C++ Demo: Next Permutation Algorithm
```

Program Listing/Output 8.5: Permutation_Generator_Client.cpp - using the next_
permutation algorithm

8.6 HANOI TOWERS

❖ An Ancient puzzle, publicised in the West by French mathematician Édouard Lucas in 1883, the Hanoi Towers puzzle has been a favourite children's (and adult's!:-)) game since

❖ It comprises 3 towers, one of them has a stack of rings, the widest at the bottom, getting progressively narrower going up

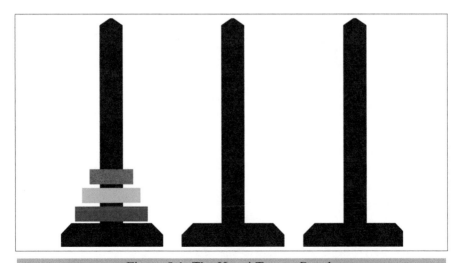

Figure 8.1: The Hanoi Towers Puzzle

❖ The objective is to transfer all the rings to another tower

> ➤ The rings can be moved only one at a time
> ➤ Only a ring at the top of a tower can be moved
> ➤ A wider ring cannot be placed on top of a narrower one

❖ One of the most common ways to solve this problem is using recursion

> ➤ The fundamental strategies remain unchanged no matter how many rings have to be moved i.e.
> ➤ See how to solve this problem using the minimal case of 3 rings
> ➤ Apply the same techniques for any larger number

❖ In the following examples:

> ➤ Ring 1 is the narrowest at the top
> ➤ The rings get wider as the numbers increase

8.6.1 STRATEGY FOR 3 RINGS

❖ Start: Rings 3, 2, 1 on Tower A

	RING	FROM TOWER	TO TOWER	TOWER (BOTTOM UP) A	B	C
				3, 2, 1		
1.	1	A	B	3, 2	1	
2.	2	A	C	3	1	2
3.	1	B	C	3		2, 1
4.	3	A	B		3	2, 1
5.	1	C	A	1	3	2
6.	2	C	B	1	3, 2	
7.	1	A	B		3, 2, 1	

8.6.2 STRATEGY FOR 4 RINGS

❖ Start: Rings 4, 3, 2, 1 on Tower A

	RING	From TOWER	TO TOWER	TOWER (BOTTOM UP)		
				A	B	C
				4, 3, 2, 1		
1.	1	A	B	4, 3, 2	1	
2.	2	A	C	4, 3	1	2
3.	1	B	C	4, 3		2, 1
4.	3	A	B	4	3	2, 1
5.	1	C	A	4, 1	3	2
6.	2	C	B	4, 1	3, 2	
7.	1	A	B	4	3, 2, 1	
8.	4	A	C		3, 2, 1	4
9.	1	B	C		3, 2	4, 1
10.	2	B	A	2	3	4,1
11.	1	C	A	2, 1	3	4
12.	3	B	C	2, 1		4, 3
13.	1	A	B	2	1	4, 3
14.	2	A	C		1	4, 3, 2
15.	1	B	C			4, 3, 2, 1

8.6.3 ITERATIVE STRATEGIES

❖ There are slightly different procedures depending on whether you have an odd or even number of disks

  ➢ For an odd number of disks

   • Start by moving the top disk to the destination tower

  ➢ For an even number of disks

   • Start by moving the top disk to the "non-destination" tower

❖ In subsequent steps

  ➢ Move the topmost ring to an empty or part filled tower

  ➢ Move the narrowest on top of this

  ➢ Move the 3rd narrowest to the vacant tower

❖ Either way, for n disks, the minimum number of moves is $2^n - 1$

8.6.4 RECURSIVE STRATEGIES

❖ There is a way to look at Hanoi Towers using Recursion

❖ Think of the stack of disks on top of the bottom one as a "sub-stack"

➤ Move this "sub-stack" to a vacant tower

➤ Now move the bottom disk to the vacant tower

➤ Now move the "sub-stack" on top of the widest disk to complete the move

❖ Clearly, it is prohibited to move more than one disk at a time, so now we focus on moving the sub-stack

➤ We look at the "sub-sub-stack" sitting on top of the bottom ring

➤ We then move this one accordingly

❖ This strategy thus exhibits recursion

➤ We move "sub-stacks" of towers, by moving sub-stacks within each sub-stack

➤ We keep on doing this till we reach the

➤ Base Case: Sub-stack of size 1

8.6.5 IMPLEMENTING RECURSION IN HANOI TOWERS

❖ The following program shows how the Recursive algorithm can be implemented in C++:

```
/*      ***********************************************************
        Hanoi Towers
        ***********************************************************
        -       Demonstrates using Recursion for Hanoi Towers
*/
#include <iostream>
#include <string>
using namespace std;

//+++++++++++++++++++++++++++++++++++++++++++++++++++++++++++++++++++
void operate_HanoiTowers( int ,char, char, char );

//+++++++++++++++++++++++++++++++++++++++++++++++++++++++++++++++++++
int main( )
{
        //»»»»»»»»»»»»»»»»»»»»»»»»»»»»»»»»»»»»»»»»»»»»»»»»»»»»»»»»»»»»»
        int num_Disks = 0;

        cout << "******* 3 Disks *******"<< endl;
        num_Disks = 3;
        operate_HanoiTowers( num_Disks, 'A', 'B', 'C' );

        cout << endl<< "******* 4 Disks *******"<< endl;
        num_Disks = 4;
        operate_HanoiTowers( num_Disks, 'A', 'B', 'C' );
```

```cpp
        cout << endl<< "****** 6 Disks ******"<< endl;
        num_Disks = 6;
        operate_HanoiTowers( num_Disks, 'A', 'B', 'C' );


        //»»»»»»»»»»»»»»»»»»»»»»»»»»»»»»»»»»»»»»»»»»»»»»»»»»»»»»»»»»»»»»»
        cout   << endl<< endl<< "==========="
               << endl
               << "END...C++ Demo: Hanoi Towers"
               << " "
               << endl;
        while(true){}
        return 0;

}//END...Function MAIN

//+++++++++++++++++++++++++++++++++++++++++++++++++++++++++++++++++++
void operate_HanoiTowers( int top,
                          char tower_From,
                          char tower_Intermediate,
                          char tower_To
                          )
{
    //»»»»»»»»»»»»»»»»»»»»»»»»»»»»»»»»»»»»»»»»»»»»»»»»»»»»»»»»»»»»»»»
    if( top == 1 )        // Base Case: 1 Disk
    {
        cout << "Disk 1 from "<< tower_From
             << " to "<< tower_To
             << endl;
    }
    //»»»»»»»»»»»»»»»»»»»»»»»»»»»»»»»»»»»»»»»»»»»»»»»»»»»»»»»»»»»»»»»
    else
    {
        // Move Sub Tower
        operate_HanoiTowers(   ( top - 1 ),
        tower_Intermediate,
                               tower_From,
                               tower_To
                          );
        cout << "Disk "<< top
             << " from "<< tower_From
             << " to "<< tower_To
             << endl;

        // Operating the Sub Tower
        operate_HanoiTowers(   ( top - 1 ),
                               tower_Intermediate,
                               tower_From,
                               tower_To
                          );
    }
}//END...Function operate_HanoiTowers
```

```
******* 3 Disks *******
Disk 1 from Tower A to Tower C
Disk 2 from Tower B to Tower C
Disk 1 from Tower A to Tower C
Disk 3 from Tower A to Tower C
Disk 1 from Tower A to Tower C
Disk 2 from Tower B to Tower C
Disk 1 from Tower A to Tower C

******* 4 Disks *******
Disk 1 from Tower B to Tower C
Disk 2 from Tower A to Tower C
Disk 1 from Tower B to Tower C
Disk 3 from Tower B to Tower C
Disk 1 from Tower B to Tower C
Disk 2 from Tower A to Tower C
Disk 1 from Tower B to Tower C
Disk 4 from Tower A to Tower C
Disk 1 from Tower B to Tower C
Disk 2 from Tower A to Tower C
Disk 1 from Tower B to Tower C
Disk 3 from Tower B to Tower C
Disk 1 from Tower B to Tower C
Disk 2 from Tower A to Tower C
Disk 1 from Tower B to Tower C

******* 6 Disks *******
Disk 1 from Tower B to Tower C
Disk 2 from Tower A to Tower C
Disk 1 from Tower B to Tower C
Disk 3 from Tower B to Tower C
Disk 1 from Tower B to Tower C
Disk 2 from Tower A to Tower C
Disk 1 from Tower B to Tower C
Disk 4 from Tower A to Tower C
Disk 1 from Tower B to Tower C
Disk 2 from Tower A to Tower C
Disk 1 from Tower B to Tower C
Disk 3 from Tower B to Tower C
Disk 1 from Tower B to Tower C
Disk 2 from Tower A to Tower C
Disk 1 from Tower B to Tower C
Disk 5 from Tower B to Tower C
Disk 1 from Tower B to Tower C
Disk 2 from Tower A to Tower C
Disk 1 from Tower B to Tower C
Disk 3 from Tower B to Tower C
Disk 1 from Tower B to Tower C
Disk 2 from Tower A to Tower C
Disk 1 from Tower B to Tower C
```

```
Disk 4 from Tower A to Tower C
Disk 1 from Tower B to Tower C
Disk 2 from Tower A to Tower C
Disk 1 from Tower B to Tower C
Disk 3 from Tower B to Tower C
Disk 1 from Tower B to Tower C
Disk 2 from Tower A to Tower C
Disk 1 from Tower B to Tower C
Disk 6 from Tower A to Tower C
Disk 1 from Tower B to Tower C
Disk 2 from Tower A to Tower C
Disk 1 from Tower B to Tower C
Disk 3 from Tower B to Tower C
Disk 1 from Tower B to Tower C
Disk 2 from Tower A to Tower C
Disk 1 from Tower B to Tower C
Disk 4 from Tower A to Tower C
Disk 1 from Tower B to Tower C
Disk 2 from Tower A to Tower C
Disk 1 from Tower B to Tower C
Disk 3 from Tower B to Tower C
Disk 1 from Tower B to Tower C
Disk 2 from Tower A to Tower C
Disk 1 from Tower B to Tower C
Disk 5 from Tower B to Tower C
Disk 1 from Tower B to Tower C
Disk 2 from Tower A to Tower C
Disk 1 from Tower B to Tower C
Disk 3 from Tower B to Tower C
Disk 1 from Tower B to Tower C
Disk 2 from Tower A to Tower C
Disk 1 from Tower B to Tower C
Disk 4 from Tower A to Tower C
Disk 1 from Tower B to Tower C
Disk 2 from Tower A to Tower C
Disk 1 from Tower B to Tower C
Disk 3 from Tower B to Tower C
Disk 1 from Tower B to Tower C
Disk 2 from Tower A to Tower C
Disk 1 from Tower B to Tower C


===========
END...C++ Demo: Hanoi Towers
```

Program Listing/Output 8.6: HanoiTowers_Client.cpp

8.6.6 TRADITIONAL LEGENDS OF HANOI TOWERS

❖ There are stories that there are 3 Hanoi towers, one of them with 64 gold
   rings

❖ When a group of workers have moved all these rings to another tower, the world will end!

❖ The stories do vary:
- ➢ Some say the venue is a temple, others say a monastery
- ➢ Some say the towers are studded with diamonds
- ➢ Several locations are cited, the most common being in
  - • India, especially Kashi Vishwanath
  - • Vietnam
- ➢ Some say the workers are monks, others say priests

❖ Even if we assume the story is true, there is no need to panic...for quite some time yet!:-)
- ➢ The minimum number of moves required is $2^{64} - 1$
- ➢ Even if we assume the workers take 1 second to move a ring...
  - • which is a sizeable estimate given that the rings are made of gold,
- ➢ ...The full move will take them ∼585 billion years!
- ➢ Most estimates for the age of Planet Earth are ∼4.6 billion years

8.7 FURTHER RECURSIVE PROBLEMS

❖ Several further world class problems can be tackled using Recursion

8.7.1 TRIANGULAR NUMBERS

❖ Triangular Numbers are ones which are generated by adding 1 more than the previous addition

❖ The first Triangular Numbers are
- ➢ 1
- ➢ 3   1 + 2
- ➢ 6   1 + 2 + 3
- ➢ 10   1 + 2 + 3 + 4
- ➢ 15   1 + 2 + 3 + 4 + 5
- ➢ ...

The Recursive Approach is to

- ➢ Add to the previous number
  - • 1 more than the previous addition
- ➢ Until you reach the Base Case of 1

8.7.2 KNAPSACK/RUCKSACK PROBLEM

❖ This comprises investigating different ways of fitting articles of different sizes into a container of a fixed size

➢ You have a fixed number of items, each with a fixed value

➢ You have a container, a Rucksack/Knapsack, of limited size

➢ What quantities of each item, if any, should be placed in the container, in order to maximise the total content value?

❖ The Recursive Approach is to look at the problem on different scales

❖ Initial Case:

➢ We have

• 1 Item
• 1 Knapsack

➢ There are thus 2 possible decisions

• We place the Item in the Knapsack
  ○ This will mean
    ■ We get a certain total value in the knapsack
    ■ We have lost some capacity in the knapsack
• We do not place the Item in the Knapsack
  ○ This will mean
    ■ We have zero total value in the knapsack
    ■ We have lost no capacity in the knapsack

❖ We can now work on this for further cases

➢ The scales will be different, we thus adjust the

• Weight
• Size of knapsack

➢ Until we reach the Base Case

• The remaining space in the Knapsack is too small for any object

8.7.3 N QUEENS PROBLEM

❖ In Chess

➢ On a standard 8 x8 board

➢ A maximum of 8 Queens can be placed so that none of them are attacking any of the others

❖ How do you place N Queens on an N x N board so that none of them are in the line of attack of any of the others?"

❖ The Recursive strategy comprises

➢ Placing 1 Queen,

  • Then looking at the squares which this one prohibits

➢ Placing another Queen in one of the allowed Squares

  • Then looking at the squares which this one prohibits

➢ Repeating procedure until the Base Case of 1 remaining allowed square is reached

DATA STRUCTURES & ALGORITHMS

❖ Data Structures and Algorithms are often described as the two fundamental building blocks of any computer language

➢ This is directly analogous to vocabulary and grammar being the two fundamental building blocks of any human language

❖ One will not get very far

➢ In a human language using only individual words,

● Sentences, paragraphs and further word structures will be required

➢ In a computer language, using individual data items

● Data Structures are required

❖ The array is the single most commonly used data structure in virtually all programming languages

➢ No further libraries need to be included in order to use it

➢ As we'll see in Cybernetics in C++ - Expert, when further data structures need to be used, libraries need to be defined, 2 of the most popular being

● STL Standard Template Library
● Boost

9. ARRAYS

9.1 WHAT ARE ARRAYS?

❖ The Array is the single most commonly used data structure

❖ It is used to hold a collection of data items

➢ e.g. a single entity, holding the cost of a ticket might defined as

```
int cost_Admission
```

➢ If several ticket costs need to be stored, a collection of such data might be defined as follows

```
int cost_Admission[5]
```

- This will mean there will be 5 data items, each of which represents a cost

❖ Arrays are

➢ Homogeneous
➢ Indexed from 0 to N-1
- N = the total number of elements in the Array
➢ Data Structures which
- Use square brackets [] as operators
- Have the data items stored in consecutive runtime memory locations

9.1.1 HOMOGENEOUS NATURE OF DATA STRUCTURES

❖ **Arrays are homogeneous** i.e. very data item must be of the same type

➢ This is the case for all data structures
- This includes linked lists, tress, graphs (covered in Cybernetics in C++ - Expert)
➢ i.e. We cannot have an array where the first element is of type int, the 2nd of type double, the third is an ADT type such as struct Employee (section 17)
➢ All elements must be of type int, or double or Employee...

9.1.2 INDEXING IN ARRAYS

❖ Arrays elements are numbered from 0 to N-1, where N is the total size; thus the

➢ 1st element is at index 0
➢ 2nd element is at index 1
➢ 3rd element is at index 2...
➢ ...m^{th} element is at index (m-1)

❖ The idea is that the index will represent the offset from the beginning
❖ The analogy with UK buildings is often used:

	UK BUILDING	ARRAY
➢	The floor at street level, the first one people entering the building encounter is the **ground** floor	The first element in an array is of index **0**
➢	The floor number is a measure of how many staircases need to be ascended	The index is a measure of the offset from the first element
➢	The top floor in an N storey building is floor (N - 1)	The last index in an array of size N is element (N - 1)

❖ Note: Arrays in other languages

➢ This convention, of numbering from 0 to N-1 is followed by the bulk of languages,

• including C, Java, C#, F#, Swift

➢ Only a few languages use the 1 to N system

• The two most common examples are Pascal and MATLAB

9.1.3 THE ARRAY OPERATOR

❖ The compiler recognises arrays using the [ ] square brackets operator[1]

➢ For this reason, [ ] are often called "Array Brackets"

9.1.4 STORAGE IN RUNTIME MEMORY

❖ The elements of an array are stored in **consecutive** locations in runtime memory

❖ For this reason, arrays are part of the family known as "contiguous data structures"

❖ Thus when we have an array of ints, each of size 4 bytes

➢ If the 2nd element is at position 0x234567A

➢ The 3rd element will be at position 0x234567E

9.2 DEFINING ARRAYS

❖ Just like all individual variables need to be defined, an array needs to be defined at the beginning

---

[1]This is the most common operator. The pointer operator * can also be used; you will meet this in section 12, then using this with arrays is covered in detail in section 18.

❖ The fundamental syntax is as follows:

```
dataType arrayName[];
```

9.3 WRITING DATA TO AN ARRAY

❖ To write data to an array,

> The square brackets [] operator is used to specify the element

> The = operator is used to assign the value, just like primitive variables

```
arrayName[index] = newValue;
```

9.4 DEFINING AND INITIALISING ARRAYS

❖ There are primarily 2 ways to do this:

> Defining and initialising separately

> Defining and initialising on the same line

❖ Initialising the values of each element is especially important

> Failure to do so will mean the array will fill with garbage

> While several modern day systems, such as MS Visual Studio and QT initialise automatically, do not assume this will happen

9.4.1 DEFINING AND INITIALISING SEPARATELY

9.4.1.1 Stand Alone Definition

❖ The definition of an array is as follows:

```
dataType arrayName[ arraySize ];
```

❖ Examples:

> An array, type int, size 5, name sizeOfTeams

```
int sizeOfTeams[5];
```

> An array, type class Employee (section 18), size 10, name office

```
Employee office[10];
```

9.4.1.2 Initialisation

❖ The most common way to initialise an array is to use a for loop:

```
for( int i = 0; i < arraySize; ++i )
{
        arrayName[i] = Null value such as 0 or " ";
}
```

➢ Example:

```
for( int i = 0; i < 5; ++i )
{
        sizeOfTeams[i] = 0;
}
```

9.4.2 DEFINING AND INITIALISING ON THE SAME LINE

❖ When defining and initialising on the same line, the form is as follows:

➢ The = operator is used

➢ The definition, on the left hand side, still used the [] operators

➢ The initial values, on the right hand side, use the braces { } operators

```
dataType arrayName[] = { values, separated by commas };
```

❖ Example:

```
int stadiaCapacities[] = {44000, 32000, 250, 100000, 25000};
```

9.5 READING DATA FROM AN ARRAY

❖ Reading data from an array is identical in style to writing data to it

➢ The square brackets [] operator is used to specify the element

```
arrayName[index];
```

9.6 USING ARRAYS

❖ The following program shows how arrays are used in C++:

```
/*      ****************************************************************
        ARRAYS IN C++
        ****************************************************************
        -       This file demonstrates using Arrays in C++
        -       DEFINING an Array: 2 methods
        (1)     Define and Initialise separately
        (2)     Define and Initialise on the same line
        -       Writing Data to an Array
        -       Reading Data from an Array
*/
#include <iostream>
#include <string>

using namespace std;

int main( )
{
        //»»»»»»»»»»»»»»»»»»»»»»»»»»»»»»»»»»»»»»»»»»»»»»»»»»»»»»»»»»»»»»»»»
```

```
      // METHOD (1) Define and Initialise separately
      int myArrInt_A[4]; // Definition
      for( int i = 0; i < 4; ++i ) // Initialising using a for loop
      {
            myArrInt_A[i] = 0;
      }
            // Cowardly method!:-) Far better to use STL Iterators

      //»»»»»»»»»»»»»»»»»»»»»»»»»»»»»»»»»»»»»»»»»»»»»»»»»»»»»»»»»»»»»»»»»»»
      // METHOD (2) Define and Initialise on the same line
      string arrStr_Greeting[] = { "Hello,", "how", "are", "you", "?" };


      //»»»»»»»»»»»»»»»»»»»»»»»»»»»»»»»»»»»»»»»»»»»»»»»»»»»»»»»»»»»»»»»»»»»
      // WRITING Data to the Array - use []
      //
      myArrInt_A[3] = 25;
      myArrInt_A[1] = 4;
      myArrInt_A[2] = 12;
      myArrInt_A[0] = 27;

      //»»»»»»»»»»»»»»»»»»»»»»»»»»»»»»»»»»»»»»»»»»»»»»»»»»»»»»»»»»»»»»»»»»»
      //    READING Data from the Array - again use []
      //
      cout  << "The 3rd element of myArrInt_A is "
            << myArrInt_A[2]<< endl;

      //===========================================================
      cout  << "Contents of myArrInt_A: [ ";
      for( int i = 0; i < 5; ++i)
      {
            cout << myArrInt_A[i]<< " ";
      }
      cout  << "]"<< endl;

      //===========================================================
      cout  << "Contents of arrStr_Greeting: [ ";
      for( int i = 0; i < 5; ++i)
      {
      cout << arrStr_Greeting[i]<< " ";
      }
      cout  << "]"<< endl;

      //»»»»»»»»»»»»»»»»»»»»»»»»»»»»»»»»»»»»»»»»»»»»»»»»»»»»»»»»»»»»»»»»»»»
      cout  << endl<< endl<< "==========="
            << endl
            << "END...C++ Demo: Arrays"
            << " "
            << endl;
      while(true){}
      return 0;

}//END...Function MAIN
```

```
The 3rd element of myArrInt_A is 12
Contents of myArrInt_A: [ 27 4 12 25 -858993460 ]
Contents of arrStr_Greeting: [ Hello, how are you ? ]
```

```
===========
END...C++ Demo: Arrays
```

Program Listing/Output 9.1: Arrays_Demo_Client.cpp

EXERCISE

❖ *Write a program which asks the user to input 10 numbers (integers will be fine), and will then display them in reverse order*

9.7 ARRAYS AS FUNCTION INPUTS

❖ Arrays can be passed as function inputs:

> ➤ The declaration of a function which takes an array as an input is as follows:

```
returnType arrayName( typeOfArray[] );
```

❖ The following program shows

> ➤ How an array can be input to a function
> ➤ The effect it has on the data being passed in:

```
/*      ************************************************************
        ARRAYS AS FUNCTION INPUTS
        ************************************************************
        -       The following program demonstrates defining and using a
                function which takes an array as its input
*/
#include <iostream>
#include <string>
using namespace std;

// Function DECLARATION
void set_FifthSixth_Elements( string[] );

int main( )
{
        //»»»»»»»»»»»»»»»»»»»»»»»»»»»»»»»»»»»»»»»»»»»»»»»»»»»»»»»»»»»»
        string arrStr_Greeting[] = { "Hello,", "how", "are", "you", "?",
                                " ", " " };
        cout   << "Contents of arrStr_Greeting: [ ";
        for( int i = 0; i < 7; ++i)
        {
                cout << arrStr_Greeting[i]<< " ";
        }
        cout   << "]"<< endl;

        //»»»»»»»»»»»»»»»»»»»»»»»»»»»»»»»»»»»»»»»»»»»»»»»»»»»»»»»»»»»»
```

```
        cout  << endl
              << "...Calling the function to set the 6th & 7th elements"
              << endl<< endl;

     set_FifthSixth_Elements( arrStr_Greeting );

     //»»»»»»»»»»»»»»»»»»»»»»»»»»»»»»»»»»»»»»»»»»»»»»»»»»»»»»»»»»»»»»»»»»»»»»»»»
     cout << "Contents of arrStr_Greeting: [ ";
     for( int i = 0; i < 7; ++i)
     {
           cout << arrStr_Greeting[i]<< " ";
     }
     cout << "]"<< endl;

     //»»»»»»»»»»»»»»»»»»»»»»»»»»»»»»»»»»»»»»»»»»»»»»»»»»»»»»»»»»»»»»»»»»»»»»»»»
     cout << endl<< endl<< "==========="
          << endl
          << "END...C++ Demo: Arrays as Function Inputs"
          << " "
          << endl;
     while(true){}
     return 0;

}//END...Function MAIN

//++++++++++++++++++++++++++++++++++++++++++++++++++++++++++++++++++++++++
// Function DEFINITION
void set_FifthSixth_Elements( string strArr_IN[] )
{
     strArr_IN[4] = "today";
     strArr_IN[5] = "?";

}//END...Function set_FifthSixth_Elements
```

```
Contents of arrStr_Greeting: [ Hello, how are you ?     ]

...Calling the function to set the 6th & 7th elements

Contents of arrStr_Greeting: [ Hello, how are you today ?   ]


===========
END...C++ Demo: Arrays as Function Inputs
```

Program Listing/Output 9.2: Arrays_asFunctionInputs_Client.cpp

❖ Thus the variable, here a data structure, itself changed

  ➢ However, as we saw in section 5.4, when we pass in individual variables, they are passed by value, so the variables themselves are unaffected

  ➢ The reason for the difference is that the array name is in fact a pointer (section 10 onwards)

- Thus whenever an array is the input to a function, it is passed as a pointer, not by value (covered in detail in section 16)

9.8 ISSUES WITH ARRAYS

❖ Arrays cannot be set equal to one another

➤ The following statement will compile fine, but will not produce the desired effect:

```
myIntArr_A = myIntArr_B;
```

❖ An array cannot be a function output

➤ Thus declarations/definitions of the following form are forbidden in C and C++

```
type[] functionName( inputs );
```

- A subtle trick to achieve this is to get an array to output a pointer (section 10 onwards)

➤ An Array cannot hold functions

- Thus declarations/definitions of the following form are forbidden in C and C++

```
outputType arrayName[ ]( inputTypes );
```

9.9 MULTIDIMENSIONAL ARRAYS

❖ It is perfectly acceptable, in fact it is a common practice, to have an array representing more than one dimension

➤ Each element of an array is defined as being another array

- Some programmers call this a "Sub-Array"

❖ In C and C++, these are called "Multidimensional Arrays"

❖ In most other languages, such as Pascal and Visual Basic, they are called "Arrays-Of-Arrays"

❖ The following example shows this in use, by modelling Flight times: Outbound & Inbound departure times

FLIGHT	1	2	3	4	5
DEPARTURE	0925	0015	0715	0932	1145
ARRIVAL	1235	1105	1215	1355	1945

```
/*          ************************************************************
            C++ MULTIDIMENSIONAL ARRAYS
            ************************************************************
            -       This programs demonstrates
                    -       Defining
                    -       Using
                    Multidimensional Arrays in C++
*/
#include <iostream>
#include <string>
using namespace std;

int main( )
{
        //»»»»»»»»»»»»»»»»»»»»»»»»»»»»»»»»»»»»»»»»»»»»»»»»»»»»»»»»»»»»»»»»»»»
        // DEFINITION \& INITIALISATION
        // - Double [][] required
        //
        double flightTime_Dept_Arr[2][5];
        for( int i = 0; i < 2; ++i)
        {
                for( int j = 0; j < 5; ++j )
                {
                                flightTime_Dept_Arr[i][j] = 0.0;
                }
        }
        //»»»»»»»»»»»»»»»»»»»»»»»»»»»»»»»»»»»»»»»»»»»»»»»»»»»»»»»»»»»»»»»»»»»
        // Populating the Multidimensional Array
        //
        flightTime_Dept_Arr[0][0] = 9.25;
        flightTime_Dept_Arr[1][0] = 12.35;

        flightTime_Dept_Arr[0][1] = 0.15;
        flightTime_Dept_Arr[1][1] = 11.15;

        flightTime_Dept_Arr[0][2] = 7.15;
        flightTime_Dept_Arr[1][2] = 12.15;

        flightTime_Dept_Arr[0][3] = 9.32;
        flightTime_Dept_Arr[1][3] = 13.55;

        flightTime_Dept_Arr[0][4] = 11.45;
        flightTime_Dept_Arr[1][4] = 19.45;


        //»»»»»»»»»»»»»»»»»»»»»»»»»»»»»»»»»»»»»»»»»»»»»»»»»»»»»»»»»»»»»»»»»»»
        // Displaying the contents of the Multidimensional Array
        //
        cout << "Flight\t\t";
        for(int i = 0; i < 5; ++i )
        {
                cout << i<< "\t";
        }
        cout << endl<< "Departure:\t";
        for(int i = 0; i < 5; ++i )
        {
```

```
        cout << flightTime_Dept_Arr[0][i]<< "\t";
        }

        cout << endl<< "Arrival:\t";
        for(int i = 0; i < 5; ++i )
        {
                cout << flightTime_Dept_Arr[1][i]<< "\t";
        }

        //»»»»»»»»»»»»»»»»»»»»»»»»»»»»»»»»»»»»»»»»»»»»»»»»»»»»»»»»»»»»»»»»»»»
        cout    << endl<< endl<< "==========="
                << endl
                << "END...C++ Demo: Multidimensional Arrays"
                << " "
                << endl;
        while(true){}
        return 0;

}//END...Function MAIN
```

```
Flight          0       1       2       3       4
Departure:      9.25    0.15    7.15    9.32    11.45
Arrival:        12.35   11.15   12.15   13.55   19.45

===========
END...C++ Demo: Multidimensional Arrays
```

Program Listing/Output 9.3: Multidimensional_Arrays_Demo_Client.cpp

- o At 1st glance, it is tempting to think that a 2-D Array is laid out as a 2-D Matrix in runtime memory
- o They are in fact laid out linearly i.e. consecutive memory locations
  - Each element of the Outer Array
  - Is each Inner Array

9.10 DYNAMIC ARRAYS

❖ A static array, the one discussed in this section, is one where the size

  ➢ Has to be predetermined at compile time
  ➢ Cannot be modified at runtime

❖ A Dynamic Array is one where the size is determined at runtime, and can change

  ➢ This is covered in Cybernetics in C++ - Expert

MEMORY & RESOURCE MANAGEMENT PROGRAMMING

❖ Memory Management is the practice of getting down to the grass roots level and manipulating the individual registers on your computer hardware

❖ Since this is the lowest level possible, it is often called "Low Level Programming" or "Firmware"

  ➢ Embedded programming, which is all about controlling individual hardware components, uses this extensively

  ➢ A major part of the work of Operating Systems is Memory Management

❖ It is required for several reasons

  ➢ In order to optimise the performance of your program

  ➢ Without memory management being performed by the programmer, all has to be dealt with "under the bonnet" by the computer

❖ C++, as well as C, supports this extensively, using the enitit(y)(ies)

  ➢ The Pointer

    • Supported by C++ and C

  ➢ The Reference

    • This is a further entity added to C++

❖ Difficult or easy?

  ➢ There's no denying that memory management is widely considered one of the major challenges of C and C++

  ➢ It is often argued that the syntax is confusing

  ➢ However, we take the view that when taught properly,

    • Not only can all the obstacles be overcome, but

    • The exciting and powerful nature of this paradigm can be unleashed

179

❖ In addition, it places immense power in the hands of the programmer

➢ For this reason, C++ has sometimes been likened to a scalpel
  • In the right hands, it's a life saving tool
  • In the wrong hands, it's a murder weapon

❖ Partly for these reasons, several modern day languages such as

➢ Java
➢ Perl
➢ Web Scripting Languages such as JavaScript, PHP
➢ Visual Basic

hide memory management from the programmer (almost) entirely

❖ The prices paid in these languages include

➢ Generally being inadequate for high performance
➢ Requiring considerably extra work when implementing several entities such as
  • Aggregate Ownership (Cybernetics in C++ - Expert)
  • The Observer Design Pattern

❖ It is largely the strong support for Memory and Resource Management which maintains C++ as the preferred language for high performance systems such as:

➢ Real time systems, such as
  • Telecommunications and Data Communications
  • Medical Applications/Biotechnology
➢ Graphics Programming
➢ Financial Applications

❖ We take the view that avoiding learning Memory & Resource Management is False Economy

➢ By investing the time and effort in the short term, you will be able to
  • Develop high performance programs
  • Implement advanced concepts with considerable more ease than when trying to avoid them

10. POINTERS

10.1 WHAT IS THE PRACTICE OF ADDRESSING?

❖ Pointers are the main entity which implements the practice of "Addressing"

❖ Addressing is a means of thinking which comprises controlling reading and writing access using the location/address of an object as opposed to its name

➢ Much of the time this can mean more efficient data usage

➢ Often, further actions are facilitated

❖ This is a very powerful means of thinking, used abundantly outside the programming world

10.1.1 ADDRESSING OUTSIDE THE PROGRAMMING WORLD

❖ Building Locations/Postal Addresses

➢ Perhaps the first thing which come to mind when hearing the word "address" is the address of a building - we use this all the time to specify locations

➢ Bjarne Stroustrup: If I wanted you to sort out books in a library,

• The equivalent of not using addressing would be sending you the entire library, getting you to do the job, then transporting it all back!

• The practice of addressing would be to send you the address

❖ Serial Numbers

➢ Identifying individual items, and their versions, by name alone will be a near impossibility

➢ Assigning a unique integer number solves this problem and is widely used

• Car registration numbering is a good example

❖ Enquiries

➢ Most companies, including Dell, Adobe and Microsoft, give a unique number to every enquiry, whether it is sales, technical support or feedback

➢ This provides a convenient means of reference for future enquiries

❖ IT

➢ Internet

- Humans feel comfortable referring to entities on the Internet by name e.g.
  ○ Websites by the URL
  ○ Servers by name
- Machines refer to entities on the Internet by their IP addresses
  ○ This is much more versatile and prevents problems such as ambiguities

➢ Databases

- Humans like to refer to database records using words e.g. we like to remember a person by their name as opposed to their ID number
  ○ Duplicate names lead to ambiguity
- It is common to create a new variable, such as a unique number or code, and assign it the status "Primary Key"
  ○ Using this for reference in programs will be more efficient and avoid ambiguities

10.1.2 ADDRESSING INSIDE THE PROGRAMMING WORLD

❖ Even for languages where memory management is hidden from the programmer, such as Java and VB, addressing is still taking place

❖ Programming is all about moving around and processing data

➢ Addressing is what makes this moving around possible

❖ Early languages, including Machine Code and Assembler based languages, were all about addressing

➢ Since this was working very close to the hardware, this is often called "low level programming"

❖ C and C++ enable this low level thinking by supporting memory and resource management

10.2 WHAT ARE POINTERS?

❖ Pointers are Variables which contain Memory Addresses

❖ They are the fundamental tool in Memory Management

➢ In programming languages which support Memory Management, such as C and C++, they are pervasive

❖ Note that since pointers are variables, like all other variables, they have a

> ➢ Name
> ➢ Type
> ➢ Value
> ➢ Address in runtime memory

❖ The value is an address

> ➢ This often looks confusing at first glance
> ➢ However, it becomes clear as soon as one acknowledges that the value is the address of a different variable
>> • This latter variable, the one being pointed to, is often called the "pointee"

10.2.1 L-VALUE VS R-VALUE

❖ The value contained in a Pointer is called an L-Value

> ➢ This is a Location (see section 2.4.1.2.2)
> ➢ Known at Compile Time

❖ The contents of the value in a Pointer is called an R-Value

> ➢ This is a Runtime Value (see section 2.4.1.2.2)
> ➢ Not known until Runtime

10.2.2 THE IMPORTANCE OF POINTERS

❖ Memory management enables high performance programming

> ➢ In programming languages where memory management is hidden from the programmer, it must take place via an interpreter
> ➢ At the very least, this will mean a significant drop in performance

❖ Pointers allow you to access large data structures/items and pass them as function parameters

> ➢ In modern times, graphics, video and medical data are becoming increasingly popular
>> • A 16 megapixel camera will generate at least 48 million numbers every time a photograph is taken
>> • When moving this data to functions, as is required in languages such as MATLAB and Basic, there will be a significant time cost overhead

- Passing a Pointer, as is strongly supported in C and C++, is very fast

❖ An entirely new kind of engineering, "Firmware engineering" where hardware is manipulated directly from a program, is enabled

➢ Bugs which occur at a hardware level can be fixed

➢ A programming language which hides memory management would disallow access to bugs of this nature

❖ At the software level, pointers offer a means of organisation and direction which would either be impossible or tedious and less efficient to implement otherwise

➢ Pointers are invaluable for implementing data structures in C and C++

- Arrays are implemented using pointers entirely (see section 16)
- Linked Lists, Trees and Graphs heavily use pointers in their implementation (Cybernetics in C++ - Expert)

➢ In Object Oriented Programming, pointers facilitate the implementation of

- Aggregate Ownership (Cybernetics in C++ - Expert)
- The Observer Design Pattern (Cybernetics in C++ - Expert)

❖ Pointers enable using more than one kind of runtime memory: the Heap and the Stack (section 14)

❖ In C++, pointers are used to implement Polymorphism (section 23)

10.3 IMPLEMENTING POINTERS IN C & C++

10.3.1 DEFINING POINTERS: THE * OPERATOR

❖ The asterisk operator * is used to define a pointer

❖ An entity cannot just be of type Pointer, it must be of type Pointer-to-DataType e.g.

➢ Pointer to primitive types

- Pointer-to-Int
- Pointer-to-Char

➢ Pointer to user defined types

- Pointer-to-Struct (section 17 onwards)
- Pointer-to-Class (section 17 onwards)

❖ TIP: The need for wisely choosing a name is more important than ever now

❖ Syntax:

```
Data Type of Pointee * pointerName
```

❖ Suppose you want to define a variable

➢ Type: pointer-to-Int
➢ Name: `ptrInt_myInt_A`

Both of the following are possible:

➢ `int* ptrInt_myInt_A;`
➢ `int *ptrInt_myInt_A;`

❖ Which one is preferable is down to personal choice

10.3.1.1 Placing the Asterisk BEFORE the space

❖ Some programmers prefer placing the asterisk before the space, next to the variable type

```
int* ptrInt_myInt_A;
```

because

➢ The variable name is `ptrInt_myInt_A`...not `*ptrInt_myInt_A`
➢ The type is Pointer-to-int i.e. int*

so this is more readable

❖ They dislike the syntax

```
int *ptrInt_myInt_A;
```

because

➢ Your eyes/brain have to

• Straddle the space between the pointee data type, in this case `int`, and the asterisk*
• Insert a logical space between the asterisk and the variable name

10.3.1.2 Placing the Asterisk AFTER the space

❖ The practice of placing the asterisk after the space, before the variable name

```
int *ptrInt_myInt_A;
```

tends to be more popular among C and C++ programmers because it eliminates potential errors:

❖ In the line

```
int* a, b, c;
```

➢ Only `a` is going to be of type `Pointer-to-int`

➢ Both `b` and `c` will be of type int

❖ Whereas, in the line

```
int *a, *b, *c;
```

➢ All of `a,` `b` and `c` will all be of type `Pointer-to-int`

10.3.2 POPULATING POINTERS: THE & OPERATOR

❖ Since the value of a pointer is the address of another variable, an operator is needed which means "the address of"

➢ The Ampersand operator & is used

❖ Suppose we have defined the variables

➢ `myInt_A` of type int, which contains the value 5

```
int myInt_A = 5;
```

➢ `ptrInt_myInt_A` of type pointer-to-int, as above

❖ In order to get `ptrInt_myInt_A` to contain the address of `myInt_A`, the following syntax is required:

```
ptrInt_myInt_A = &myInt_A;
```

❖ The following is now created in runtime memory:

MEMORY ADDRESS	LAYOUT IN RUNTIME MEMORY	NAME	
0x002C27	5	myInt_A	
0x0FC234	0x002C27	ptrInt_myInt_A	

Figure 10.1   Pointers in Runtime Memory

❖ The definition and population have been shown separately above for demonstration purposes

❖ Like with all variables, it is advisable to define and initialise pointer variables on the same line:

```
int *ptrInt_myInt_A = &myInt_A;
```

❖ Without this, here too, there is the problem, that for a moment at least, the pointer variable will be filled with garbage

10.3.3 DECODING POINTERS: THE * OPERATOR

❖ The question now is how do we read/write the value of the contents of a pointer

   ➢ The asterisk * operator is used again

   ➢ The statement

```
*pointerName
```

means the Value of the Contents of `pointerName`

❖ In our case, we want to access `myInt_A` via a pointer to it, here `ptrInt_myInt_A`

   ➢ We require the statement `*ptrInt_myInt_A`

❖ Thus we now have two ways of accessing the variable

   ➢ Directly

   ➢ Via the Pointer

10.3.4 DEMONSTRATION: CREATING & USING POINTERS

❖ The following program demonstrates defining, populating and decoding pointers

```
/*      ************************************************************
        C++ POINTERS
        ************************************************************
        - This program demonstrates
                    - Defining Pointers
                    - Populating Pointers
                    - Decoding Pointers
*/
#include <iostream>
```

```cpp
using namespace std;


int main()
{
    // »»»»»»»»»»»»»»»»»»»»»»»»»»»»»»»»»»»»»»»»»»»»»»»»»»»»»»»»»»»»»»»»»»
    int myInt_A = 5;
    double myDbl_B = 7.2;

    /* »»»»»»»»»»»»»»»»»»»»»»»»»»»»»»»»»»»»»»»»»»»»»»»»»»»»»»»»»»»»»»»»»»
            DEFINING A POINTER
            -       The Asterisk operator * is used
    */
    int *ptrInt_myInt_A;        // Defining ptrInt_myInt_A as type
                                // Pointer-to-int using int *

    /* »»»»»»»»»»»»»»»»»»»»»»»»»»»»»»»»»»»»»»»»»»»»»»»»»»»»»»»»»»»»»»»»»»
            POPULATING i.e. Initialising A POINTER
            -       The VALUE held in ptrInt_myInt_A will be
                    the hexadecimal ADDRESS of myInt_A
            -       The Ampersand & operator is used
    */
    ptrInt_myInt_A = &myInt_A;  // Take the hexadecimal address of
                                // myInt_A
                                // - Put it into ptrInt_myInt_A

    //================================================================
    // Better to define and populate on the same line:
    double *ptrDbl_myDbl_B = &myDbl_B;

    //================================================================
    //    At this stage, we can display the hexadecimal address
    //    - this is only sometimes needed7
    cout  << "========= DISPLAYING ADDRESSES ========="<< endl;
    cout  << "Hexadecimal address of myInt_A: ptrInt_myInt_A = "
          << ptrInt_myInt_A
          << endl;
    cout  << "Hexadecimal address of myDbl_B: ptrDbl_myDbl_B = "
          << ptrDbl_myDbl_B
          << endl;

    /* »»»»»»»»»»»»»»»»»»»»»»»»»»»»»»»»»»»»»»»»»»»»»»»»»»»»»»»»»»»»»»»»»»
            DECODING THE CONTENTS OF A POINTER
            -       i.e. Extracting the value of 5, held in myInt_A,
                    using ptrInt_myInt_A
            -       The asterisk operator * is used again
    */
    cout  << endl<< "========= DECODING POINTERS ========="<< endl;
    cout  << "Accessing via Pointers: "<< endl;
    cout  << "*ptrInt_myInt_A = "<< *ptrInt_myInt_A<< endl;
    cout  << "*ptrDbl_myDbl_B = "<< *ptrDbl_myDbl_B<< endl;

    cout  << endl<< "Accessing directly: "<< endl;
```

```
    cout   << "myInt_A = "<< myInt_A<< endl;
    cout   << "myDbl_B = "<< myDbl_B<< endl;

    //»»»»»»»»»»»»»»»»»»»»»»»»»»»»»»»»»»»»»»»»»»»»»»»»»»»»»»»»»»»»»
    cout   << endl<< "==========="
           << endl<< "END...C++ Pointers Demo"
           << endl;
    while(true){}
    return 0;

}//END...function main


========= DISPLAYING ADDRESSES =========
Hexadecimal address of myInt_A: ptrInt_myInt_A = 001FFDE8
Hexadecimal address of myDbl_B: ptrDbl_myDbl_B = 001FFDD8

========= DECODING POINTERS =========
Accessing via Pointers:
*ptrInt_myInt_A = 5
*ptrDbl_myDbl_B = 7.2

Accessing directly:
myInt_A = 5
myDbl_B = 7.2
===========
END...C++ Pointers Demo
```

Program Listing/Output 10.1: Pointers_Demo_Client.cpp Using Pointers in C and C++

10.3.5 POTENTIAL PROBLEM: TYPE MISMATCH

❖ An essential point to bear in mind is that the type of the pointer must be compatible with its proposed contents

➢ A pointer-to-int can contain the address of only an integer

➢ A pointer-to- double can contain the address of only an double

❖ Any attempt to place an incompatible type, such as a double in a pointer-to-int will result in a compile-time error

10.3.6 EQUATING POINTERS

❖ Pointers are themselves variables, and so their values, which are memory addresses of other variables, can be, and very frequently are, varied

❖ This means that the pointer is a modifiable l-value

➢ It can be placed on the left hand side of an assignment operator

❖ The following program demonstrates this:

```
/*      ************************************************************
        EQUATING POINTERS
        ************************************************************
        -       This program shows how to make one pointer contain the value
                -       i.e. address of another variable
                -       of another pointer
*/
#include <iostream>
using namespace std;

int main( )
{


        //»»»»»»»»»»»»»»»»»»»»»»»»»»»»»»»»»»»»»»»»»»»»»»»»»»»»»»»»»»»»»»
        int myInt_A = 5, myInt_B = 17;
        int *ptr_Int_01 = &myInt_A;
        int *ptr_Int_02 = &myInt_B;

        cout    << endl<< "========= INITIALLY... ========="<< endl;
        cout    << "myInt_A = "<< myInt_A<< endl;
        cout    << "myInt_B = "<< myInt_B<< endl;
        cout    << " ptr_Int_01 points to = myInt_A, so *ptr_Int_01"
                << *ptr_Int_01<< endl;
        cout    << " ptr_Int_02 points to = myInt_B, so *ptr_Int_02"
                << *ptr_Int_02<< endl;


        //»»»»»»»»»»»»»»»»»»»»»»»»»»»»»»»»»»»»»»»»»»»»»»»»»»»»»»»»»»»»»»
        cout    << endl<< "========= EQUATING... ========="<< endl;
        ptr_Int_02 = ptr_Int_01;
        cout    << "...saying ptr_Int_02 = ptr_Int_01"<< endl;

        //»»»»»»»»»»»»»»»»»»»»»»»»»»»»»»»»»»»»»»»»»»»»»»»»»»»»»»»»»»»»»»
        cout    << endl<< "========= NEW SETTINGS... ========="<< endl;
        cout    << "myInt_A = "<< myInt_A<< endl;
        cout    << "myInt_B = "<< myInt_B<< endl;
        cout    << " ptr_Int_01 still points to = myInt_A, so"
                << " *ptr_Int_01 = "<< *ptr_Int_01<< endl;
        cout    << " ptr_Int_02 now also points to = myInt_A, so"
                << " *ptr_Int_02 = "<< *ptr_Int_02<< endl;


        //»»»»»»»»»»»»»»»»»»»»»»»»»»»»»»»»»»»»»»»»»»»»»»»»»»»»»»»»»»»»»»
        cout    << endl<< endl<< "==========="
                << endl
                << "END...C++ Demo: Equating Pointers"
                << " "
                << endl;
        while(true){}
        return 0;
```

```
}//END...Function MAIN


========= INITIALLY... =========
myInt_A = 5
myInt_B = 17
 ptr_Int_01 points to = myInt_A, so *ptr_Int_015
 ptr_Int_02 points to = myInt_B, so *ptr_Int_0217

========= EQUATING... =========
...saying ptr_Int_02 = ptr_Int_01

========= NEW SETTINGS... =========
myInt_A = 5
myInt_B = 17
ptr_Int_01 still points to = myInt_A, so *ptr_Int_01 = 5
ptr_Int_02 now also points to = myInt_A, so *ptr_Int_02 = 5


===========
END...C++ Demo: Equating Pointers
```

Program Listing/Output 10.2: Equating_Pointers_Client.cpp

10.3.7 REVIEW: 2 USES OF THE ASTERISK OPERATOR

❖ A major point to note is that the asterisk symbol is used for two different operations:

➢ Defining as type Pointer-to-DataType
➢ Decoding the Pointee

❖ (Q) How do we distinguish between the two?

(A) We look at what is to the left of the asterisk

➢ If the asterisk **follows a data type**, then it means defining as type Pointer-to data type e.g.
- `int *a`   Defines a of type Pointer-to-int
- `double *ptrDbl_myDbl_B` Defines `ptrDbl_myDbl_B of type` Pointer-to-double
- `char* c`   Defines c of type Pointer-to-char

➢ If the asterisk **does not have a data type before it**, only the variable name after it, then it is decoding the pointee

10.4 THE NULL POINTER

10.4.1 WHAT IS THE NULL POINTER?

❖ A question now is "Is there an identity value for a pointer?"

  ➢ For numbers:

   • Addition: Integer, it is zero, for a double it is 0.0
   • Multiplication: Integer, it is 1, for a double it is 1.0

  ➢ For a matrix, it is the Identity matrix, 1s down the leading diagonal, 0s elsewhere

  ➢ For a character, it is the blank character ' '

❖ For the pointer, it is the NULL pointer

  ➢ It is often denoted using the symbol ∅

   • It means that for the time being, the pointer contains nothing

  ➢ It is a good idea, in fact essential, to initialise your pointer variables with this, when you don't know till runtime which variables they will be pointing to

   • Common mistake: to try initialising a pointer with a hexadecimal zero 0x000...

    ○ Since we do not know/think about the actual addresses of the runtime memory, this is often an actual address which contains something else

10.4.1.1 Do not Confuse NULL Pointers and NUL Characters

❖ There is something else which sounds identical in spoken English

  ➢ It is the NUL character ' \0' which is used to indicate the end of a string

❖ In written English, the difference is clearer: the spelling

❖ The following "poem" may help!

  ➢ The one L NUL ends an ASCII string
  ➢ The two L NULL points to nothing

10.4.2 DEFINING THE NULL POINTER

❖ The keyword NULL is used to place the NULL value in a pointer

```
int *ptr_myInt_A = NULL;
```

10.4.3 APPLICATIONS OF THE NULL POINTER

❖ Apart from initialising pointer variables, the NULL pointer underpins several applications

❖ When using the heap runtime memory (section 14 onwards), often the same pointer will need to point to different parts during runtime

> ➢ The NULL pointer is used whenever there is a period when the pointee is unknown, to prevent the dangling pointer bug (section 14.5 and Cybernetics in C++ - Expert)

❖ Data Structures

> ➢ Linked Lists (Cybernetics in C++ - Expert) are implemented using the NULL pointer to indicate the end of the list

> ➢ Trees (Cybernetics in C++ - Expert), the most commonly used being Binary Search Trees, use NULL pointers to implement leaf nodes

10.5 THE VOID POINTER

❖ The void pointer is a pointer which can point to any type

❖ The syntax is as follows:

```
void *pointerName = &pointeeVariableName
```

❖ The void pointer cannot be decoded directly using the asterisk * operator

❖ It must be cast into type pointer-to-variableType first

❖ The following program shows how it can be used with various primitive types:

```
#include <iostream>
using namespace std;

int main( )
{
        //»»»»»»»»»»»»»»»»»»»»»»»»»»»»»»»»»»»»»»»»»»»»»»»»»»»»»»»»»»»»»»
        double myDbl_A = 7.35;
        int myInt_A = 9;

        // A void pointer can point to different types
        void *ptrVoid_myDbl_A = &myDbl_A;
        void *ptrVoid_myInt_A = &myInt_A;

        //»»»»»»»»»»»»»»»»»»»»»»»»»»»»»»»»»»»»»»»»»»»»»»»»»»»»»»»»»»»»»»
        //      Cannot dereference void pointers directly...
        //      - The following are illegal:
```

```
        /*
                cout << "The value of the pointee of ptrVoid_myDbl_A is "
                    << *ptrVoid_myDbl_A<< endl;

                cout << "The value of the pointee of ptrVoid_myInt_A is "
                    << *ptrVoid_myInt_A<< endl;
        */
        //»»»»»»»»»»»»»»»»»»»»»»»»»»»»»»»»»»»»»»»»»»»»»»»»»»»»»»»»»»»»»»»»
        // ...They need to be cast
        //
        cout   << "The value of the pointee of ptrVoid_myDbl_A is "
               << *(static_cast<double*>(ptrVoid_myDbl_A))
               << endl;

        cout   << "The value of the pointee of ptrVoid_myInt_A is "
               << *(static_cast<int*>(ptrVoid_myInt_A))
               << endl;


        //»»»»»»»»»»»»»»»»»»»»»»»»»»»»»»»»»»»»»»»»»»»»»»»»»»»»»»»»»»»»»»»»
        cout   << endl<< "============"
               << endl<< "END...C++ Void Pointers Demo"
               << endl;
        while(true){}
        return 0;

}//END...function main
```

```
The value of the pointee of ptrVoid_myDbl_A is 7.35
The value of the pointee of ptrVoid_myInt_A is 9

============
END...C++ Void Pointers Demo
```

Program Listing/Output 10.3: File VoidPointer_Demo_Client.cpp Using the Void pointer

10.5.1 ISSUES WITH THE VOID POINTER

❖ At first glance, the void pointer seems to offer a desirable level of flexibility; after all, it can point to anything

➢ It may be tempting to use this all the time

❖ Problem: There is no way to keep track of the type of the variable it is pointing to

❖ In C programming, this was undesirable but not an immense issue

❖ In C++ programming, this is a major problem, since type flexibility is strongly supported through **Polymorphism**:

➢ Dynamic Polymorphism in object oriented programming (section 23 onwards)

➢ Static Polymorphism in generic programming (section 27 onwards)

❖ **Rule of Thumb in C++**: aVOID the Void pointer! :-)

10.6 POINTERS AS FUNCTION INPUTS

10.6.1 PROBLEMS: LACK OF ADDRESSING

❖ The importance of addressing, even when dealing with small data items, often becomes apparent when we look at the sorts of bugs that arise when there is a lack of it

❖ **Common Interview Question:** What is the flaw in the following program?

```
include <iostream>
using namespace std;

// Function DECLARATIONS
void swap( int int_IN_A, int int_IN_B );


#int main()
{
        //»»»»»»»»»»»»»»»»»»»»»»»»»»»»»»»»»»»»»»»»»»»»»»»»»»»»»»»»»
        int myInt_A = 5, myInt_B = 7;
        cout   << "Original values: "
               << "myInt_A = "<< myInt_A
               << " and myInt_B = "<< myInt_B
               << endl;

        //»»»»»»»»»»»»»»»»»»»»»»»»»»»»»»»»»»»»»»»»»»»»»»»»»»»»»»»»»
        //     Attempting the swap
        swap( myInt_A, myInt_B );

        //»»»»»»»»»»»»»»»»»»»»»»»»»»»»»»»»»»»»»»»»»»»»»»»»»»»»»»»»»
        cout   << "Values after attempted swap: "
               << "myInt_A = "<< myInt_A
               << " and myInt_B = "<< myInt_B
               << endl;

        //»»»»»»»»»»»»»»»»»»»»»»»»»»»»»»»»»»»»»»»»»»»»»»»»»»»»»»»»»
        cout   << endl<< "=========="
               << endl<< "End...Demo, problems, lack of Memory Management"
               << endl;
        while(true){}
        return 0;

}//END...Function Main
```

```
void swap( int int_IN_A, int int_IN_B )
{
        int temp = int_IN_A;
        int_IN_A = int_IN_B;
        int_IN_B = temp;

}//END...Function swap
```

```
Original values: myInt_A = 5 and myInt_B = 7
Values after attempted swap: myInt_A = 5 and myInt_B = 7

==========
End...Demo, problems, lack of Memory Management
```

Program Listing/Output 10.4: Attempting a swap - with a fundamental flaw

❖ The problem is that the default behaviour is to pass actual parameters BY VALUE

➤ Thus in this example, all that was passed to the function were the values 5 and 7

➤ These were swapped over but since these values were not stored any-where, this was a meaningless operation - we didn't even detect it at runtime!

➤ The variables `myInt_A` and `myInt_B` were untouched

10.6.2 PASSING FUNCTION INPUTS AS POINTERS

❖ What we need to pass **is** the addresses of the variables?

❖ Since the values of pointers are addresses, a function which will perform the desired swap operation can be implemented using inputs as pointers

10.6.2.1 Formal Parameters as Pointers

❖ Like all other variables, pointers need to have their types specified in the formal parameters

➤ Revision: the asterisk operator and the data type of the pointee are used

❖ Thus a function with

➤ The name myFunc

➤ 2 inputs:

• 1, called ptrInt_In_A, of type pointer-to-int

- 1, called ptrDbl_In_B, of type pointer-to-double

➤ No output

will have the following prototype:

```
void myFunc( int *ptrInt_In_A, double *ptrDbl_In_B )
{
    // Functionalities of myFunc
}
```

❖ The signature of the function, used in the declaration, will be as follows:

```
void myFunc( int *, double * );
```

10.6.2.2 Actual Parameters as Pointers

❖ When invoking a function, the inputs, the actual parameters, need to be the values that the input variables take on

❖ In the case of pointers, these "values" are memory addresses

➤ Since the ampersand & operator is used to mean "the address of" for pointers, these are mostly seen for pointer actual parameters

❖ So for our example, a call to myFunc will be as follows:

```
myFunc( &myInt_A, &myDbl_B );
```

➤ where

- myInt_A is a previously defined variable of type int
- myDbl_B is a previously defined variable of type double

10.6.3 FIXING THE PROBLEMS USING POINTER INPUTS

❖ The following program shows how the swap can work properly when the inputs are pointers

```
#include <iostream>
using namespace std;

void swap_usingPointers( int *, int * );

int main()
{
    //»»»»»»»»»»»»»»»»»»»»»»»»»»»»»»»»»»»»»»»»»»»»»»»»»»»»»»»»»»»»»»»»
    int myInt_A = 7, myInt_B = 9;
```

```
        cout   << "Before the attempted swap myInt_A = "<< myInt_A
               << " and myInt_B = "<< myInt_B<< endl;

        //»»»»»»»»»»»»»»»»»»»»»»»»»»»»»»»»»»»»»»»»»»»»»»»»»»»»»»»»»»»»
        /* INVOKING swap_usingPointers - inputs Pointers
                  -      Want the inputs to be
                  -       "The ADDRESS of myInt_A",
                  -       "The ADDRESS of myInt_B"
                  -      We say &myInt_A and &myInt_B, since
                  -         the operator to say "The address of a variable
                            pointed to by a pointer": AMPERSAND &
        */
        swap_usingPointers( &myInt_A, &myInt_B );

        //»»»»»»»»»»»»»»»»»»»»»»»»»»»»»»»»»»»»»»»»»»»»»»»»»»»»»»»»»»»»
        cout   << "After the attempted swap myInt_A = "<< myInt_A
               << " and myInt_B = "<< myInt_B<< endl;

        //»»»»»»»»»»»»»»»»»»»»»»»»»»»»»»»»»»»»»»»»»»»»»»»»»»»»»»»»»»»»
        cout   << endl<< "==========="
               << endl<< "END...C++ Pointers as Function Inputs Demo"
               << endl;
        while(true){}
        return 0;

}//END...function main


/*     ++++++++++++++++++++++++++++++++++++++++++++++++++++++++++++
       Function DEFINITION swap_usingPointers
       -        We are now going to pass the variables themselves,
                not their values
       -        When a variable is defined, (for the best part)
                it stays in the SAME memory location until it goes
                out of scope
       -        Thus passing this memory ADDRESS to the function
                is the way we ensure the variable itself
                is passed
       -        Here we achieve this using Pointers
*/
void swap_usingPointers( int *ptrInt_In_A, int *ptrInt_In_B )
{
        int temp = *ptrInt_In_B;
        *ptrInt_In_B = *ptrInt_In_A;
        *ptrInt_In_A = temp;

}// END...function swap_usingPointers
```

```
Before the attempted swap myInt_A = 7 and myInt_B = 9
After the attempted swap myInt_A = 9 and myInt_B = 7

===========
```

```
END...C++ Pointers as Function Inputs Demo
```

Program Listing/Output 10.5: Pointers_Demo_Client.cpp Fixing the swap problem using pointers

❖ Now, since the addresses of `myInt_A` and `myInt_B` are being passed, it is the contents of these memory addresses, i.e. the variables themselves, which get manipulated

10.6.4 THE IMPORTANCE OF POINTER INPUTS

❖ Even a simple operation like swapping two variables, both of which occupy very little memory, illustrates the importance of memory management

❖ However, the benefits are fully felt when dealing with large data structures:

❖ The performance of an executable program is determined largely by the speed with which it can process and move around data

❖ A package which allows picture editing will require the photo data to be communicated throughout the program

> If there is no memory management, the entire picture will need to be moved between the different functions

• Given that a 15MP colour photo may be ~7MB in size, to have to keep moving this around will significantly degrade performance

> If there is memory management, instead pointers can be moved around, with the photo being kept in one place

❖ It is largely due to memory management that C and C++ programs are generally considered faster than those built in a language which does not support it, such as Java and MATLAB

10.7 THE POINTER AS THE FUNCTION OUTPUT

❖ The prototype of a function which outputs a pointer is as follows

```
DataType* functionName( inputType(s) inputName(s) )
```

> Thus a function, name myFunc which

• Takes as inputs an integer and a double
• Returns a pointer to an integer

will have a prototype as follows:

```
int *myFunc( int myIntIn, doublemyDblIn )
```

10.7.1 PROSPECTIVE DANGLING POINTER PROBLEM

❖ A common mistake is to return a pointer to a non-static local variable

➢ This kind of mistake will not cause a compile time error

❖ However, during runtime, after the program exits from the function, the local variable will have gone out of scope

➢ The address of this will remain valid, but it will now contain garbage

❖ Thus the output of the function will be a pointer, the contents of which have gone out of scope

➢ This is called a dangling pointer (see section 14.5.3)

11. ADVANCED POINTERS

11.1 POINTER TO POINTER HIERARCHIES

❖ Pointers contain the address of variables

❖ However, the pointer is itself a variable

❖ Thus it is perfectly possible, in fact a common practice, to have a pointer containing the address of another pointer

❖ The address of this can in turn be contained in another pointer

❖ Thus a hierarchy of Pointer-to-Pointer-to... can be developed

11.1.1 ONE LEVEL: POINTER TO A POINTER

❖ In order to implement one level i.e. Pointer-to-Pointer-to-DataType

➢ Definition: Two asterisk operators ** are required
➢ Population: The address of a pre-defined Pointer is required:

```
int **ptrPtrInt_myInt_A = &ptrInt_myInt_A
```

❖ Decoding: Two asterisk operators ** are required to give the value of the contents of the contents

❖ The following program shows how a Pointer-to-Pointer-to-int and a Pointer-to-Pointer-to-double can be defined and used

```cpp
/*      ***********************************************************
        POINTER-POINTER HIERARCHIES
        ***********************************************************
        -       This program shows how we can implement and use a
                -       Pointer-to-Pointer-to-int
                -       Pointer-to-Pointer-to-double
*/
#include <iostream>
#include <string>
using namespace std;

int main( )
{
        //»»»»»»»»»»»»»»»»»»»»»»»»»»»»»»»»»»»»»»»»»»»»»»»»»»»»»»»»»»»»»
        //      Implementing
        //      -       Variables
        //      -       Pointers to them
        //      -       Pointers to these Pointers
        //
        int myInt_A = 7;
        double myDbl_B = 8.2;

        //=========================================================
        //      Pointer-to-int, Pointer-to-double
        //
        int *ptrInt_myInt_A = &myInt_A;
        double *ptrDbl_myDbl_B = &myDbl_B;


        //=========================================================
        //      Pointer-to-Pointer-to-int, Pointer-to-Pointer-to-double
        //
        int **ptrPtrInt_ptrInt_myInt_A = &ptrInt_myInt_A;
        double **ptrPtrDbl_ptrDbl_myDbl_B = &ptrDbl_myDbl_B;

        //»»»»»»»»»»»»»»»»»»»»»»»»»»»»»»»»»»»»»»»»»»»»»»»»»»»»»»»»»»»»»
        //      Decoding Pointers
        //
        cout    << "myInt_A = "<< myInt_A<< endl;
        cout    << "*ptrInt_myInt_A = "<< *ptrInt_myInt_A<< endl;
        cout    << "**ptrPtrInt_ptrInt_myInt_A = "
                        << **ptrPtrInt_ptrInt_myInt_A<< endl;
        cout    << endl;
        cout    << "myDbl_B = "<< myDbl_B<< endl;
        cout    << "*ptrDbl_myDbl_B = "<< *ptrDbl_myDbl_B<< endl;
        cout    << "**ptrPtrDbl_ptrDbl_myDbl_B = "
                        << **ptrPtrDbl_ptrDbl_myDbl_B<< endl;

        //»»»»»»»»»»»»»»»»»»»»»»»»»»»»»»»»»»»»»»»»»»»»»»»»»»»»»»»»»»»»»
        cout    << endl<< endl<< "============"
                << endl
                << "END...C++ Demo: Pointer-Pointer Hierarchies"
```

```
                << " "
                << endl;
        while(true){}
        return 0;

}//END...Function MAIN
```

```
myInt_A = 7
*ptrInt_myInt_A = 7
**ptrPtrInt_ptrInt_myInt_A = 7

myDbl_B = 8.2
*ptrDbl_myDbl_B = 8.2
**ptrPtrDbl_ptrDbl_myDbl_B = 8.2



===========
END...C++ Demo: Pointer-Pointer Hierarchies
```

Program Listing/Output 11.1: PointerHierarchies_Client.cpp

11.1.2 MULTIPLE LEVEL POINTERS

❖ In order to implement further levels, it is simply a case of extending the definitions and uses practised with one level

❖ The following program shows multiple level pointers in use:

```
/*      *****************************************************************
        MULTIPLE POINTER-POINTER HIERARCHIES
        *****************************************************************
        -       This program shows how we can implement and use a
                -       Pointer-to-Pointer-to-int
                -       Pointer-to-Pointer-to-double
*/
#include <iostream>
#include <string>
using namespace std;

int main( )
{
        //»»»»»»»»»»»»»»»»»»»»»»»»»»»»»»»»»»»»»»»»»»»»»»»»»»»»»»»»»»»»»»»»»»»»»»
        //      Implementing
        //      -    Variables
        //      -    Pointers to them
        //      -    Pointers to these Pointers
        //
        int myInt_A = 7;
        double myDbl_B = 8.2;
```

```
//==============================================================
//      Pointer-to-int, Pointer-to-double
//
int *ptrInt_myInt_A = &myInt_A;
double *ptrDbl_myDbl_B = &myDbl_B;


//==============================================================
//      Pointer-to-Pointer-to-int, Pointer-to-Pointer-to-double
//
int **ptrPtrInt_myInt_A = &ptrInt_myInt_A;
double **ptrPtrDbl_myDbl_B = &ptrDbl_myDbl_B;


//==============================================================
//    Pointer-to-Pointer-to-Pointer-to-int,
//    Pointer-to-Pointer-to-Pointer-to-double
//
int ***ptrPtrPtrInt_myInt_A = \&ptrPtrInt_myInt_A;
double ***ptrPtrPtrDbl_myDbl_B = \&ptrPtrDbl_myDbl_B;


//»»»»»»»»»»»»»»»»»»»»»»»»»»»»»»»»»»»»»»»»»»»»»»»»»»»»»»»»»»»»»»»»»»
//    Decoding Pointers
//
cout   << "myInt_A = "<< myInt_A<< endl;
cout   << "*ptrInt_myInt_A = "<< *ptrInt_myInt_A<< endl;
cout   << "**ptrPtrInt_myInt_A = "
                    << **ptrPtrInt_myInt_A<< endl;

cout   << "***ptrPtrPtrInt_myInt_A = "
                    << ***ptrPtrPtrInt_myInt_A<< endl;

cout   << endl;

cout   << "myDbl_B = "<< myDbl_B<< endl;
cout   << "*ptrDbl_myDbl_B = "<< *ptrDbl_myDbl_B<< endl;
cout   << "**ptrPtrDbl_myDbl_B = "
                    << **ptrPtrDbl_myDbl_B<< endl;

cout   << "***ptrPtrPtrDbl_myDbl_B = "
                    << ***ptrPtrPtrDbl_myDbl_B<< endl;


//»»»»»»»»»»»»»»»»»»»»»»»»»»»»»»»»»»»»»»»»»»»»»»»»»»»»»»»»»»»»»»»»»»
cout   << endl<< endl<< "==========="
        << endl
        << "END...C++ Demo: Mulitple Pointer-Pointer Hierarchies"
        << " "
        << endl;
    while(true){}
    return 0;

}//END...Function MAIN
```

```
myInt_A = 7
```

```
*ptrInt_myInt_A = 7
**ptrPtrInt_myInt_A = 7
***ptrPtrPtrInt_myInt_A = 7

myDbl_B = 8.2
*ptrDbl_myDbl_B = 8.2
**ptrPtrDbl_myDbl_B = 8.2
***ptrPtrPtrDbl_myDbl_B = 8.2


===========
END...C++ Demo: Mulitple Pointer-Pointer Hierarchies
```

Program Listing/Output 11.2: PointerHierarchies_MultipleLevels_Client.cpp

11.2 READ-ONLY POINTERS USING CONST

❖ Revision: the keyword `const` is used to make a variable read-only i.e. no longer a variable

❖ As we have been discussing, when dealing with pointers, we are looking at two items of data:

➢ The pointer itself
➢ The data being pointed to, the pointee

❖ The question now is how do we use `const` in order to make

➢ Only the pointer read-only, not the data
➢ Only the data being pointed to read only, not the pointer
➢ Both the pointer and the data being pointed to read only

❖ The answer depends on where the word const is placed, in the definition, in relation to the asterisk operator*

➢ You might want to use BP TAP
  • If `const` comes <u>B</u>EFORE the `asterisk` *, the data <u>P</u>OINTED-<u>T</u>O is read only
  • If `const` comes <u>A</u>FTER the `asterisk` *, the <u>P</u>OINTER is read only

❖ Note that when we say the data being pointed to is read only

➢ It means the value of the pointee cannot be written to by the pointer
➢ The value can still be written to
  • Directly

- Via another Pointer
- Via a Reference (section 12 onwards)

❖ The following program demonstrates this concept:

```
/*      ************************************************************
        READ ONLY POINTERS WITH CONST
        ************************************************************
        -       This Program demonstrates how to assign READ-ONLY status,
                using const, to
                -       The Pointer
                -       The Pointee (data being [pointed to)
                -       Both the Pointer and Pointee
        -       In the definition, look at where the const keyword is in
                relation to the asterisk * operator and use BPTAP
                -       Before Pointer-To (i.e. pointee )
                -       After Pointer
*/
#include <iostream>
using namespace std;

int main( )
{
        //»»»»»»»»»»»»»»»»»»»»»»»»»»»»»»»»»»»»»»»»»»»»»»»»»»»»»»»»»»»»»»»
        int myInt_A = 14, myInt_B = 17, myInt_C = 23, myInt_D = 27,
                myInt_E = 35, myInt_F = 44, myInt_G = 53, myInt_H = 67;


        //»»»»»»»»»»»»»»»»»»»»»»»»»»»»»»»»»»»»»»»»»»»»»»»»»»»»»»»»»»»»»»»
        int *ptrInt_myInt_A = &myInt_A;


        //»»»»»»»»»»»»»»»»»»»»»»»»»»»»»»»»»»»»»»»»»»»»»»»»»»»»»»»»»»»»»»»
        // The const keyword is BEFORE the asterisk *
        //
        const int *ptrInt_myInt_B = &myInt_B;
        // *ptrInt_myInt_B = 79; // ILLEGAL - contents of ptrInt_myInt_B
                                //          are read-only ( via the pointer )
        ptrInt_myInt_B = &myInt_F; // OK - ptrInt_myInt_B is writeable

        myInt_B = 78;      // OK - "Outside" the pointer,
                           // myInt_B is still writeable


        //»»»»»»»»»»»»»»»»»»»»»»»»»»»»»»»»»»»»»»»»»»»»»»»»»»»»»»»»»»»»»»»
        // The const keyword is AFTER the asterisk *
        //
        int * const ptrInt_myInt_C = &myInt_C;
        // ptrInt_myInt_C = &myInt_G;  // ILLEGAL - ptrInt_myInt_C
                                     // is read-only

           *ptrInt_myInt_C = 25;         // OK - contents of ptrInt_myInt_C
                                     // are writeable


        //»»»»»»»»»»»»»»»»»»»»»»»»»»»»»»»»»»»»»»»»»»»»»»»»»»»»»»»»»»»»»»»
        // The const keyword is both BEFORE and AFTER the asterisk *
```

```
      //
      const int * const ptrInt_myInt_D = &myInt_D;

      // ptrInt_myInt_D = &myInt_H; // ILLEGAL - ptrInt_myInt_D
                                    // is read-only

      // *ptrInt_myInt_D = 29;      // ILLEGAL - contents of
                                    // ptrInt_myInt_D are read-only
                                    // ( via the pointer )

      myInt_D = 30;   // OK - "Outside" the pointer, myInt_D is still
                      // writeable

      //»»»»»»»»»»»»»»»»»»»»»»»»»»»»»»»»»»»»»»»»»»»»»»»»»»»»»»»»»»»»
      cout << endl<< endl<< "==========="<< endl
           << "END...C++ Demo Const with Pointers"<< endl;
      while(true){}
      return 0;

}//END...Function MAIN
```

Program Listing/Output 11.3: Const_With_Pointers_Client.cpp Demonstration: using const with Pointers

11.3 POINTERS TO FUNCTIONS

❖ A function is, like data items, an entity which is stored in runtime memory

❖ We can thus have a pointer to this

❖ Function Pointers are widely used in C programming (covered in Cybernetics in C++ - Expert)

> It is possible to have a data structure of these

> Then, functions can be called in sequence

❖ In C++, the Function Object/Functor has many advantages over Function Pointers, and is largely used in its place (covered in detail in Cybernetics in C++ - Expert)

11.4 ARRAYS OF POINTERS

❖ Since Pointers are items of data, they can be, and much of the time are, stored in data structures

> This section looks at storing Pointers in the most common data structure, the array

❖ The data type of the array must be of type Pointer-to-requiredDataType

```
dataType* arrayName[size];
```

> ➤ The array brackets operator [] takes precedence over the pointer definition operator *
>   - Thus this is an array of type Pointer-to-dataType
>   - Not a pointer to an array of dataType

❖ This is never a problem, because arrays are themselves implemented using Pointers (section 16)

❖ An array of type Pointer-to-dataType is simply a data item of type Pointer-to-Pointer-to-dataType

11.4.1 ARRAY OF POINTERS TO PRIMITIVE VARIABLES

❖ Let us start by defining an array of Pointers-to-int

❖ The definition will be as follows:

```
int   *capacities_London2012_Venues[]
```

❖ Each element will need to be an address of an int

❖ The following program shows this in practice:

```
/*    ***************************************************************
      ARRAY OF POINTERS
      ***************************************************************
      -     Defining
      -     Populating
      -     Decoding
      an array of Pointers-to-int
*/
#include <iostream>
#include <string>
using namespace std;

int main( )
{
        //»»»»»»»»»»»»»»»»»»»»»»»»»»»»»»»»»»»»»»»»»»»»»»»»»»»»»»»»»»»»»»
        int capacity_London2012_Stadium = 80000;
        int capacity_London2012_Velodrome = 6000;
        int capacity_London2012_Aquatics = 17500;

        //»»»»»»»»»»»»»»»»»»»»»»»»»»»»»»»»»»»»»»»»»»»»»»»»»»»»»»»»»»»»»»
        // Defining and Populating an Array of type Pointer-to-int
        //
        int *capacities_London2012_Venues[]
            = {   &capacity_London2012_Stadium,
                  &capacity_London2012_Velodrome,
```

```
                        &capacity_London2012_Aquatics
                };

        //===========================================================
        cout << "The London 2012 venue capacities: "
             << "Olympic Stadium, Velodrome and Aquatics Centre"
             << " are as follows: "
             << endl;
        for( int i = 0; i < 3; ++i )
        {
        cout << *capacities_London2012_Venues[i]<< " ";
        }

        //»»»»»»»»»»»»»»»»»»»»»»»»»»»»»»»»»»»»»»»»»»»»»»»»»»»»»»»»»»»
        cout << endl<< endl<< "==========="
             << endl
             << "END...C++ Demo: Array Of Pointers"
             << " "
             << endl;
        while(true){}
        return 0;

}//END...Function MAIN
```

```
The London 2012 venue capacities: Olympic Stadium, Velodrome and
Aquatics Centre are as follows:
80000 6000 17500

===========
END...C++ Demo: Array Of Pointers
```

Program Listing/Output 11.4: ArrayOfPointers_Client.cpp

11.4.2 ARRAY OF POINTERS TO FUNCTIONS

❖ In C programming, a common practice is to store Pointers to Functions in an array (Cybernetics in C++ - Expert)

➢ This can then be used to call functions in a specified order

❖ However, in C++

➢ This is a less common practice in C++

➢ Since the Function Object/Functor (discussed in detail in Cybernetics in C++ - Expert) is generally preferred over the Function Pointer,

12. REFERENCES

12.1 WHAT ARE REFERENCES?

❖ References are also individual containers which hold memory addresses

➢ In this sense, they are similar to pointers

➢ However, their syntax is different, as are their capabilities and usage

❖ References were introduced into C++ largely due to operator overloading (section 21)

➢ A major application which makes use of this is the Smart Pointer (Cybernetics in C++ - Expert)

❖ References have fewer capabilities than pointers

➢ A common problem with pointers is that they are often too powerful

❖ However, there are several situations where references are preferable over pointers when both will do (section 15)

12.1.1 L-VALUE VS R-VALUE

❖ The concept of L-Values and R-Values for References is identical to that of Pointers

❖ The value contained in a Reference is called an L-Value

➢ This is a Location (see section 2.4.1.2.2)

➢ Known at Compile Time

❖ The contents of the value in a Reference is called an R-Value

➢ This is a Runtime Value (see section 2.4.1.2.2)

➢ Not known until Runtime

12.1.2 RESTRICTIONS WITH REFERENCES

❖ References must be initialised in the same line as their definition

❖ References cannot be edited after initialisation

➢ i.e. References are always read-only

• Effectively, it is as though the `const` keyword has been used

❖ A Reference cannot contain the address of another Reference

➢ Consequently, you cannot implement a "Hierarchy of References" as you can with Pointers

12.2 IMPLEMENTING REFERENCES IN C++

12.2.1 DEFINING REFERENCES

❖ Like with pointers, an entity must be of type Reference-to-dataType e.g.

➢ `Reference-to-int`

➢ `Reference-to-double`

➢ `Reference-to-class` (OOP concept, section 17 onwards)

❖ Syntax: In order to define an entity of type reference, the & operator is used

`Data Type of Referee &`   `&referenceName`

➢ `int& ref_a`	Defines `ref_a` of type Reference-to-int
➢ `double &refDbl_myDbl_B`	Defines `refbl_myDbl_B` of type Reference-to-double
➢ `char& refChar_c`	Defines `refChar_c` of type Reference-to-char

❖ As this shows, the ampersand can be placed:

➢ Before the space, next to the data type

➢ After the space, before the reference name

This idea of syntax is the same as that encountered in the definition of pointers with the placement of the asterisk * operator. (section 10.3.1)

❖ **Note:** Do not confuse this with the ampersand used to populate pointers (see section 15 for details)

12.2.2 POPULATING REFERENCES

❖ References MUST be initialised in the same line as the definition

➢ With pointers, this is advisable but not compulsory

❖ The operator used to say "the address of" for a reference: NOTHING

❖ Thus definitions and initialisations look as follows:

➢ `int& ref_a = a`	Defines `ref_a` of type Reference-to-int to contain the address of the variable a

➢ `double`
`&refDbl_myDbl_B =`
`myDbl_B`
	Defines `refDbl_myDbl_B` of type Reference-to-double, to contain the address of the variable `myDbl_B`

➢ `char& ref_c = c`	Defines `ref_c` of type Reference-to-char to contain the address of the variable `c`

❖ Recall that for pointers, the ampersand operator & is used to mean "the address of" (see section 15 for details)

➢ For references, NOTHING is used

❖ Consequently, if we have the following lines of code in a program:

```
int myInt_A = 7;
int& ref_myInt_A = myInt_A;
```

the layout in runtime memory will be of the following form:

MEMORY ADDRESS	LAYOUT IN RUNTIME MEMORY	NAME	
0x002D39	7	myInt_A	
0x0ED93C	0x002D39	ref_myInt_A	

Figure 12.1 References in Runtime Memory

12.2.3 DECODING REFERENCES

❖ The operator used to say "the value of the contents of" for a reference: NOTHING

❖ Thus once we have populated a reference, stating its name will be an alternative to stating the name of the variable of its referee e.g.

➢ If double &refDbl_myDbl_B = myDbl_B

- Then saying refDbl_myDbl_B thereafter will
  o Give the value of the contents of refDbl_myDbl_B
  o This is the an alternative to saying myDbl_B

❖ This does indeed mean, that unlike pointers, there is no way to display the hexadecimal address, using cout, for a reference; the line

```
cout << refDbl_myDbl_B;
```

will display the value of the contents of `refDbl_myDbl_B`, not the hexadecimal address

❖ Recall that for pointers, the asterisk operator * is reused to mean "the address of" (see section 15 for details)

➢ For references, NOTHING is reused

12.3 USING REFERENCES

❖ The following program puts the above into practice

```
#include <iostream>
using namespace std;

int main()
{
    //»»»»»»»»»»»»»»»»»»»»»»»»»»»»»»»»»»»»»»»»»»»»»»»»»»»»»»»»»»»»»»»»»»»
    int myInt_A = 7;
    double myDbl_B = 9;

    //»»»»»»»»»»»»»»»»»»»»»»»»»»»»»»»»»»»»»»»»»»»»»»»»»»»»»»»»»»»»»»»»»»»
    //      DEFINING & Populating References
    //
    int& ref_myInt_A = myInt_A;
    double& ref_myDbl_B = myDbl_B;

    //»»»»»»»»»»»»»»»»»»»»»»»»»»»»»»»»»»»»»»»»»»»»»»»»»»»»»»»»»»»»»»»»»»»
    //      DECODING References
    //
    cout  << "The value of the Integer referred to by ref_myInt_A: "
          << ref_myInt_A<< endl;

    cout  << "The value of the Double referred to by ref_myDbl_B: "
          << ref_myDbl_B<< endl;

    cout  << "myInt_A = "<< myInt_A<< endl;
    cout  << "myDbl_B = "<< myDbl_B<< endl;

    //»»»»»»»»»»»»»»»»»»»»»»»»»»»»»»»»»»»»»»»»»»»»»»»»»»»»»»»»»»»»»»»»»»»
    cout  << endl<< "=============="
          << endl<< "END...C++ Demo: What are References?"<< endl;
    while(true){}
    return 0;
}//END...Function: main
```

```
The value of the Integer referred to by ref_myInt_A: 7
The value of the Double referred to by ref_myDbl_B: 9
myInt_A = 7
myDbl_B = 9
```

```
===============
END...C++ Demo: What are References?
```

Program Listing/Output 12.1: WhatAreReferences_Client.cpp Demonstrates the usage of C++ References

12.4 REFERENCES AS FUNCTION INPUTS

12.4.1 PROBLEMS: LACK OF ADDRESSING

❖ The importance of addressing, which we saw with pointers, can also be implemented using references

❖ To start with. let's remind ourselves of the swap program, and the problems with lack of addressing

```cpp
#include <iostream>
using namespace std;

// Function DECLARATIONS
void swap( int int_IN_A, int int_IN_B );


int main()
{
    //»»»»»»»»»»»»»»»»»»»»»»»»»»»»»»»»»»»»»»»»»»»»»»»»»»»»»»»»»»»»»»»»»»»»
    int myInt_A = 5, myInt_B = 7;
    cout  << "Original values: "
          << "myInt_A = "<< myInt_A
          << " and myInt_B = "<< myInt_B
          << endl;

    //»»»»»»»»»»»»»»»»»»»»»»»»»»»»»»»»»»»»»»»»»»»»»»»»»»»»»»»»»»»»»»»»»»»»
    //    Attempting the swap
    swap( myInt_A, myInt_B );

    //»»»»»»»»»»»»»»»»»»»»»»»»»»»»»»»»»»»»»»»»»»»»»»»»»»»»»»»»»»»»»»»»»»»»
    cout  << "Values after attempted swap: "
          << "myInt_A = "<< myInt_A
          << " and myInt_B = "<< myInt_B
          << endl;

    //»»»»»»»»»»»»»»»»»»»»»»»»»»»»»»»»»»»»»»»»»»»»»»»»»»»»»»»»»»»»»»»»»»»»
    cout  << endl<< "=========="
          << endl<< "End...Demo, problems, lack of Memory Management"
          << endl;
    while(true){}
    return 0;

}//END...Function Main
```

```
void swap( int int_IN_A, int int_IN_B )
{
        int temp = int_IN_A;
        int_IN_A = int_IN_B;
        int_IN_B = temp;

}//END...Function swap
```

```
Original values: myInt_A = 5 and myInt_B = 7
Values after attempted swap: myInt_A = 5 and myInt_B = 7

==========
End...Demo, problems, lack of Memory Management
```

Program Listing/Output 12.2: Attempting a swap - with a fundamental flaw

❖ Revision: The problem in this program is the inputs are being passed by VALUE, so only two numbers, not the variables are being swapped...

12.4.2 PASSING FUNCTION INPUTS AS REFERENCES

❖ ...In order to fix this problem, the variables themselves need to be passed: this is possible by passing their addresses

➤ One possibility is to pass the inputs as pointers, section 12.4.3

➤ We will now see an alternative: pass them by reference

12.4.2.1 Formal Parameters as References

❖ Like all other variables, references need to have their types specified in the formal parameters

➤ Revision: the ampersand & operator and the data type of the pointee are used

❖ Thus a function with

➤ The name myFunc

➤ 2 inputs:
  - 1, called refInt_In_A, of type pointer-to-int
  - 1, called refDbl_In_B, of type pointer-to-double

➤ No output

will have the following prototype:

```
void myFunc( int &refInt_In_A, double &refDbl_In_B )
{
        // Functionalities of myFunc
}
```

❖ The signature of the function, used in the declaration, will be as follows:

```
void myFunc( int&, double & );
```

12.4.2.2 Actual Parameters as References

❖ When invoking a function, the inputs, the actual parameters, need to be the values that the input variables take on

❖ In the case of references, these "values" are memory addresses

  ➢ Since NOTHING is used to mean "the address of" for references, this is seen for reference actual parameters

❖ So for our example, a call to myFunc will be as follows:

```
myFunc( myInt_A, myDbl_B );
```

  ➢ where
    • myInt_A is a previously defined variable of type int
    • myDbl_B is a previously defined variable of type double
  ➢ The inputs are saying "The Address of" the variables; the first is saying:
    • The address of (myInt_A)
    • NOTHING myInt_A

12.4.3 FIXING THE PROBLEMS USING REFERENCES

❖ The following program shows how the swap can work properly when the inputs are References

```
#include <iostream>
using namespace std;

void swap_usingReferences( int&, int& );


int main()
{
    //»»»»»»»»»»»»»»»»»»»»»»»»»»»»»»»»»»»»»»»»»»»»»»»»»»»»»»»»»»»
    int myInt_A = 11, myInt_B = 13;

    cout   << "Before the attempted swap myInt_A = "<< myInt_A
           << " and myInt_B = "<< myInt_B
           << endl;

    //»»»»»»»»»»»»»»»»»»»»»»»»»»»»»»»»»»»»»»»»»»»»»»»»»»»»»»»»»»»
    /*     INVOKING swap_usingReferences - inputs References
           -      Want the inputs to be
```

```
                        -        "The ADDRESS of myInt_A",
                        -        "The ADDRESS of myInt_B"
                -       We say myInt_A and myInt_B, since
                        -        the operator to say "The address of a variable
                                 referred to by a reference": NOTHING
        */
        swap_usingReferences( myInt_A, myInt_B  );

        //»»»»»»»»»»»»»»»»»»»»»»»»»»»»»»»»»»»»»»»»»»»»»»»»»»»»»»»»»»»»»»
        cout  << "After the attempted swap myInt_A = "<< myInt_A
              << " and myInt_B = "<< myInt_B
              << endl;

        //»»»»»»»»»»»»»»»»»»»»»»»»»»»»»»»»»»»»»»»»»»»»»»»»»»»»»»»»»»»»»»
        cout  << endl<< "==========="
              << endl<< "END...C++ References as Function Inputs Demo"
              << endl;
        while(true){}
        return 0;

}//END...function main


/*      ++++++++++++++++++++++++++++++++++++++++++++++++++++++++++++++
        Function DEFINITION swap_usingReferences
        -       We are now going to pass the variables themselves,
                not their values
        -       When a variable is defined, (for the best part) it stays
                in the SAME memory location until it goes out of scope
        -       Thus passing this memory ADDRESS to the function is the
                way we ensure the variable itself is passed
                -       Here we achieve this using References
*/
void swap_usingReferences( int& ref_Int_In_A, int& ref_Int_In_B  )
{
        int temp = ref_Int_In_B; // temp = CONTENTS OF ref_Int_In_B
        ref_Int_In_B = ref_Int_In_A;
        ref_Int_In_A = temp;

}// END...function swap_usingReferences
```

```
Before the attempted swap myInt_A = 11 and myInt_B = 13
After the attempted swap myInt_A = 13 and myInt_B = 11


===========
END...C++ References as Function Inputs Demo
```

Program Listing/Output 12.3: swap_usingReferences.cpp Fixing the swap problem using references

❖ Now, since the addresses of my Int_A and my Int_B are being passed, it is the contents of these memory addresses, i.e. the variables themselves, which get manipulated

13. ADVANCED REFERENCES

13.1 READ-ONLY REFERENCES USING CONST

❖ Revision: the keyword const is used to make a variable read-only i.e. no longer a variable

➢ This section is similar to making pointers read only, but simpler,

❖ As we have been discussing, when dealing with references, just like with pointers, we are looking at two items of data:

➢ The reference itself
➢ The data being referred to

❖ Revision: a reference is always read only

❖ Again, the question is how do we use const in order to make

➢ Only the data being referred to read only,

❖ The answer depends on where the word const is placed, in the definition, relation to the ampersand operator &

➢ You might want to use BR TAR
   • If const comes BEFORE the ampersand &, the data REFERRED-TO is read only
   • If const comes AFTER the ampersand &, the REFERENCE is read only
     ○ This is the case anyway, so this placement of const is redundant

❖ As you are seeing, this is a similar story to using const with pointers

❖ The following program demonstrates this concept:

```
#include <iostream>
using namespace std;

int main( )
{
        //»»»»»»»»»»»»»»»»»»»»»»»»»»»»»»»»»»»»»»»»»»»»»»»»»»»»»»»»»»»»»»»»»»»
        int myInt_A = 14, myInt_B = 17, myInt_C = 23, myInt_D = 27,
            myInt_E = 35, myInt_F = 44, myInt_G = 53, myInt_H = 67;
```

```
        //»»»»»»»»»»»»»»»»»»»»»»»»»»»»»»»»»»»»»»»»»»»»»»»»»»»»»»»»»»»»»»»
        int &refInt_myInt_A = myInt_A;

        //»»»»»»»»»»»»»»»»»»»»»»»»»»»»»»»»»»»»»»»»»»»»»»»»»»»»»»»»»»»»»»»
        // The const keyword is BEFORE the ampersand &
        // -          Contents of the Reference are Read-only
        //
        const int &refInt_myInt_B = myInt_B;
//      refInt_myInt_B = 79;        // ILLEGAL - contents of refInt_myInt_B
                                    // are read-only ( via the reference )

        myInt_B = 78;        // OK - "Outside" the reference,
                             // myInt_B is still writeable


        //»»»»»»»»»»»»»»»»»»»»»»»»»»»»»»»»»»»»»»»»»»»»»»»»»»»»»»»»»»»»»»»
        //    The const keyword is AFTER the ampersand & - REDUNDANT
        //    -     The Reference is Read-only
        //          -     It is anyway, so this is redundant!
        //
        int& const refInt_myInt_C = myInt_C;

        refInt_myInt_C = 25;        //    OK - contents of refInt_myInt_C
                                    //    are writeable


        //»»»»»»»»»»»»»»»»»»»»»»»»»»»»»»»»»»»»»»»»»»»»»»»»»»»»»»»»»»»»»»»
        // The const keyword is both BEFORE and AFTER the ampersand &
        //
        const int & const refInt_myInt_D = myInt_D;

//      refInt_myInt_D = myInt_H;       // ILLEGAL - refInt_myInt_D
                                        // is read-only

//      refInt_myInt_D = 29;        // ILLEGAL - contents of
                                    // refInt_myInt_D are read-only
                              .     // ( via the pointer )

        myInt_D = 30;        //    OK - "Outside" the pointer, myInt_D is
                             //    still writeable


        //»»»»»»»»»»»»»»»»»»»»»»»»»»»»»»»»»»»»»»»»»»»»»»»»»»»»»»»»»»»»»»»
        cout   << endl<< endl<< "==========="<< endl
               << "END...C++ Demo: Const with References"<< endl;
        while(true){}
        return 0;

}//END...Function MAIN
```

Program Listing/Output 13.1: constReferences_Client.cpp Making references and referees read only

13.2 POTENTIAL AMBIGUITIES IN REFERENCE NOTATION

❖ The notation used to define and decode references can lead to some potentially ambiguous code

➢ The Line

```
ref_myInt_A = myInt_C;
```

might look like we are trying to edit `ref_myInt_A` to refer to the address of `myInt_C`

- It seems plausible since `myInt_C` does also mean the address of `myInt_C` when dealing with references
- However, references are read-only after initialisation
- What this is actually doing is taking the value of `myInt_C` and placing it into the contents of `ref_myInt_A`
  - This is fine because `ref_myInt_A` also means "the value of the contents of `ref_myInt_A` "
  - i.e. This line has the same effect as

  ```
  myInt_A = myInt_C;
  ```

  - where myInt_A is the data being referred to by `ref_myInt_A`

13.3 REFERENCE TO A FUNCTION

❖ It is possible to have a reference to a function, just like it is possible to have a Pointer to a function

❖ However, since references cannot be stored in arrays (and most other data structures)[1], this is a rare practice

14. RUNTIME MEMORY MANAGEMENT

14.1 GARBAGE COLLECTION

❖ Once you have finished with data, it is **Garbage**

❖ It is imperative the garbage is discarded, otherwise, in very little time, your runtime memory will fill with garbage, causing your program to crash

➢ We have all visited offices and houses where people never throw away their garbage! :-)

---

[1]This is the case with most modern day compilers. However, even if a compiler allows it, this practice would mean mixing together pointers and references in the same hierarchy which is highly inadvisable

➤ This will have disastrous consequences in computer science

❖ Managing the discarding of garbage data is called "Garbage Collection"

❖ There are primarily two kinds of runtime memory in C and C++

   ➤ The **RUNTIME STACK**:
      • The Garbage Collection is carried out automatically by the computer
   ➤ The **RUNTIME HEAP**:
      • The Garbage Collection is managed by the programmer
      (i.e. YOU! :-))
      • Terminology Note: Some programmers use the following variation:
        ◦ The Heap to mean the runtime memory where primitive variables, and ADTS can be allocated, used in C programming
        ◦ The Free Store to mean the runtime memory where all the C programming entities, as well as the new C++ OOP entity, objects can be allocated
        ◦ In this book, we stick to the term "Heap" to mean the runtime memory for objects, as well as ADTs and primitive variables

❖ Note that the whole of this chapter is looking at **runtime** memory

   ➤ This can be accessed only while the program is running
      • As soon as it ends, all the data is erased
   ➤ In order to save data, so it can be retrieved at a later date "Streaming" is required, to save the data to disk in a file
      • We will cover this in section 25

14.2 THE RUNTIME STACK

❖ The Runtime Stack is the memory you have been using so far

   ➤ The computer takes care of your garbage collection

14.2.1 LIFO: LAST IN FIRST OUT NATURE

❖ A stack is a structure which follows a LIFO: Last In First Out nature

   ➤ Outside the programming world: The analogy of a stack of dinner plates is often used:
      • This can be accessed from nowhere but the top
   ➤ In C++ programming, a stack is used for

➤ The runtime stack

➤ The specialist Stack data structure (Cybernetics in C++ - Expert)

❖ The runtime stack works with **blocks** of data

➤ A block is any entity which is surrounded by braces { }, examples include:

- Conditional statements
- Loops
- Functions
- ADTs - structs, enums and unions (section 17 onwards)
- Classes (section 17 onwards)

❖ As soon as a block where a variable was defined is exited, that variable goes **out of scope**

➤ It is then discarded from memory

- What actually happens is the computer allocates this memory for other resources

❖ In C++ programming, having blocks within blocks is ubiquitous

➤ The later a variable is defined (i.e. comes into scope), the sooner it goes out of scope

➤ Thus the automated management of variables follows that of a stack - a LIFO Last In First Out system

❖ The following pseudo code program shows the workings of the runtime stack

```
#include <iostream>
#include <string>
using namespace std;

// Demo only, don't try to build and run this

// Global Function
int myFunc( int myInt_IN );

// Global Variable
int myGlobalInt = 8;      // AVOID global variables

int main()  // Variable myGlobalInt comes into scope, placed on runtime
            // stack
{
```

```
        int myMainInt = 7; // Variable myMainInt comes into scope, placed
                           // on runtime stack

        int myresult = myFunc( myMainInt ); // Variable myresult comes
                                            // into scope, placed on
                                            // runtime stack

                                            // Program now jumps to the
                                            // function myFunc

        return 0;

}

//  End of main() - myresult, myMainInt go OUT OF SCOPE - discarded from
//  the runtime stack

// End of Program: myLocalStaticInt, myGlobalInt go OUT OF SCOPE -
// discarded from the runtime stack

// Function Definition
int myFunc( int myInt_IN )      // Variable myMainInt comes into scope,
                                // placed on runtime stack
{
        int myLocalInt = 7;             // Variable myLocalInt comes into
                                        // scope, placed on runtime stack
        static int myLocalStaticInt = 16;   // Variable myLocalStaticInt
                                            // comes into scope, placed
                                            // on runtime stack

        for( int i = 0; i < 10; ++i  ) // Variable i comes into scope,
                                        // placed on runtime stack
        {
             int myForInt = 19; // Variable myForInt comes into scope,
                                // placed on runtime stack

             if( myForInt == 10 )
             {
                  // Conditional actions
                  double myIfInt = 9; // Variable myIfInt comes into
                                      // scope, placed on runtime stack

             }// END...if statement...myIfInt goes OUT OF SCOPE -
              // discarded from the runtime stack

             // myIfInt = 9;            // ERROR: myIfInt has gone OUT OF
             // SCOPE - i.e. no longer exists

        }// END...for loop...myForInt and i go OUT OF SCOPE - discarded
         //from the runtime stack

        return myLocalInt;
```

```
}// END...function myFunc...myLocalInt and myInt_IN go OUT OF SCOPE -
    discarded from the runtime stack
// NB. myLocalStaticInt stays in scope for the duration of the program
//   -    that's why it can retain its value between function calls

// Jump back to the part of the program after the function call
```

Program Listing/Output 14.1: Pseudo Code: Demonstration: Scope of Variables and the Runtime Stack

14.2.2 STACK OVERFLOW

❖ A very common runtime bug is when the stack runs out of space

❖ Common reasons include:

➢ Infinite function calls; these occur in
  • Circular references (section 5.2.4.1)
  • Recursion when a base case has been missed/not defined altogether (section 8.1.2)
➢ Trying to place too much data on the stack
  • This commonly happens when there is a long chain of function calls
  • Many of them comprise having large data structures as inputs or local variables

❖ This is such a common problem, it is considered a part of software culture

➢ There is a web forum, for computer programmers, called www.stackove rflow.com

14.3 THE RUNTIME HEAP / FREE STORE

❖ The Runtime Heap is the runtime memory where the programmer (i.e. You! :-)) control(s) the scope of variable

❖ It can only ever be accessed using pointers

➢ Never directly or via references

14.3.1 THE NEW OPERATOR

❖ The `new` operator returns a pointer to the first memory address where the data is stored

❖ The following syntax is required:

```
dataType *pointerName = new dataType
```

❖ In order to allocate an int on the heap, use the following

```
int *ptr_Int_A = new int;
```

➢ It is advisable, but not compulsory, to initialise it on the same line using the parentheses operator():

```
int *ptr_Int_A = new int(0);
```

➢ After this, it can be manipulated in the usual way

```
*ptr_Int_A = 5
```

❖ Note that the pointer itself is still on the runtime stack

➢ Its contents are on the runtime heap

❖ As we will see in section 18.5.5, further functionalities take place when allocating objects on the heap

14.3.2 THE DELETE OPERATOR

❖ It is essential we do not forget to delete the variable we have allocated i.e. force it to go out of scope

❖ We achieve this as follows:

```
delete ptr_Int_A;
```

❖ As we will see in section 18.5.5, further functionalities take place when deleting objects on the heap

14.3.2.1 Dangling Pointer

❖ It is important to note that the `delete` command

➢ Deletes/Forces out of scope the data being pointed to, on the Heap
➢ The Pointer itself remains in scope, (usually) on the Runtime Stack

❖ We now have issue of the Pointer being a **dangling** pointer

➢ A pointer, the contents of which, have gone out of scope

❖ Any attempt to decode the contents of this pointer will result in reading garbage

❖ It is thus good practice to set this pointer to NULL

```
ptr_Int_A = NULL;
```

❖ **Rules of Thumb:** When finished with data on the Runtime Heap

➢ Free it up using the `delete` operator
➢ Set the pointer to NULL

14.4 USING THE RUNTIME HEAP

14.4.1 DIRECT USAGE FROM THE STACK

❖ The following program demonstrates the usage of the heap

➢ Note again that the pointers ptr_numPeople and ptr_totalCost are on the runtime stack

➢ Thus the heap is being controlled from the stack:

```
/*     *************************************************************
       HEAP SIMPLE USAGE
       *************************************************************
       -      This program places data on the Runtime Heap
       -      Performs calculations on this data
              -      using Pointers which are themselves on the Stack
              -      Stores the result on the Runtime Stack
       -      Frees up the Heap Memory
       -      As a safety measure, ensures the used Pointers are set to
              NULL
*/
#include <iostream>
#include <string>
using namespace std;

int main( )
{
    //»»»»»»»»»»»»»»»»»»»»»»»»»»»»»»»»»»»»»»»»»»»»»»»»»»»»»»»»»»»»»»»»»
    //     Placing data on the Heap
    //
    int *ptr_numPeople = new int(5);
    double *ptr_totalCost = new double(85.5);

    cout  << "The number of people, in the Runtime Heap, pointed to"
          << " by ptr_numPeople is "<< *ptr_numPeople
          << endl;
    cout  << "The total cost, in the Runtime Heap, pointed to by "
          << "ptr_totalCost is "<< *ptr_totalCost
          << endl;

    //    Using the data on the Heap
    //    -      The result is being stored on the Runtime Stack
    double pricePerPerson = *ptr_totalCost /

    static_cast<double>(*ptr_numPeople);
    cout  << "The price per person is "<< pricePerPerson<< endl;

    //»»»»»»»»»»»»»»»»»»»»»»»»»»»»»»»»»»»»»»»»»»»»»»»»»»»»»»»»»»»»»»»»»
    // Freeing up the Heap - do not forget this! :-)
    //
    delete ptr_numPeople;
    delete ptr_totalCost;
```

```
        // Good Practice: Set the now Dangling Pointers to NULL
        ptr_numPeople = NULL;
        ptr_totalCost = NULL;

        //»»»»»»»»»»»»»»»»»»»»»»»»»»»»»»»»»»»»»»»»»»»»»»»»»»»»»»»»»»»»»»»»»»»»»»
        cout   << endl<< endl<< "==========="
               << endl
               << "END...C++ Demo: Heap Simple Usage"
               << " "
               << endl;
        while(true){}
        return 0;

}//END...Function MAIN
```

```
The number of people, in the Runtime Heap, pointed to by ptr_numPeople
    is 5
The total cost, in the Runtime Heap, pointed to by ptr_totalCost is
    85.5
The price per person is 17.1


===========
END...C++ Demo: Heap Simple Usage
```

Program Listing/Output 14.2: Heap_SimpleUsage_Client.cpp

14.4.2 SHARING THE RUNTIME HEAP BETWEEN FUNCTION CALLS

❖ One of the major strengths of using the runtime heap is it enables the programmer (i.e. you! :-)) to control the scope of data between function calls

➢ When using the runtime stack, the control is almost entirely down to the computer

➢ The one small amount of control you have is making a local variable static

  • However, this means the data stays in scope for the rest of the duration of the program

  • This may well be wasteful

❖ The following program shows this in practice:

```
/*      ****************************************************************
        FUNCTION USING THE RUNTIME HEAP
        ****************************************************************
        -       Function places data on the Heap Runtime Memory
                -       Calculates Total Ticket price
                -       Returns a Pointer to this
        -       This Pointer is then used by function calling it
                -       In this case main
*/
#include <iostream>
#include <string>
using namespace std;

//++++++++++++++++++++++++++++++++++++++++++++++++++++++++++++++++++++
double* calc_TotalCost( int numTickets, double pricePerTicket );

//++++++++++++++++++++++++++++++++++++++++++++++++++++++++++++++++++++
int main( )
{
        //»»»»»»»»»»»»»»»»»»»»»»»»»»»»»»»»»»»»»»»»»»»»»»»»»»»»»»»»»»»»»»»
        int num_London2012_CyclingTickets = 23;
        double cost_London2012_Ticket_Cycling = 75.0;

        double *ptr_Cost_Cycling
                = calc_TotalCost( num_London2012_CyclingTickets,

        cost_London2012_Ticket_Cycling
                                                );

        cout   << "The cost of taking "
               << num_London2012_CyclingTickets<< " people, "
               << "each ticket costing "<< cost_London2012_Ticket_Cycling
               << " will be "<< *ptr_Cost_Cycling
               << endl;

        // Do not forget to free up the Heap Memory
        delete ptr_Cost_Cycling;
        ptr_Cost_Cycling = NULL;

        //»»»»»»»»»»»»»»»»»»»»»»»»»»»»»»»»»»»»»»»»»»»»»»»»»»»»»»»»»»»»»»»
        cout   << endl<< endl<< "==========="
               << endl
               << "END...C++ Demo: Function using the Runtime Heap"
               << " "
               << endl;
        while(true){}
        return 0;

}//END...Function MAIN

/*++++++++++++++++++++++++++++++++++++++++++++++++++++++++++++++++++++
```

```
      FUNCTION: calc_TotalCost
         -      Calculates total ticket price
         -          Inputs: Number of Tickets, Price per Ticket
         -          Places the result on the heap
                -      Returns a Pointer to it
*/
double* calc_TotalCost( int numTickets, double pricePerTicket )
{
      double *ptr_totalCost = new double(0.0);

      *ptr_totalCost = static_cast<double>(numTickets) * pricePerTicket;

      return ptr_totalCost;

}//END...Function calc_TotalCost
```

```
The cost of taking 23 people, each ticket costing 75 will be 1725

===========
END...C++ Demo: Function using the Runtime Heap
```

Program Listing/Output 14.3: HeapFunction_client.cpp

14.5 PROSPECTIVE BUGS WHEN USING THE RUNTIME HEAP

❖ As we are seeing when we use the runtime heap, we have in our hands great power...

➢ We are controlling the scope of variables

❖ ...with which comes great responsibility

➢ We must make sure the allocated memory is freed appropriately

- Doing this too late or forgetting to do so altogether causes a **memory leak**
- Doing so too early causes **memory corruption**

14.5.1 MEMORY LEAKS

❖ A memory leak is freeing up the heap memory too late or forgetting altogether

❖ As we have seen, heap memory can be accessed only using a pointer

❖ We have completely lost control of the allocated heap memory if

➢ We lose access to that pointer

➤ That pointer loses access to that part of the memory

❖ As the following example shows, these can happen very easily

```
/*      ************************************************************
        HEAP MEMORY LEAK
        ************************************************************
        -       Demonstrates Memory Leaks
                -       Simply reallocation of a Pointer to a Heap
                        -       Lost all access to this part of the Heap
        -       Leaky Loop
                -       Reallocates the same pointer to different parts of
                        the Heap per iteration
                        -       Each  iteration - it loses the pointer it
                                allocated on the last iteration
*/
#include <iostream>
#include <string>
using namespace std;

int main( )
{
        //»»»»»»»»»»»»»»»»»»»»»»»»»»»»»»»»»»»»»»»»»»»»»»»»»»»»»»»»»»»»»»»»
        int myInt_Stack_A = 7;

        // Creating an int on the Heap, accessing it using a Pointer
        int *ptr_myInt_Heap_A = new int(13);

        // POTENTIAL DISASTER: Reallocating the Pointer to a Stack
        // Variable
        ptr_myInt_Heap_A = &myInt_Stack_A;

        // ===== MEMORY LEAK! ===== All access to the integer on the
        // Heap has now been lost!

        //»»»»»»»»»»»»»»»»»»»»»»»»»»»»»»»»»»»»»»»»»»»»»»»»»»»»»»»»»»»»»»»»
        //      LEAKY LOOP!
        //      -       Each iteration - it loses the pointer it allocated
        //              on the last iteration!
        //
        for( int i = 0; i < 7; ++i )
        {
                double *ptr_Dbl = new double(i);
        }


        //»»»»»»»»»»»»»»»»»»»»»»»»»»»»»»»»»»»»»»»»»»»»»»»»»»»»»»»»»»»»»»»»
        cout  << endl<< endl<< "==========="
                << endl
                << "END...C++ Demo: Heap Memory Leak"
                << " "
                << endl;
        while(true){}
```

```
        return 0;

}//END...Function MAIN
```

Program Listing/Output 14.4: HeapMemory_Leak_Client.cpp

❖ What this means is that for the duration of this program, the (expensive!) parts of the runtime memory are no longer accessible

➢ At best , this is a waste

➢ Too much of this will cause the program to crash

❖ Only when the program is ended will the problem be tackled

➢ In some cases, this is a problem, but not an unbearable one

• For something like office software, e.g. a word processor, this is expected to be closed and reopened at least a few times a day

• An operating system often requires a reboot at least every few days

➢ For a program which should be rebooted infrequently, this can be catastrophic; examples include

• Software which is embedded in a microprocessor, located under the sea bed, to signal to passing submarines

14.5.2 MEMORY CORRUPTION

❖ Memory corruption is freeing up memory while it is still in use

❖ This can also happen very easily

```
/*      ************************************************************
        HEAP MEMORY CORRUPTION
        ************************************************************
        -       Demonstrates when Heap Memory is freed too early
                -       Garbage is output when attempting to access the
                        part of the heap which has been freed
*/
#include <iostream>
#include <string>
using namespace std;

int main( )
{
        //»»»»»»»»»»»»»»»»»»»»»»»»»»»»»»»»»»»»»»»»»»»»»»»»»»»»»»»»»»»»»»»»»»»»»
        int myInt_Stack_A = 7;

        // Creating an int on the Heap, accessing it using a Pointer
        int *ptr_myInt_Heap_A = new int(17);
```

```
                // Some operations using ptr_myInt_Heap_A...

        delete ptr_myInt_Heap_A;

        cout   << "The integer on the Heap, pointed to by ptr_myInt_Heap_A"
               << " is "<< *ptr_myInt_Heap_A
               << endl;
               // Will display garbage

        //»»»»»»»»»»»»»»»»»»»»»»»»»»»»»»»»»»»»»»»»»»»»»»»»»»»»»»»»»»»»»»»»»»»»»»
        cout   << endl<< endl<< "==========="
               << endl
               << "END...C++ Demo: Heap Memory Corruption"
               << " "
               << endl;
        while(true){}
        return 0;

}//END...Function MAIN
```

```
The integer on the Heap, pointed to by ptr_myInt_Heap_A is -17891602


===========
END...C++ Demo: Heap Memory Corruption
```

Program Listing/Output 14.5: HeapMemory_Corruption_Client.cpp

❖ After any runtime memory has been freed up, it is reallocated for further use

❖ The issue here is that the program has displayed the value, now on the heap, which is garbage

❖ However, this is not noticeable until runtime and even then, can be difficult to spot

14.5.3 DANGLING POINTERS

❖ A dangling pointer is a pointer, the contents of which have gone out of scope

❖ Since the pointer itself remains perfectly valid and in scope, often the only indication that a pointer is dangling is when garbage is encountered at runtime

❖ This is a very common, particularly problematic bug

➢ It is commonly encountered when using the heap

❖ Smart Pointers are a special entity which mimic pointers, but are able to prevent much of these bugs

  ➢ These are covered in Cybernetics in C++ - Expert

14.6 MALLOC() AND FREE() FROM C PROGRAMMING

❖ C programmers use the functions

  ➢ `malloc()` to allocate memory on the heap

    • This is a portmanteau for <u>mem</u>ory <u>alloc</u>ation

  ➢ `free()` to free this memory

❖ There are several reasons C++ programmers do not use these

  ➢ `malloc()` and `free()` cannot deal with objects (section 17 onwards)

    • `malloc()` will not call constructors (section 18.9)
    • `free()` will not call destructors (section 18.10)

  ➢ In addition, there is automatic error checking in `new()`

    • `malloc()` will not give any visible indication if memory has not been successfully allocated
      ○ It will simply return a NULL pointer
      ○ Consequently, C programmers are always advised to check if the returned pointer from a `malloc()` call is NULL, before attempting to allocate anything on the heap
    • The `new()` operator will give an automatic error message if the heap is full
      ○ i.e. It performs the checking for you

❖ **Rule of Thumb:** Use `malloc()` and `free()` only in situations where backward compatibility with C programming is required, otherwise stick to `new()` and `delete()`

14.7 USING BOTH THE RUNTIME STACK AND HEAP

❖ Putting together the above discussions, when we want to place variables in runtime memory then use them, there are no less than 4 different ways of doing so

  ➢ On the runtime **stack**, accessing **directly**
  ➢ On the runtime **stack**, accessing using **pointers**

➤ On the runtime **stack**, accessing using **references**

➤ On the runtime **heap**, accessing using **pointers**

❖ The following program, which performs the simple task of adding together two integers, shows the 4 different ways this can be accomplished

```
/*      »»»»»»»»»»»»»»»»»»»»»»»»»»»»»»»»»»»»»»»»»»»»»»»»»»»»»»
        SIMPLE TASK: Adding 2 numbers
        No less than 4 different ways to do this
        (1) Using the runtime STACK
                (1.1) Accessing directly
                (1.2) Accessing using Pointers
                (1.3) Accessing using References
        (2) Using the runtime HEAP
                (2.1) Accessing using Pointers
*/
#include <iostream>
#include <string>

using namespace std;

int main()
{
        //»»»»»»»»»»»»»»»»»»»»»»»»»»»»»»»»»»»»»»»»»»»»»»»»»»»
        //      (1)     Using the runtime STACK
        //              (1.1) Accessing directly
        cout    << endl<< "======="<< endl<< "Using the Runtime Stack, "
                                        << "accessing Directly"
                                    << endl;
        int myInt_1_1_A = 5, myInt_1_1_B = 6, myInt_1_1_C = 0;
        myInt_1_1_C = myInt_1_1_A + myInt_1_1_B;
        cout    << "The total of "<< myInt_1_1_A<< " and "<< myInt_1_1_B
                << " is "<< myInt_1_1_C<< endl;

        //-------------------------------------------------
        //      (1.2) Accessing using Pointers
        cout    << endl<< "======="<< endl<< "Using the Runtime Stack, "
                                        "accessing using Pointers"
                                    << endl;
        int myInt_1_2_A = 7, myInt_1_2_B = 8, myInt_1_2_C;

        int *ptr_myInt_1_2_A = &myInt_1_2_A;
        int *ptr_myInt_1_2_B = &myInt_1_2_B;
        int *ptr_myInt_1_2_C = &myInt_1_2_C;

        *ptr_myInt_1_2_C = *ptr_myInt_1_2_A + *ptr_myInt_1_2_B;
        cout    << "The total of " << *ptr_myInt_1_2_A<< " and "
                                << *ptr_myInt_1_2_B<< " is "
                                << *ptr_myInt_1_2_C
                                << endl;
```

```
    //---------------------------------------------------
    //    (1.3) Accessing using References
    cout << endl<< "======="<< endl<< "Using the Runtime Stack, "
                              << "accessing using References"
         << endl;
    int myInt_1_3_A = 9, myInt_1_3_B = 10, myInt_1_3_C;

    int& ref_myInt_1_3_A = myInt_1_3_A;
    int& ref_myInt_1_3_B = myInt_1_3_B;
    int& ref_myInt_1_3_C = myInt_1_3_C;

  ref_myInt_1_3_C = ref_myInt_1_3_A + ref_myInt_1_3_B;

  cout << "The total of "<< ref_myInt_1_3_A<< " and "
                         << ref_myInt_1_3_B<< " is "
                         << ref_myInt_1_3_C
                         << endl;

//»»»»»»»»»»»»»»»»»»»»»»»»»»»»»»»»»»»»»»»»»»»»»»»»»»»»»»
//    (2) Using the runtime HEAP
//        - Accessing using Pointers
    cout << endl<< "======="<< endl<< "Using the Runtime Heap, "
                              << "accessing using Pointers"
         << endl;
    int *ptr_myInt_2_1_A = new int;
    int *ptr_myInt_2_1_B = new int;
    int *ptr_myInt_2_1_C = new int;

    *ptr_myInt_2_1_A = 11;
    *ptr_myInt_2_1_B = 12;

    *ptr_myInt_2_1_C = *ptr_myInt_2_1_A + *ptr_myInt_2_1_B;

  cout << "The total of " << *ptr_myInt_2_1_A<< " and "
                          << *ptr_myInt_2_1_B<< " is "
                          << *ptr_myInt_2_1_C
                          << endl;

  // Do not forget to discard these!!!
  delete ptr_myInt_2_1_A;
  delete ptr_myInt_2_1_B;
  delete ptr_myInt_2_1_C;

//»»»»»»»»»»»»»»»»»»»»»»»»»»»»»»»»»»»»»»»»»»»»»»»»»»»»»»
  cout << endl<< "========="
       << endl<< "End...C++ Demo: Adding numbers using "
              << "the Runtime Stack And Heap"
       << endl;
  while(true){}
  return 0;

}//END...Function MAIN
```

```
=======
Using the Runtime Stack, accessing Directly
The total of 5 and 6 is 11

=======
Using the Runtime Stack, accessing using Pointers
The total of 7 and 8 is 15

=======
Using the Runtime Stack, accessing using References
The total of 9 and 10 is 19

=======
Using the Runtime Heap, accessing using Pointers
The total of 11 and 12 is 23

=========
End...C++ Demo: Adding numbers using the Runtime Stack And Heap
```

Program Listing/Output 14.6: Using the Runtime Stack (3 ways) and the Runtime Heap

14.7.1 WHEN TO USE WHICH RUNTIME MEMORY

❖ For the same tasks, using the runtime stack is actually faster than the runtime heap

➢ For the heap, since everything has to be read using a pointer, at least 2 fetch instructions are required to read or write data

➢ There is a lower overhead with the runtime stack

❖ This is often pleasantly surprising news! :-)

➢ It would be as if getting someone else to clean your house would be cheaper than doing it yourself! :-)

❖ **Rule of Thumb:** Use the runtime stack whenever possible; use the heap only when you need to control the scope of a variable

➢ Often jokingly said: "Stick to the Stack"

14.8 DYNAMIC ARRAYS

❖ A dynamic array is one which can change its size during runtime

❖ If you want to implement this yourself, you must use the runtime heap (covered in Cybernetics in C++ - Expert)

❖ The STL provides libraries for this; you are advised to use these whenever possible

14.8.1 THE NEW[] AND DELETE[] OPERATORS

❖ There are modified operators `new[]` and `delete[]` which are used whenever arrays need to be allocated on the heap

❖ Such arrays can be, and usually are dynamic: the size is determined and altered at runtime

❖ Creating your own dynamic arrays

➢ Is a common practice in C Programming

➢ Is less common in C++ (covered in Cybernetics in C++ - Expert)

• You are advised to use the STL Libraries vector and deque whenever possible (covered in Cybernetics in C++ - Expert)

❖ In summary, for allocation and deletion from the heap runtime memory

➢ `new[]` and `delete[]`are used for arrays

➢ `new` and `delete` are used for single entities

15. POINTERS VS REFERENCES

❖ As we have been seeing, Pointers and References are very similar, but their differences are particularly important

❖ The aim of this section is to

➢ Revise the concepts of Pointers and References

➢ Compare and contrast their similarities and differences, especially regarding their

• Concepts
• Syntax

15.1 CONTRAST: DEFINITIONS AND USAGE

OPERATOR FOR...	POINTERS	REFERENCES
❖ Definition	Asterisk *	Ampersand &
❖ The "Address of"	Ampersand &	NOTHING
❖ The "Value of the Contents of"	Asterisk *	NOTHING

OPERATION

	POINTERS	REFERENCES
	```#include <iostream>``` ``` using namespace std;``` ``` int main ( )``` ``` {``` ``` int myInt_A = 5;``` ``` int myInt_B = 16;```	```#include <iostream>``` ``` using namespace std;``` ``` int main ( )``` ``` {``` ``` int myInt_C = 7;``` ``` int myInt_D = 48;```
Definition & Initial Assignment	```//»»»»»»»»»»»»»»»»»»»»»»»»»``` ```// Definition & Initial Assignment``` ```// - Can be performed on different``` ```// lines for Pointers``` ```//``` ```int *ptrInt myInt_A;``` ```ptrInt_myInt_A = &myInt_A;``` ```//=========================``` ```// - Better to perform on the same line``` ```//``` ```int *ptrInt_myInt_B = &myInt_B;```	```//»»»»»»»»»»»»»»»»»»»»»»»»»``` ```// Definition & Initial Assignment``` ```// - Must be performed in the same``` ```// line for References``` ```//``` ```int& refInt_myInt_C = myInt_C;```
Decoding	```//»»»»»»»»»»»»»»»»»»»»»»»»»``` ```// DECODING``` ```//``` ```cout << "Value, myInt A: "``` ``` << myInt_A<< endl;``` ```cout << "Value, "``` ``` << "contents of ptrInt myInt A: "```	```//»»»»»»»»»»»»»»»»»»»»»»»»»``` ```// DECODING``` ```//``` ```cout << "Value, myInt C: "<< myInt_C<< endl;``` ```cout << "Value, contents of refInt_myInt_C: "``` ``` << refInt_myInt_C<< endl;```

OPERATOR FOR...	POINTERS	REFERENCES

Further Assignment	`<< *ptrInt_myInt_A<< endl;`
	
	`// >>>>>>>>>>>>>>>>>>>>>>>>>>>>>>>>>>>>>>`
	`//   FURTHER ASSIGNMENT`
	`//   - allowed for Pointers`
	`//`
	
	`ptrInt_myInt_A = &myInt_B;`
	
	`// >>>>>>>>>>>>>>>>>>>>>>>>>>>>>>>>>>>>>>`
	`cout  << endl<< endl<< "==========="`
	`     << endl<< "END...C++ Demo: Pointers"`
	`     << endl;`
	`while(true){}`
	`return 0;`
	
	`}//END...Function MAIN`
		`// >>>>>>>>>>>>>>>>>>>>>>>>>>>>>>>>>>>>>>`
		`//   FURTHER ASSIGNMENT - NOT allowed`
		`//                    for References`
		`//`
		
		`// refInt_myInt_C = myInt_D`
		
		`//    - Does not reassign refInt_myInt_C`
		`//    - Instead, it assigns`
		`//       the value of myInt_D`
		`//       to the contents of`
		`//       refInt_myInt_C`
		`//       - i.e. myInt_C`
		
		`// >>>>>>>>>>>>>>>>>>>>>>>>>>>>>>>>>>>>>>`
		`cout  << endl<< endl<< "==========="`
		`     << endl<< "END...C++ Demo: References"`
		`     << endl;`
		`while(true){}`
		`return 0;`
		
		`}//END...Function MAIN`

	`value, myInt_A: 5`
	`value, contents of ptrInt_myInt_A: 5`
	
	`===========`
	`END...C++ Demo: Pointers`
		`value, myInt_C: 7`
		`value, contents of refInt_myInt_C: 7`
		
		`===========`
		`END...C++ Demo: References`

	Program Listing/Output 15.1 Pointers_Demo_Client.cpp	**Program Listing/Output 15.2** References_Demo_Client.cpp

❖ As a piece of revision, let us look again at how the program with the function which swaps two integers, can be implemented using both pointers and references:

	POINTERS	REFERENCES
| Function Declaration | ```cpp
#include <iostream>
using namespace std;

void swap_usingPointers( int *, int * );
``` | ```cpp
#include <iostream>
using namespace std;

void swap_usingReferences( int&, int& );
``` |
| Performing the swap | ```cpp
int main()
{
    // >>>>>>>>>>>>>>>>>>>>>>>>>>>>>>>>>>>
    int myInt_A = 7, myInt_B = 9;

    cout << "Before the attempted swap myInt_A = "
         << myInt_A
         << " and myInt_B = "
         << myInt_B
         << endl;

    // >>>>>>>>>>>>>>>>>>>>>>>>>>>>>>>>>>>
    /*  INVOKING swap_usingPointers - inputs
        Pointers
        -   Want the inputs to be
            -   "The ADDRESS of myInt_A",
            -   "The ADDRESS of myInt_B"
        -   We say &myInt_A and &myInt_B, since
                the operator to say "The address
                of a variable pointed
                to by a pointer":
                AMPERSAND &
    */
    swap_usingPointers( &myInt_A, &myInt_B );

    // >>>>>>>>>>>>>>>>>>>>>>>>>>>>>>>>>>>
    cout << "After the attempted swap myInt_A = "
         << myInt_A
         << " and myInt_B = "
         << myInt_B
         << endl;
``` | ```cpp
int main()
{
    // >>>>>>>>>>>>>>>>>>>>>>>>>>>>>>>>>>>
    int myInt_A = 11, myInt_B = 13;

    cout << "Before the attempted swap myInt_A = "
         << myInt_A
         << " and myInt_B = "
         << myInt_B
         << endl;

    // >>>>>>>>>>>>>>>>>>>>>>>>>>>>>>>>>>>
    /*  INVOKING swap_usingReferences - inputs
        References
        -   Want the inputs to be
            -   "The ADDRESS of myInt_A",
            -   "The ADDRESS of myInt_B"
        -   We say myInt_A and myInt_B, since
                the operator to say "The address
                of a variable
                referred to by a reference":
                NOTHING
    */
    swap_usingReferences( myInt_A, myInt_B );

    // >>>>>>>>>>>>>>>>>>>>>>>>>>>>>>>>>>>
    cout << "After the attempted swap myInt_A = "
         << myInt_A
         << " and myInt_B = "
         << myInt_B
         << endl;
``` |

POINTERS	REFERENCES
| ```
//»»»»»»»»»»»»»»»»»»»»»»»»»»»»»»»»»»»»»
cout  << endl<< "============"<< endl
      << "END...C++ Pointers as Function Inputs
         Demo"
      << endl;
while(true){}
return 0;

}//...function main
``` | ```
//»»»»»»»»»»»»»»»»»»»»»»»»»»»»»»»»»»»»»
cout  << endl<< "============"
      << endl<< "END...C++ References as Function
                 Inputs Demo"
      << endl;
while(true){}
return 0;

}//END...function main
``` |

	POINTERS	REFERENCES
| Function Definition | ```
void swap_usingPointers( int *ptrInt_In_A,
                         int *ptrInt_In_B )
{

    int temp = *ptrInt_In_B;
    *ptrInt_In_B = *ptrInt_In_A;
    *ptrInt_In_A = temp;

}// END...function swap_usingPointers
``` | ```
void swap_usingReferences( int& ref_Int_In_A,
                           int& ref_Int_In_B )
{

    int temp = ref_Int_In_B;
    ref_Int_In_B = ref_Int_In_A;
    ref_Int_In_A = temp;

}// END...function swap_usingReferences
``` |
| | ```
Before the attempted swap myInt_A = 7 and myInt_B = 9
After the attempted swap myInt_A = 9 and myInt_B = 7

============
END...C++ Pointers as Function Inputs Demo
``` | ```
Before the attempted swap myInt_A = 11 and myInt_B = 13
After the attempted swap myInt_A = 13 and myInt_B = 11

============
END...C++ References as Function Inputs Demo
``` |
| | **Program Listing/Output 15.3**   swap_Pointers_Client.cpp | **Program Listing/Output 15.4**   swap_usingReferences.cpp |

15.2 RULES OF THUMB: WHEN TO USE POINTERS & REFERENCES

❖ You must use a pointer if

➢ There is a chance the contents will need to vary

- A reference must be initialised on the same line as its definition and is read-only thereafter
- A pointer does not have to be initialised on the same line as the definition, though this is advisable

➢ There is a chance the contents of your memory address will be nothing even temporarily

- There is no such thing as a NULL reference - it must always refer to something

➢ You want to use the Runtime Heap

- This is accessible only by pointers, not by references or directly

➢ You want to create a data structure, as opposed to have single entities

- A data structure of references will generally mean attempting to mix together pointers and references
  - This is best avoided
  - Even when compilers allow this, messy programs which are difficult to maintain, are the result
- The bulk of C++ data structures (covered in Cybernetics in C++ - Expert) either
  - Are Implemented as pointers
  - Heavily use Pointers in their implementation
- Having data structures contain pointers is fine
  - Implementing hierarchies of pointers is no problem and a common practice

➢ Your memory address is to contain another memory address

- It is not possible to have a hierarchy of references, only one level

❖ You must use a reference if

➢ You want an overloaded operator to return a memory address (section 21 and Cybernetics in C++ - Expert)

- This is the main reason references were created in C++

➢ You are implementing the Copy Constructor (section 20.6 onwards)

❖ There will be several cases when both will do

➢ e.g. When you need to pass the addresses of the variables into a function, as opposed to their values, they can be passed as either pointers or references

➢ In these cases, give preference to the reference

➢ Since references are less powerful than pointers, there are far fewer opportunities to go wrong with them

16. POINTERS VS ARRAYS: MOSTLY INTERCHANGEABLE, SOMETIMES NOT

❖ Comments such as "Arrays and Pointers are the same thing" and "Arrays and Pointers are interchangeable" are common in C and C++ programming

❖ However, they are half-truths!

➢ While it is very important to understand how their natures overlap

➢ It is equally important to understand the less common but essential cases when they do not

16.1 THE NATURE OF POINTERS & ARRAYS

16.1.1 PLACEMENT IN RUNTIME MEMORY

❖ Let's start by looking at what happens in runtime memory when an array is defined:

MEMORY ADDRESS - DIRECT	LAYOUT IN RUNTIME MEMORY	MEMORY ADDRESS - POINTER		PROGRAM CODE
0 x 002304	H	← greeting		
0 x 002305	e	← greeting +	1	
0 x 002306	l	← greeting +	2	
0 x 002307	l	← greeting +	3	#include <iostream>
0 x 002308	o	← greeting +	4	using namespace std;
0 x 002309	,	← greeting +	5	int main()
0 x 00230A		← greeting +	6	{
0 x 00230B	h	← greeting +	7	
0 x 00230C	o	← greeting +	8	char greeting[] =
0 x 00230D	w	← greeting +	9	"Hello, how are
0 x 00230E		← greeting +	10	you?";
0 x 00230F	a	← greeting +	11	
0 x 002310	r	← greeting +	12	return 0;
0 x 002311	e	← greeting +	13	}//END...Function MAIN
0 x 002312		← greeting +	14	
0 x 002313	y	← greeting +	15	
0 x 002314	o	← greeting +	16	
0 x 002315	u	← greeting +	17	
0 x 002316	?	← greeting +	18	

Figure 16.1   How an Array is laid out in Runtime Memory

❖ **Revision:** Arrays are data structures where the elements are laid out in consecutive memory locations:

➢ Whenever we access individual elements, we are accessing individual memory addresses

❖ The **array name is a pointer to the first address in runtime memory where the array is stored**

➢ Accessing elements means using this pointer, then adding the appropriate, possibly 0, offset to it

❖ Consequently, when we want to access the letter "w", there are (at least) 2 ways to do it

➢ Using "Array Operations"   `greeting[9]`
➢ Using "Pointer Operations"   `*( greeting + 9 )`

❖ We can now see why most languages, including C, C++, Java and Swift, index arrays of size N from 0 to (N-1) as opposed to 1 to N

➢ The index represents the offset from the first address in memory where the array is stored

16.1.2 ARRAY AND POINTER OPERATORS

❖ As far as the compiler is concerned, it always deals with arrays as pointers

➢ Whenever it sees a statement or part statement like

```
greeting[5]
```

it will be translated to

```
*( greeting + 5 )
```

❖ **Common Interview Questions**

(Q) Which **two** operators can be used to define a pointer?

(A) The asterisk * and the square brackets []

(Q) Which **two** operators can be used to decode a pointer i.e. say "the value of the contents of"?

(A) The asterisk * and the square brackets []

(Q) Which **two** operators are used to perform two different operations with pointers?

(A) Both the asterisk * and the square brackets [] are used for both definition and decoding

➤ It is common for programmers to associate

- Only the asterisk * with pointers
- Only the square brackets [] with arrays

➤ However, as we are seeing, since arrays and pointers are mostly interchangeable, both can be used for both

- However, as we will see, not every combination gives optimum performance

❖ **Hacker's Note:** You may be thinking the following:

➤ Is it possible to write a statement like `5[greeting]` ?

➤ The unexpected answer is Yes:

- From our rules above, the compiler will decode this to

```
*(5 + greeting)
```

  ○ Since addition is commutative, this makes perfect sense to the computer at runtime!

➤ However, any programmer who writes code like this has clearly forgotten that there are 2 audiences to keep happy: your fellow humans as well as the computer!

- Conceptually, this statement does not make any sense
- TIP: Avoid writing statements like this - a program of this nature will be an unnecessary nightmare to maintain and debug

❖ **Programming Guideline:** The usual tip of ensuring your program is readable for a human is especially important at an advanced level and when dealing with concepts which can easily cause confusion such as memory management

16.2 WHERE ARRAYS AND POINTERS ARE INTERCHANGEABLE

16.2.1 MIXING POINTER AND ARRAY SYNTAXES IN DEFINITIONS AND ACCESSING

❖ It follows from the above discussion that a data structure, where the elements are contiguous in memory can be

➤ Defined as a(n)

- Array
- Pointer

➢ Accessed as a(n)

- Array
- Pointer

❖ The following program demonstrates how array and pointer operations can be mixed together,

➢ This is no surprise since they are both interpreted by the compiler as pointers

➢ Note that this program is for demonstration only

- Mixing together definitions does create an unnecessarily unclear program for humans, including yourself, who will be maintaining your program

```
#include <iostream>
using namespace std;

int main()
{
    //»»»»»»»»»»»»»»»»»»»»»»»»»»»»»»»»»»»»»»»»»»»»»»»»»»»»»»»»»»»»»»»»»»»
    cout  << endl<< "======================="<< endl
          << "Definition as an ARRAY, access as a POINTER: "
          << endl<< "----------"
          << endl<< endl;

    cout  << "Definition as an ARRAY: "<< endl
          << "greeting_English[]  = Hello, how are you?"
          << endl<< endl;
    char greeting_English[] = "Hello, how are you?";

    cout  << "Access as a POINTER: "<< endl
          << "The 4th element of greeting_English,"
          << " *( greeting_English + 3 ) = "
          << *( greeting_English + 3)
          << endl<< endl;

    //»»»»»»»»»»»»»»»»»»»»»»»»»»»»»»»»»»»»»»»»»»»»»»»»»»»»»»»»»»»»»»»»»»»
    cout  << endl<< "======================="<< endl
          << "Definition as a POINTER, access as an ARRAY: "
          << endl<< "----------"
          << endl<< endl;

    cout  << "Definition as a POINTER: "<< endl
          << "*greeting_French = Bonjour, comment allez vous?"
          << endl<< endl;
    char *greeting_French = "Bonjour, comment allez vous?";
```

```
        cout  << "Access as an ARRAY: "<< endl
              << "The 5th element of greeting_French,"
              << " greeting_French[4] = "
              << greeting_French[4]
              << endl<< endl;

        //»»»»»»»»»»»»»»»»»»»»»»»»»»»»»»»»»»»»»»»»»»»»»»»»»»»»»»»»»»»»»»»»»»»»»
        cout << endl<< endl<< "========================"
             << endl
             << "END...Demo, \t Define as Array, access as Pointer, "
             << endl<< "\t\t Define as Pointer, access as Array"
             << endl;
        while(true){}
        return 0;

}//END...Function MAIN


========================
Definition as an ARRAY, access as a POINTER:
----------

Definition as an ARRAY:
greeting_English[]  = Hello, how are you?

Access as a POINTER:
The 4th element of greeting_English, *( greeting_English + 3 ) = l


========================
Definition as a POINTER, access as an ARRAY:
----------

Definition as a POINTER:
*greeting_French = Bonjour, comment allez vous?

Access as an ARRAY:
The 5th element of greeting_French, greeting_French[4] = o



========================
END...Demo,     Define as Array, access as Pointer,
                Define as Pointer, access as Array
```

Program Listing/Output 16.1: Array_Pointer_Declarations_Client.cpp

16.2.1.1 Define as Pointer, accessed as Array

❖ From above, the runtime steps when we define as a Pointer, access as an array are as follows:

➢ The program must first go to where the Pointer variable is stored
➢ It must next read the value, i.e. memory address here, then go there
➢ Now it is at the address in memory where the first array element is stored

16.2.1.2 Array of Pointers

❖ There is no problem with having an array of pointers

➢ Since the compiler translates an array as a pointer
➢ An array of Pointers will be translated as a Pointer-to-a-Pointer

16.2.2 INPUT TO A FUNCTION

❖ Whenever an input to a function is an array

➢ It is always decoded by the compiler as a pointer...
➢ ...to the first element in memory where the array is stored

❖ i.e. The following declaration

```
void myFunc( int myArr_IN[] )
```

will be rewritten by the compiler as the following:

```
void myFunc( int *myArr_IN )
```

❖ This is why there is no immediate way to check the size of an input array inside a function,

➢ If the `sizeof` operator is used (see section 16.3.4 for a full discussion of this), it will not return the size of the array
➢ All that gets passed is the pointer (to the first element of the array in runtime memory)
  • The size of only this individual pointer will be reported

❖ Thus when an array is being input to a function, some programmers write it as a pointer

➢ Argument(s) in favour:
  • This will be carried out by the compiler anyway, so this saves a step
➢ Argument(s) against:
  • The one step it saves is at compile time
    ○ Programs are almost always judged on their performance at runtime, not compile time

- It is more programmer friendly to humans to write the input(s) as (an) array(s)
    - An array indicates that a group of data is being dealt with
    - A pointer could be dealing with a single entity or a group of data

16.3 WHERE ARRAYS AND POINTERS ARE NOT INTERCHANGEABLE

16.3.1 DEFINITION AS ARRAY, DECLARATION AS A POINTER FROM A SEPARATE FILE

❖ Perhaps the greatest difference is that arrays and pointers cannot be mixed together in definitions and declarations

➢ If it is defined as an array, it needs to be declared as an array

➢ If it is defined as a pointer, it needs to be declared as an pointer

❖ Recall the following, when we define as a Pointer, access as an array

➢ The program must first go to where the Pointer variable is stored

➢ It must next read the value, i.e. memory address here, then go there

➢ Now it is at the address in memory where the first array element is stored

❖ Consider the following project where a variable has been

➢ Defined as an array in one file

➢ Declared as a pointer in another

```
#ifndef MULTILINGUALGREETINGS_H
#define MULTILINGUALGREETINGS_H

char greeting_French[] = "Bonjour, comment allez vous?";
char greeting_English[] = "Hello, how are you?";

#endif
```

Program Listing/Output 16.2: MultiLingualGreetings.h

```
#include <iostream>
#include <string>
using namespace std;

#include "MultiLingualGreetings.h"

extern char *greeting_French; // ERRONEOUS Declaration
```

```
extern char greeting_English[]; // GOOD Declaration


int main()
{
        //»»»»»»»»»»»»»»»»»»»»»»»»»»»»»»»»»»»»»»»»»»»»»»»»»»»»»»»»»»»»»»»»»»»»»
        greeting_French[ 27 ] = '?';
        greeting_English[ 18 ] = '?';



        //»»»»»»»»»»»»»»»»»»»»»»»»»»»»»»»»»»»»»»»»»»»»»»»»»»»»»»»»»»»»»»»»»»»»»
        cout   << endl<< endl<< "========================"
               << endl
               << "END...Demo, Define as Array,"
               << " Extern Declaration a Pointer"
               << endl;
        while(true){}
        return 0;

}//END...Function MAIN
```

Program Listing/Output 16.3: Declaration_Ptr_Arr_Client.cpp

❖ Here's what the compiler and linker are attempting to do:

➢ First, in the file `Declaration_Ptr_Arr_Client.cpp` where `greeting_French` has been defined as a pointer-to-char, it goes to the contents of this
  • This will be a memory address
➢ It now goes to this memory address and reads its contents
  • These will be of type char
  • However, if they are pointing to an array, they need to be a hexadecimal address
  • What happens at this stage depends on the system
    ○ Several modern day systems, including Microsoft Visual Studio, will give a compiler/linker error
    ○ Be cautious - older systems may well attempt to interpret ASCII characters as memory addresses at runtime, with potentially catastrophic consequences

❖ The remedy to this somewhat complex problem, is very simple: ensure declarations match definitions

➢ If a variable is defined as an array, declare it as an array
  • This is generally preferable when dealing with a group of data

- In our example `greeting_English` has been defined and declared as an array, so does not cause any errors
➢ If a variable is defined as a pointer, declare it as a pointer

16.3.2 OUTPUT OF A FUNCTION

❖ As we have seen, a function cannot return an array,

❖ Consequently, you will never see declarations of the form

```
int myFunction()[]
```

❖ However, the following trick can be applied to effectively return an array from a function since

➢ An array name is a pointer
➢ A Pointer can be, and frequently is, a function output

❖ Define the function with a return type as pointer-to-typeOfTheArray

❖ Inside the function, deal with the array as normal

❖ Return the name of the array

➢ Since this is a pointer, there is no type mismatch

❖ **CAUTION: Dangling Pointers:**

➢ A crucial point to bear in mind when applying this trick, is not to return a pointer to a non-static local variable (section 10.7.1)
➢ When the program exits the function, the pointer it returns will be a dangling pointer

- A pointer, the contents of which have gone out of scope

❖ The following program shows this in action:

```
#include <iostream>
#include <string>
using namespace std;

char *greet_DifferentLanguages(char);

int main()
{
        //»»»»»»»»»»»»»»»»»»»»»»»»»»»»»»»»»»»»»»»»»»»»»»»»»»»»»»»»»»»»»»»»»»»»»»»
        cout  << greet_DifferentLanguages('F');
        cout  << endl;
        cout  << greet_DifferentLanguages('G');
```

```
        cout  << endl;
        cout  << greet_DifferentLanguages('P');
        cout  << endl;

        //»»»»»»»»»»»»»»»»»»»»»»»»»»»»»»»»»»»»»»»»»»»»»»»»»»»»»»»»»»»»»»»»
        cout << endl << endl << "========================"
            << endl
            << "END...Demo, Function effectively returning an array"
            << endl;
        while (true) {}
        return 0;

}//END...Function MAIN


/*    FUNCTION:
-     Effectively returns an array, by returning a pointer
-     Returns a different greeting ( array of chars )
depending on the user input
*/
char *greet_DifferentLanguages(char selectedLanguage_IN)
{
        static char greeting_English[] = { "Hello, how are you?" };
        static char greeting_French[] = { "Bonjour, comment allez vous?"
};
        static char greeting_German[] = { "Guten Tag, wie geht es Ihnen?"
};

        if (selectedLanguage_IN == 'F')
        {
                return greeting_French;
        }
        else if (selectedLanguage_IN == 'G')
        {
        return greeting_German;
        }
        else
        {
        return greeting_English;
        }
}//END...Function greeting_DifferentLanguages
```

```
Bonjour, comment allez vous?
Guten Tag, wie geht es Ihnen?
Hello, how are you?


========================
END...Demo, Function effectively returning an array
```

Program Listing/Output 16.4: FunctionReturnArray_Client.cpp

❖ Declaration Note:

➤ When we have the declaration:

```
char *greet_DifferentLanguages(char)
```

- greet_DifferentLanguages is a Function, which returns a Pointer-to-char
- NOT a Pointer-to-a Function which returns a Char
- The reason is that the parentheses operator () for a function takes precedence over the asterisk operator * for the pointer

➤ Unscrambling declarations is discussed in detail in Cybernetics in C++ - Expert

16.3.3 USING THE ASSIGNMENT OPERATOR

❖ **Revision:** The assignment operator cannot be used with an array name on the left hand side

➤ The array name is a non-modifiable l-value

❖ However, a pointer name is a modifiable l-value, so can be, and very frequently is, on the left hand side of an assignment operator

❖ The following program demonstrates these concepts:

```
#include <iostream>
#include <string>
using namespace std;

int main()
{
        //»»»»»»»»»»»»»»»»»»»»»»»»»»»»»»»»»»»»»»»»»»»»»»»»»»»»»»»»»»
        int myInt_A = 5;
        int *ptrInt_myInt_A = &myInt_A;
        int myArrInt_A[] = { 4, 34, 23, 86, 23, 7 };

        cout  << "Contents of myArrInt_A: [ " ;
        for( int i = 0; i < 6; ++i )
        {
               cout << myArrInt_A[i]<< " ";
        }
        cout  << "]"<< endl<< endl;

        //»»»»»»»»»»»»»»»»»»»»»»»»»»»»»»»»»»»»»»»»»»»»»»»»»»»»»»»»»»
        // ASSIGNMENT
        //
//      myArrInt_A = myArrInt_B;      // ILLEGAL - cannot assign to
//      myArrInt_A = ptrInt_myInt_A;  // ILLEGAL  an array
```

```
        ptrInt_myInt_A = myArrInt_A;  // OK - can assign to a pointer

        cout << "Contents of ptrInt_myInt_A: [" ;
        for( int i = 0; i < 6; ++i )
        {
            cout << ptrInt_myInt_A[i]<< " ";
        }
        cout << "]"<< endl;

        //»»»»»»»»»»»»»»»»»»»»»»»»»»»»»»»»»»»»»»»»»»»»»»»»»»»»»»»»»»»»»»»»
        cout << endl<< endl<< "======================="
            << endl
            << "END...Demo, Using the Assignment Operator on Arrays"
            << " & Pointers"
            << endl;
        while(true){}
        return 0;

}//END...Function MAIN
```

```
Contents of myArrInt_A: [ 4 34 23 86 23 7 ]

Contents of ptrInt_myInt_A: [4 34 23 86 23 7 ]


=======================
END...Demo, Using the Assignment Operator on Arrays & Pointers
```

Program Listing/Output 16.5: AssignmentOperator_Array_Pointer_Client.cpp

16.3.4 USING THE SIZEOF OPERATOR

❖ A subtle but often essential difference is observed when using the `sizeof` operator, as the following program shows:

```
#include <iostream>
#include <string>
using namespace std;

int main()
{
        //»»»»»»»»»»»»»»»»»»»»»»»»»»»»»»»»»»»»»»»»»»»»»»»»»»»»»»»»»»»»»»»»
        int myInt_A = 5;
        int *ptrInt_myInt_A = &myInt_A;
        int myArrInt_A[] = { 4, 34, 23, 86, 23, 7 };

        //»»»»»»»»»»»»»»»»»»»»»»»»»»»»»»»»»»»»»»»»»»»»»»»»»»»»»»»»»»»»»»»»
        // OPERANDS TO SIZEOF
        //
```

```
        int size_myInt_A = sizeof( myInt_A );
        int size_ptrInt_myInt_A = sizeof( ptrInt_myInt_A );
        int size_myArrInt_A = sizeof( myArrInt_A );

        cout   << "The size of myInt_A is "
               << size_myInt_A<< " bytes."<< endl;

        // Size of an INDIVIDUAL POINTER
        cout   << "The size of ptrInt_myInt_A is "
               << size_ptrInt_myInt_A<< " bytes."<< endl;

        // Size of an ENTIRE ARRAY
        cout   << "The size of myArrInt_A is "
               << size_myArrInt_A<< " bytes."<< endl;

        //»»»»»»»»»»»»»»»»»»»»»»»»»»»»»»»»»»»»»»»»»»»»»»»»»»»»»»»»»»»»»»»
        cout   << endl<< endl<< "========================"
               << endl
               << "END...Demo, Using sizeof Operator"
               << endl;
        while(true){}
        return 0;

}//END...Function MAIN
```

```
The size of myInt_A is 4 bytes.
The size of ptrInt_myInt_A is 4 bytes.
The size of myArrInt_A is 24 bytes.


========================
END...Demo, Using sizeof Operator
```

Program Listing/Output 16.6: SizeOf_Operation_Client.cpp

❖ Thus when using the `sizeof` operator

➢ When the input is an array, the return value is the size of the **entire** array

➢ When the input is a pointer, the return value is the size of the **individual** pointer

16.3.5 USING THE CONTENTS OF "*" OPERATOR

❖ Revision: The * operator means "the contents of"

❖ When the operand is

➢ A pointer, it means the contents of the **individual** pointer

➢ An array, it means the contents of the **entire** array

- On its own, it means the contents of the first element of the array, like all array names

❖ This is not a common occurrence but nonetheless an important operation which behaves differently when the input is an array or a pointer

❖ The following program shows this:

```cpp
#include <iostream>
#include <string>
using namespace std;

int main()
{
    //»»»»»»»»»»»»»»»»»»»»»»»»»»»»»»»»»»»»»»»»»»»»»»»»»»»»»»»»»»»»»»»»»»»»»»»»»
    int myInt_A = 5;
    int *ptrInt_myInt_A = &myInt_A;
    int myArrInt_A[] = { 4, 34, 23, 86, 23, 7 };

    //»»»»»»»»»»»»»»»»»»»»»»»»»»»»»»»»»»»»»»»»»»»»»»»»»»»»»»»»»»»»»»»»»»»»»»»»»
    //      OPERANDS TO "CONTENTS OF" & OPERATOR
    //

    // ptr_ptrInt_myInt_A is a Pointer-to-Pointer-to-Int
    int **ptr_ptrInt_myInt_A = &ptrInt_myInt_A;

    cout  << "The contents of the contents of ptr_ptrInt_myInt_A: "
          << **ptr_ptrInt_myInt_A
          << endl;
    cout  << "...i.e. an integer"
          << endl<< endl;

    // ptr_myArrInt_A is a pointer to an ENTIRE array of ints
    int (*ptr_myArrInt_A)[6] = &myArrInt_A;

    cout  << "The contents of the contents of ptr_myArrInt_A: "
          << **ptr_myArrInt_A
          << endl;
    cout  << "...i.e. the 1st element of an array of integers"
          << endl;

    //»»»»»»»»»»»»»»»»»»»»»»»»»»»»»»»»»»»»»»»»»»»»»»»»»»»»»»»»»»»»»»»»»»»»»»»»»
    cout  << endl<< endl<< "======================="
          << endl
          << "END...Demo, Using Contents of Operator & "
          << "on Arrays and Pointers."
          << endl;
    while(true){}
    return 0;

}//END...Function MAIN
```

```
The contents of the contents of ptr_ptrInt_myInt_A: 5
...i.e. an integer

The contents of the contents of ptr_myArrInt_A: 4
...i.e. the 1st element of an array of integers


========================
END...Demo, Using Contents of Operator & on Arrays and Pointers.
```

Program Listing/Output 16.7: ContentsOfOperation_Client.cpp

16.4 ARRAYS & REFERENCES: NOT INTERCHANGEABLE

❖ As we have been seeing, the Reference is similar to the Pointer in that it also contains a memory address

❖ However, the differences between them and the Pointer mean that

➢ References and Arrays have little relationship

16.5 SUMMARY: INTERCHANGEABILITY OF ARRAYS AND POINTERS

❖ The C and C++ Standards say the following about the equivalence of arrays and pointers:

➢ An array name is treated by the compiler as a pointer containing the address of the first element

An array subscript is equivalent to the offset from this first address

➢ An array name in the declaration of a function parameter is treated by the compiler as a pointer to the first element of that array

❖ The statement that "arrays and pointers are interchangeable" is valid most of the time

➢ The majority of array usage is in expressions and as function inputs

❖ It is important not to overlook the exceptions:

➢ Ensure array and pointer names are not mixed up in definitions and declarations

➢ A pointer name is a modifiable l-value, an array name is non-modifiable

- Consequently the assignment operator cannot be used for array names, but can be used for pointers

➢ Certain operators give different results for array name and pointer operands

- The `sizeof` operator
  - ○ For an array returns the size of the entire array
  - ○ For a pointer returns the size of the individual pointer
- The contents of & operator
  - ○ For an array returns the contents of the entire array
  - ○ For a pointer returns the contents of the individual pointer

➢ A function output cannot be an array, but can be a pointer

- Returning a pointer is a common trick for returning an array from a function
  - ○ Be cautious: ensure that the returned pointer is not a dangling pointer

❖ **Common question**:

➢ (Q)"If Arrays and Pointers are interchangeable in C and C++
  - Why are arrays pervasive in almost all programming languages,
  - yet pointers are unheard of in the majority?"

➢ (A) The most important difference is to the human: the concept:
  - An array is a data structure
    - ○ This will be required in all but the simplest programming languages
  - A pointer is a variable which contains a memory address
    - ○ This will be hidden from the programmer in all languages which do not support memory management programming
      - ■ i.e. Most languages, including Java, C#, Web languages such as PHP, JavaScript

OBJECT ORIENTED PROGRAMMING

❖ While Procedural Programming has made, and continues to make, a significant contribution to the world, the fact remains that it alone cannot cope with the demands of modern times

➤ The programmer (i.e. You! :-)) must remember each function and its operation

  • Modern day programs can comprise several thousand, if not millions, of functions

➤ There are only two levels of scope, local and global, and we have discussed why the latter needs to be avoided whenever possible

  • This limitation causes programmers to keep on thinking "Wouldn't it be great if there was data which only specified groups of functions can access"

❖ Object Oriented Programming, which addresses these issues and provides several further facilities, is a major, some argue the single most important, paradigm in modern day programming

❖ It comprises 4 major building blocks, A PIE

Here, they will be covered in the order

➤ **A**   Abstraction   ➤ Abstraction
➤ **P**   Polymorphism   ➤ Encapsulation
➤ **I**   Inheritance   ➤ Inheritance
➤ **E**   Encapsulation   ➤ Polymorphism

❖ The several things made possible by OOP include

➤ Dealing with Complexities

➤ Further Scope of Variables

➤ Reusability

➤ Direct Representation of the real world

❖ Object Oriented Programming Languages include

- ➤ C++
- ➤ Java
- ➤ C#
- ➤ Swift
- ➤ Objective-C
- ➤ Python
- ➤ Visual Basic
- ➤ Web Programming Languages including
  - JavaScript
  - PHP

17. ABSTRACT DATA TYPES

17.1 WHAT IS ABSTRACTION?

❖ Abstraction is the practice of focusing on the main picture (the forest) as opposed to the details (the trees)

❖ This powerful means of thinking is used abundantly outside the programming world, as well as within it

17.1.1 ABSTRACTION OUTSIDE THE PROGRAMMING WORLD

❖ Navigation: You are lost in a city

- ➤ Looking at details of the roads, buildings and surroundings will not help
- ➤ Looking at a MAP, which gives the BIGGER PICTURE is required

❖ Summaries

- ➤ Most Research Papers start with a paragraph, the "Abstract", which summarises the contents
- ➤ News programmes typically begin with Headlines, summarising the content

❖ Putting Ideas into Perspective

"A Small Step for Man, a Giant Leap for Mankind"

Neil Armstrong, as he set foot on the Moon

- ➤ The "Small Step" was the task in the short term
- ➤ The "Giant Leap" is the Abstract view, its significance in the long term

❖ Leadership & Management

"Strategy without tactics is the slowest route to victory. Tactics without strategy is the noise before defeat"

Sun Yung Tzu,
Author/Main Contributor - "The Art of War"

➢ Any form of decision making: The more abstract (higher up) an idea is, the more important it is
➢ Company Management
  • The Chairperson takes an abstract view and focuses on the long term
  • The Managing Director/CEO is in charge of the day to day running of the company

❖ IT

➢ In the early days, when computers were used very little by very few, it was fine to keep all your files in the "root of your drive"
  • It was no problem to remember the names of all your files
➢ Today, computers are used for countless activities: nearly everyone has thousands/millions of files
  • DIRECTORIES/FOLDERS are used: Abstract means of thinking
  • You do not remember the names of all your files, or even folders
  • You focus on the BIGGER PICTURE, remembering roughly the IT hierarchy

❖ Entrepreneurship/Technical management

➢ The most prosperous companies are those that focus on the long term
  • "The Art of Japanese Management" credits Japanese companies for their long term approach
➢ Until as recently as 1994, Usenet was larger than the Internet;
  • However, the Internet took a long term approach and is now the most successful computer network

❖ Mathematics

➢ Classical algebra, focusing on individual variables, is not suitable in several advanced problems
➢ Abstract Algebra is used

- Ask: does the entity satisfy the axioms to be a Group, Vector Space, Ring, Field...?
- If so, the things we know about the abstract entity can be applied
- Fermat's Last Theorem was proven this way

❖ Physics

➢ Classical Physics comprises looking at entities separately such as Forces, Energy, Power

- Individual Forces such as Friction and Centripetal Force, Forces within the Nucleus of an Atom

➢ Quantum Physics comprises taking a broader view by looking at relationships and similarities between them

- All forces can now be classified under one of the four possible categories Gravitational, Electromagnetic, Strong, Weak
- A major area of research in modern times is finding an overall category for all of them

❖ Science/Engineering: Feedback Control Systems

➢ Having a system where the input has information fed back from the output, is required across the board

- Mechanical systems
  ○ Controlling the speed of a Motor
- Electronic Systems
  ○ Controlling the Gain of an Operational Amplifier circuit
  ○ Implementation of Digital Filters

➢ There are several principles of Feedback Control Systems which apply to them all

- Representations/Stability Tests: Root Locus Diagrams, Nyquist Plots, Bode Plots

❖ Philosophy

➢ Abstract Logic is used abundantly

❖ Biology/Consciousness

Individual consciousness is only a manifestation of a unitary consciousness pervading the universe
Erwin Schrodinger, What is Life

❖ Medicine

  ➢ Focusing on and trying to tackle individual symptoms is of little use

  ➢ Looking at all the symptoms, and diagnosing the illness is required for effective treatment

❖ Linguistics, Literature, History...any written content

  ➢ It is essential to put all sentences and/or paragraphs into context

  ➢ Don't just not look at them as isolated fragments

❖ Music

  ➢ Individual notes are initially played, and often form a major part of an introduction or bridging riff

  ➢ In order to be an effective accompaniment, chords need to be played

    • The chords, i.e. combinations of notes as opposed to individual notes, are the focus of the musician

❖ Time/Team Management

  ➢ Individual days and weeks may seem unproductive

  ➢ It is the long term progress that is important

    • Often, knowing what to avoid is essential, learning this may be mistaken for unproductivity

EXERCISE

❖ *This list can be extended almost indefinitely; see if you can come up with 3 further examples of Abstraction*

17.1.2 PROBLEMS: LACK OF ABSTRACTION

❖ Individual data items are useful only to a limited extent

❖ To view the problem of relying on them too heavily, i.e. having only details, and no "broader" data type, consider the following problem:

❖ We want to model an office of employees

  ➢ Here's how it would look, only at the beginning, using only individual data items

```cpp
#include <iostream>
#include <string>

using namespace std;

int main()
{

    //»»»»»»»»»»»»»»»»»»»»»»»»»»»»»»»»»»»»»»»»»»»»»»»»»»»»»»»»»»»»»»»»»»»»
    //    Lack of abstraction can be the recipe for disaster
    //    Data for 3 out of 1000+ employees
    string joeBloggs_Name = "Joe Bloggs";
    double joeBloggs_Salary_Gross = 20500.0;
    double joeBloggs_Tax_Rate = 0.3;
    double joeBloggs_Tax_Amount = joeBloggs_Salary_Gross *
                                  joeBloggs_Tax_Rate;
    double joeBloggs_Salary_Net = joeBloggs_Salary_Gross -
                                  joeBloggs_Tax_Amount;

    string johnSmith_Name = "John Smith";
    double johnSmith_Salary_Gross = 22000.0;
    double johnSmith_Tax_Rate = 0.3;
    double johnSmith_Tax_Amount = johnSmith_Salary_Gross *
                                  johnSmith_Tax_Rate;
    double johnSmith_Salary_Net = johnSmith_Salary_Gross -
                                  johnSmith_Tax_Amount;

    string amandaTaylor_Name = "Amanda Taylor";
    double amandaTaylor_Salary_Gross = 25000.0;
    double amandaTaylor_Tax_Rate = 0.4;
    double amandaTaylor_Tax_Amount = amandaTaylor_Salary_Gross *
                                     amandaTaylor_Tax_Rate;
    double amandaTaylor_Salary_Net = amandaTaylor_Salary_Gross -
                                     amandaTaylor_Tax_Amount;

    //»»»»»»»»»»»»»»»»»»»»»»»»»»»»»»»»»»»»»»»»»»»»»»»»»»»»»»»»»»»»»»»»»»»»
    cout   << endl<< endl<< "========================"
           << endl
           << "END...C++ Demo, Lack of Abstraction "
           << endl;
    while(true){}
    return 0;

}//END...Function MAIN
```

Program Listing/Output 17.1: Problems_LackAbstraction_Client.cpp

❖ As we can see, this approach quickly becomes cumbersome

➢ We have modelled only 3 employees and a few data items for each one

➢ A large company will have several thousand employees, with several data items for each

17.1.3 INTRODUCING ABSTRACTION

❖ The practice of abstraction is to look at the data and ask "Can patterns be noticed"

❖ In this case we will realise:

➤ Joe Bloggs, John Smith and Amanda Taylor are each employees

➤ Each has the data items:
  • Name

  • NI Number

  • Gross Salary

❖ Is it possible to create an abstract entity, which will contain this data and then use it to implement various employees?...

❖ Abstraction is implemented

➤ In C and C++ using
  • Structures (commonly referred to as Structs)

  • Unions

  • Enumerated Typs (commonly referred to as Enums)

➤ In C++ also using
  • Classes
    ○ These incorporate Encapsulation as well

17.2 WHAT ARE ABSTRACT DATA TYPES?

❖ ADTs are Data Types defined by the programmer (i.e. YOU :-)), to contain composite data

➤ Primitive data types are known as primitive since they can hold only a single value at a time

➤ Composite data types can hold several values

❖ They are created using the line of thinking Abstraction

17.3 STRUCTURES

❖ The structure, known commonly by its keyword `struct`, is used here

❖ Note that the entity struct has different amounts of usage in C and C++

➤ It is widely used in C programming

➤ It is still a common sight in C++ largely because

• A lot of C++ programming comprises legacy C code

• A frequent requirement in C++ is ensuring backward compatibility with C

➤ However, in C++, the entity "class" is more commonly used instead; this will be covered in detail, section 18 onwards

17.3.1 DEFINING STRUCTURES

❖ A structure is defined as follows:

```
struct Name_Struct
  {
        dataType name_1stMember;
        dataType name_2ndMember;
        dataType name_3rdMember;
        dataType name_4thMember;
              .
              .
              .
  };
```

➤ Note that the types of the members can be, and almost always are, different

❖ In this case, the name "Employee" is convenient, and it is defined as follows:

```
struct Employee
  {
        string name;
        string nINumber;
        double salary_Gross;
        double taxRate;
        double taxAmount;
        double salary_Net;
  };
```

❖ It is very important to note that a `struct` is a data TYPE

➢ Its definition does not consume any runtime memory

➢ Only when we create an instance of it, as follows, do we write anything to memory

17.3.2 INSTANTIATING STRUCTURES

❖ In C++, the struct can now be instantiated in a similar manner to primitive variables:

```
Employee joeBloggs;
```

❖ **C Note:** In C programming, the keyword `struct` would be required again in the instantiation:

```
struct Employee joeBloggs;
```

➢ In C++, this has been eliminated

17.3.3 ACCESSING MEMBER DATA USING THE DOT . OPERATORS

❖ The Dot Operator . is used to access individual members

```
joeBloggs.name = "Joseph Bloggs";
```

❖ This is used for both

➢ Reading

➢ Writing data

17.3.4 USING STRUCTURES

❖ The following program demonstrates how the Office can be implemented more efficiently using a struct:

```
/*      ************************************************************
        Structures in C++
        ************************************************************
        -       Defining the ADT Struct
        -       Instantiating and Using Structs
*/
#include <iostream>
#include <string>
using namespace std;


struct Employee
{
        string name;
        string nINumber;
        double salary_Gross;
        double taxRate;
        double taxAmount;
        double salary_Net;
};

int main( )
{
        //»»»»»»»»»»»»»»»»»»»»»»»»»»»»»»»»»»»»»»»»»»»»»»»»»»»»»»»»»»»»»»»
        Employee joeBloggs;

        /*    To ACCESS individual members of the struct, both
              -       setting and
              -       retrieving
              use the dot/period operator .
        */

        // Setting joeBloggs's data
        joeBloggs.name = "Joseph Bloggs";
        joeBloggs.salary_Gross = 20000.0;
        joeBloggs.taxRate = 0.3;

        joeBloggs.taxAmount = joeBloggs.taxRate *
                        joeBloggs.salary_Gross;

        joeBloggs.salary_Net = joeBloggs.salary_Gross -
                                        joeBloggs.taxAmount;

        // Retrieving joeBloggs's data
        cout    << "Hello, my name is "<< joeBloggs.name
                << " & my Net Salary is "<< joeBloggs.salary_Net
                << endl;

        Employee johnSmith;
        Employee amandaTaylor;

        //»»»»»»»»»»»»»»»»»»»»»»»»»»»»»»»»»»»»»»»»»»»»»»»»»»»»»»»»»»»»»»»
```

```
        cout    << endl<< endl<< "==========="
                << endl
                << "END...C++ Demo: ADT: Struct"
                << " "
                << endl;
        while(true){}
        return 0;

}//END...Function MAIN
```

```
Hello, my name is Joseph Bloggs & my Net Salary is 14000


===========
END...C++ Demo: ADT: Struct
```

Program Listing/Output 17.2: StructDemo_Client.cpp

17.3.5 POSSIBLE NOTATION USING TYPEDEF

❖ It is possible to define a struct using a typedef alias

```
typedef struct Employee
  {
        string name;
        string nINumber;
        double salary_Gross;
        double taxRate;
        double taxAmount;
        double salary_Net;
  };
```

❖ In C programming, this practice is fairly common since it means being able to instantiate a struct as follows:

```
Employee joeBloggs;
```

instead of

```
struct Employee joeBloggs;
```

❖ In C++, since the former can be done anyway, the need for the `typedef` `alias` has been eliminated

❖ Several C programmers argue that even in C, it is not needed

  ➢ Its only benefit is eliminating the need to duplicate the keyword `struct`

17.3.6 LIMITATIONS OF STRUCTURES

❖ While structures are the most commonly used Abstract data type in C programming, their limitations are quickly becoming apparent:

➢ There is no access safety

- All the data items are accessible anywhere the ADT Employee is available-
  - ○ This could have potentially serious consequences with errors (will be covered in section 5.7.1)

➢ There are operations, such as tax calculations, which still have to be done in the main body of the program

- This is not strongly supported in structures

❖ These problems are have been tackled in classes

17.4 UNIONS

❖ Unions are similar to structs but place the restriction that only ONE of the member variables can hold a value at any one time

❖ The following program shows one of its uses: Logging the same piece of data with different levels of precision

➢ For any one task at hand, only one of the values will be useful

```
/*      ****************************************************************
        C STYLE UNIONS
        ****************************************************************
        -       This program demonstrates creating and using the ADT: a
                UNION
*/
#include <iostream>
#include <string>
using namespace std;


union Timing
{
        int crudeTiming;
        double preciseTiming;
};

int main( )
{
        //»»»»»»»»»»»»»»»»»»»»»»»»»»»»»»»»»»»»»»»»»»»»»»»»»»»»»»»»»»»»»»»»»»
        Timing marathonTime;
```

```
        marathonTime.crudeTimimg = 5;
        cout << "Crudely speaking, the marathon took "
             << marathonTime.crudeTimimg<< " hours."
             << endl;

        marathonTime.preciseTiming = 4.75;
        cout << "More Precisely, the marathon took "
             << marathonTime.preciseTiming<< " hours."
             << endl;

        //»»»»»»»»»»»»»»»»»»»»»»»»»»»»»»»»»»»»»»»»»»»»»»»»»»»»»»
        cout   << endl
               << "A union can hold only ONE value at a time:"<< endl;
        cout   << "marathonTime.crudeTimimg = "
               << marathonTime.crudeTimimg<< endl;
        cout   << "marathonTime.preciseTiming = "
               << marathonTime.preciseTiming<< endl;

        //»»»»»»»»»»»»»»»»»»»»»»»»»»»»»»»»»»»»»»»»»»»»»»»»»»»»»»
        cout   << endl<< endl<< "==========="
               << endl
               << "END...C++ Demo: ADTs, C Style Unions"
               << " "
               << endl;
    while(true){}
    return 0;

}//END...Function MAIN
```

```
Crudely speaking, the marathon took 5 hours.
More Precisely, the marathon took 4.75 hours.

A union can hold only ONE value at a time:
marathonTime.crudeTimimg = 0
marathonTime.preciseTiming = 4.75


===========
END...C++ Demo: ADTs, C Style Unions
```

Program Listing/Output 17.3: UnionDemo_Client.cpp

❖ Note: Unions are sizeably less versatile than structs and so

  ➢ Not used much in C programming

  ➢ Used very little in C++ programming, where the OOP Class is available

❖ However, their assurance that only one variable can contain a value at any one time does make them useful in a few situations

➤ In Microsoft Excel, the XLOPER data type, used when communicating with your C++ program, contains unions as its members

  • This is covered in Cybernetics in C++ - Expert

17.5 ENUMERATIONS / ENUMS

❖ The Enum/Enumerated Type is an ADT where restrictions are placed on the values that a variable can take on can be used instead

❖ With structs, there is no restriction on what values the individual member data items can take

❖ In a few occasions, it is convenient to be able to place this restriction; common examples include

  ➤ Days of the week

    • If you are asking your user to enter a day of the week, you want it possible answers to be restricted to the 7 possibilities

  ➤ Months of the year

  ➤ The 4 suits in a pack of playing cards

  ➤ The possible characters in a game of Chess

17.5.1 DECLARING ENUMS

❖ The syntax for declaring enums is as follows:

```
enum name_Of_Enum
    {
        Enum_Member_0,
        Enum_Member_1,
        Enum_Member_2,
        Enum_Member_3,
                .
                .
                .

    };
```

17.5.2 USING ENUMS

❖ When instantiating the enumerated type, the word `enum` needs to be used again in both C and C++

❖ Each member of the enum can be accessed in one of 2 possible ways

  ➤ By typing the name of the member

  ➤ By using its Index

❖ The following convention is used in the addressing:

➢ The elements

➢ The 1st element is at position 0

➢ The 2nd element is at position 1

➢ The 3rd element is at position 2

➢ ...

➢ The nth element is at position (n-1)

➢ i.e. The position index can be thought of as the offset from the beginning

❖ This is the same counting system that is used for Arrays

```
/*       ****************************************************************
         C STYLE ENUMS
         ****************************************************************
   -     This program demonstrates creating and using the ADT: a ENUM
*/
#include <iostream>
#include <string>
using namespace std;
enum day_Of_Week
{
     Sunday,
     Monday,
     Tuesday,
     Wednesday,
     Thursday,
     Friday,
     Saturday
};

int main( )
{

     //»»»»»»»»»»»»»»»»»»»»»»»»»»»»»»»»»»»»»»»»»»»»»»»»»»»»»»»»»»»»»»
     enum day_Of_Week xmasDay_2012 =  day_Of_Week(2);
     enum day_Of_Week boxingDay_2012 =  Wednesday;


     // enum day_Of_Week newYearsDay_2013 = fred; // ILLEGAL
          //    fred is not an acceptable value

     cout   << "Christmas Day 2012 will fall on a "
            << xmasDay_2012
            << endl;

     cout   << "Boxing Day 2012 will fall on a "
            << boxingDay_2012
```

```
                  << endl;



   //»»»»»»»»»»»»»»»»»»»»»»»»»»»»»»»»»»»»»»»»»»»»»»»»»»»»»»»»»»
   cout     << endl<< endl<< "============"
            << endl
            << "END...C++ Demo: ADTs, C Style Enums"
            << " "
            << endl;
     while(true){}
     return 0;
}//END...Function MAIN
```

Program Listing/Output 17.4: EnumDemo_Client.cpp

17.5.2.1 Popularity of Enum Use

❖ Enums are used sporadically in C++

➢ They are mostly used for the very common cases discussed in section 17.5

❖ In most other cases, they can potentially cause more problems than they solve

➢ Once enums have placed a restriction on the acceptable values, there is no flexibility

➢ If there is any doubt whatsoever that any new values might be needed any time in future, they are best avoided

➢ In addition, with Object Oriented Programming, Classes can provide much of the required functionalities, and a lot more

❖ When GUI programming in C, it is a common practice to place in an enum, the options which appear in

➢ List Boxes

➢ Radio Button Lists

❖ **Rule of Thumb:** Use Enums only when you are sure they are the best solution, and no new values will be added in future

➢ If in doubt, avoid Enums

17.6 POINTERS TO ADTS

❖ Pointers can, and much of the time are, used to access individual members of user defined types

❖ This section looks at implementing a Pointer-to-Struct

❖ The most common practice in C++ is to use a Pointer-to-Class (section 18.5.3, where a full discussion is given, onwards)

17.6.1 DEFINITION AND POPULATION

❖ Revision: Let us look again at our struct Employee, section 17.3:

```
struct Employee
{
    string name;
    string nINumber;
    double salary_Gross;
    double taxRate;
    double taxAmount;
    double salary_Net;
};
```

❖ In order to define a Pointer to this

   ➢ The pointer must be of type Pointer-to-struct

   ➢ It will contain the address of an instance

❖ Thus in this case, a pointer to Employee joeBloggs will take the following form

```
Employee *ptr_joeBloggs = &joeBloggs;
```

   ➢ The name of the pointer is ptr_joeBloggs

   ➢ Employee * ensures that it is of type Pointer-to-Employee

   ➢ &joeBloggs ensures that it contains the address of the object joeBloggs

17.6.2 THE "ARROW" OPERATOR

❖ The most powerful and common way to access the individual elements of a struct via a pointer is using the "arrow" operator, made up of the characters - and >

❖ Thus in this case, to access the member variable name, the following will be used:

```
ptr_joeBloggs -> name = "Joseph Bloggs";
```

❖ The following program shows this in use:

```
#pragma once

#include <iostream>
#include <string>
using namespace std;

struct Employee
{
        string name;
        string nINumber;
        double salary_Gross;
        double taxRate;
        double taxAmount;
        double salary_Net;
};
```

Program Listing/Output 17.5: Employee.h

```
/*      ***********************************************************
        POINTER TO STRUCT
        ***********************************************************
        -       Defining the ADT Struct
        -       Instantiating and Using Structs
        -       Defining a Pointer-to-Struct,
                -       using it to access individual Data Members
*/
#include <iostream>
#include <string>
using namespace std;

#include "Employee.h"

int main( )
{
        //»»»»»»»»»»»»»»»»»»»»»»»»»»»»»»»»»»»»»»»»»»»»»»»»»»»»»»»»»»»»
        // Defining instance joeBloggs and populating Data Members
        //
        Employee joeBloggs;

        //»»»»»»»»»»»»»»»»»»»»»»»»»»»»»»»»»»»»»»»»»»»»»»»»»»»»»»»»»»»»
        //      Pointer to Struct
        //
        Employee *ptr_joeBloggs = &joeBloggs;

        //==========================================================
        //      Accessing Data Members using the -> Operator
        //
        ptr_joeBloggs -> name = "Joseph Bloggs";
        ptr_joeBloggs -> salary_Gross = 20000.0;
        ptr_joeBloggs -> taxRate = 0.3;
```

```
        cout    << "The Gross Salary of Joseph Bloggs is "
                << ptr_joeBloggs -> salary_Gross
                << endl;

        // ...Further population via the Pointer

        //»»»»»»»»»»»»»»»»»»»»»»»»»»»»»»»»»»»»»»»»»»»»»»»»»»»»»»»»»»»»»»»»»»»»
        cout    << endl<< endl<< "==========="
                << endl
            << "END...C++ Demo: Pointers to Struct"
                << " "
                << endl;
    while(true){}
    return 0;

}//END...Function MAIN
```

```
The Gross Salary of Joseph Bloggs is 20000


===========
END...C++ Demo: Pointers to Struct
```

Program Listing/Output 17.6: PointerToStruct_Client.cpp

17.6.2.1 Preference over the dot operator

❖ It is also possible to use decode using the * operator, along with the dot operator

```
(*ptr_joeBloggs).greet();
```

➢ The parentheses are essential

❖ This approach is widely considered more cumbersome and less safe than the arrow operator

❖ **Rule of Thumb:** Use the arrow operator when accessing the elements of any composite type using pointers

17.7 REFERENCES TO ADTS

❖ References can also be used to access individual members of user defined types

❖ This section looks at implementing a Reference-to-Struct

➢ Since references are generally not used with data structures (section 15.2), this is a sizeably less common practice than using pointers, especially pointers to classes

17.7.1 DEFINITION AND POPULATION

❖ Revision: Let us look again at our struct Employee, section 17.3:

```
struct Employee
{
        string name;
        string nINumber;
        double salary_Gross;
        double taxRate;
        double taxAmount;
        double salary_Net;
};
```

❖ In order to define a Reference to this

➢ The pointer must be of type Reference-to-struct
➢ It will contain the address of an instance

❖ Thus in this case, a reference to Employee joeBloggs will take the following form

```
Employee &ref_joeBloggs = joeBloggs;
```

➢ The name of the reference is `ref_joeBloggs`
➢ `Employee` & ensures that it is of type Reference-to-Employee
➢ `joeBloggs` ensures that it contains the address of the object joeBloggs

17.7.2 ACCESS USING THE DOT OPERATOR

❖ In order to access the individual elements of a struct via a reference, again the dot operator is used

❖ Thus in this case, to access the member variable `name`, the following will be used:

```
ptr_joeBloggs -> name = "Joseph Bloggs";
```

❖ The following program shows this in use:

```
#pragma once

#include <iostream>
#include <string>
```

```
using namespace std;

struct Employee
{
        string name;
        string nINumber;
        double salary_Gross;
        double taxRate;
        double taxAmount;
        double salary_Net;
};
```

Program Listing/Output 17.7: Employee.h

```
/*      ****************************************************************
        REFERENCE TO STRUCT
        ****************************************************************
        -       Defining the ADT Struct
        -       Instantiating and Using Structs
        -       Defining a Reference-to-Struct,
        -       using it to access individual Data Members
*/
#include <iostream>
#include <string>
using namespace std;

#include "Employee.h"

int main()
{
        //»»»»»»»»»»»»»»»»»»»»»»»»»»»»»»»»»»»»»»»»»»»»»»»»»»»»»»»»»»»»»»»»»»
        //      Defining instance joeBloggs and populating Data Members
        //
        Employee joeBloggs;

        //»»»»»»»»»»»»»»»»»»»»»»»»»»»»»»»»»»»»»»»»»»»»»»»»»»»»»»»»»»»»»»»»»»
        //      Reference to Struct
        //
        Employee &ref_joeBloggs = joeBloggs;

        //===========================================================
        //      Accessing Data Members using the . Operator
        //
        ref_joeBloggs.name = "Joseph Bloggs";
        ref_joeBloggs.salary_Gross = 20000.0;
        ref_joeBloggs.taxRate = 0.3;

        cout    << "The Gross Salary of Joseph Bloggs is "
                << ref_joeBloggs.salary_Gross
                << endl;

        // ...Further population via the Reference
```

```
//»»»»»»»»»»»»»»»»»»»»»»»»»»»»»»»»»»»»»»»»»»»»»»»»»»»»»»»»»»»»»»»»»»»»»
cout   << endl << endl << "==========="
       << endl
       << "END...C++ Demo: References to Struct"
       << " "
       << endl;
    while (true) {}
    return 0;

}//END...Function MAIN
```

```
The Gross Salary of Joseph Bloggs is 20000


===========


END...C++ Demo: References to Struct
```
Program Listing/Output 17.8: ReferenceToStruct_Client.cpp

18. C++ CLASSES & OBJECTS

❖ C++ Classes incorporate Encapsulation as well as Abstraction

18.1 WHAT IS ENCAPSULATION?

❖ Encapsulation is the practice of hiding details

  ➤ It comes from the world "capsule"

    • An entity, with an outer membrane which encloses the inner details, enabling the outside world to think of it as a single "object"
    • Only when one penetrates the outer membrane does one look at the details

18.1.1 ENCAPSULATION OUTSIDE THE PROGRAMMING WORLD

❖ Medical Capsule

  ➤ Can have several medications in one capsule
  ➤ You take it as one entity, swallow it
  ➤ Once the outer membrane is digested in stomach, the "delving into details" begins: the different medications start work

❖ Time Capsule

   ➢ You put your things in a box/capsule which gives a picture of your life today

   ➢ In a few 1000 years' time, someone will find it and draw conclusions about human life today

❖ The Human Body

   ➢ Countless physiological activities are going on inside everyone's body

   ➢ Each is wrapped up/encapsulated in skin

      • Humans interact with each other without thinking about all these details

   ➢ Only once in a while, a surgeon may need to penetrate this membrane and delve inside

❖ Compartmental Capsule

   ➢ Rocket: astronauts' living quarters

   ➢ Chilean miners: capsule used as "lift car"

❖ IT

   ➢ Different users have accounts,

      • Network accounts

      • Email accounts

   ➢ The password protection ensures each account remains "encapsulated"

❖ All accounts

   ➢ Bank accounts

   ➢ Insurance policies

❖ Refrigerator

   ➢ Early days: each individual food item had to be cooled separately: cumbersome & expensive

   ➢ Victorians invented the refrigerator:

   ➢ You think of it as a single "capsule," a "box" which stays in your kitchen

   ➢ Only when you need to eat/drink something, you open the door and delve into details

❖ Driving a Car

➢ Lots of operations going on

• Hidden i.e. encapsulated under the bonnet

➢ You only have to worry about steering, accelerator, brake, gears (even this is encapsulated in an automatic gearbox)

➢ Only when you lift the bonnet and look inside do you delve into details

❖ Modern day electronics:

➢ Any device: TV, Computer

➢ Microchips

❖ In the modern world, as devices get increasingly sophisticated, the need for encapsulation is fast increasing

EXERCISE

❖ *This list can be also extended almost indefinitely; see if you can come up with 3 further examples of Encapsulation*

18.2 IMPLEMENTING ABSTRACTION & ENCAPSULATION IN C++

❖ The problems of lack of abstraction have been encountered in section 17.1.2

➢ C programming has gone some of the way to tackling this using ADTs

➢ However, as we saw, this has limited suitability in modern times

• All members are public, meaning they have the same exposure problems as global variables

➢ C++ provides a further means which is on the whole much more powerful:...

❖ Abstraction is implemented in C++ primarily using the entity **Class**

➢ The Class is the data type

➢ The instance of a class is called an **Object**

➢ Note that this is the first time a different word is being used for the data type and the instance

• For primitive data types e.g. int, double, the same word is used

• Also, for ADTs, struct, enum and union are used for both

❖ Encapsulation is implemented using 3 access specifiers

> private   Accessible only within that class

> protected

> public   Accessible anywhere

> The meaning of protected will make sense when we look at Inheritance, section 22 onwards

> Note that in C++ these are not full descriptions; these will be expanded when looking at

    • Friend classes and Friend Functions (section 20.13)

    • Inheritance (section 22)

18.2.1 TERMINOLOGY - NATIVE CLASS

❖ The following terminology is often used in the software industry:

❖ "**Native** Class" is used to mean a class created in ANSI (or "pure") C++

❖ Vendor specific classes have their own names:

> Microsoft calls its classes

    • "Managed Classes" - classes built in .NET

    • "Unmanaged Classes" - classes built using a Microsoft Library other than .NET such as MFC

18.2.2 BUILDING ON ADTS

❖ The work on Abstract Data Types, especially the struct, is an excellent precursor to the work on classes

> It is the strongest support C programming has for abstraction

❖ Going forwards, the use of ADTs should be minimised in C++, due to their limitations

> A major problem is that all members are publicly available

    • Consequently members of globally defined ADTs suffer from similar problems to global variables

18.3 STRUCTURE OF A CLASS

❖ To a first approximation, in OOP, a class does to data and functionalities

> What directories/folders in IT do to files

> It provides a "container" for several of them, enabling, amongst other things, organisation

❖ In C++, a class comprises 4 entities:

> Data Members

• These are the data which **all** objects need to have

> Member Functions

• These are functionalities which **all** objects need to be capable of

> Inner Class

• This is a means of further categorising data members and member functions

> Friend Declarations

• These are entities which have privileged access to the class

❖ The Abstraction is illustrated by the fact that we are focusing on

> All objects
> Not just a few specific ones

18.3.1 KNOWLEDGE & SKILL ANALOGY

❖ Note that not every class must possess all 4

> These are the 4 major categories, some may be "left blank"

❖ Data Members and Member Functions are the 2 most common

❖ They can be likened to Knowledge and Skill

> Data Members: The Data/Knowledge possessed by the objects
> Member Functions: Their Skill, ability to do things with their knowledge

18.4 IMPLEMENTING A C++ CLASS - METHOD 1 OF 2

❖ Let us now look at a fuller, more sophisticated means of modelling the Employee example

> This will now contain
• Data Members: Data which ALL employees require
• Member Functions: Functionalities which ALL employees require

➢ The first means of implementing the class comprises having the declaration and definition in one, is as follows:

```
class Employee
  {
  private:
        // Data Member(s)
        string name;
        string nI_Number;
        double salary_Gross;
        double tax_Rate;
        double tax_Amount;
        double salary_Net;

  public:
        // Member Function(s)
        void greet()
        {
            cout << "Hello, I'm an Employee"<< endl;
        }
}; // END...class Employee
```

Program Listing/Output 18.1: Implementation: Class Employee

❖ Note that

➢ The class is a block, thus surrounded by the braces { }

➢ The terminator ; is required at the end of the declaration

➢ The keywords `private` and `public`

- are followed by the colon operator
- can be used once to refer to several data members/member functions, as here, or several times
- **Java Note:** In Java, the private and public keywords must be specified separately at the beginning of each member

❖ This method, one of 2, is widely used

➢ **Java & C# Note:** In these languages, this is the only one allowed

18.4.1 SCOPE OF A VARIABLE (2)

❖ Note that one of the many essential things the class has provided is a new scope of variable

➢ Data members are accessible (primarily) by members of the class

❖ The problem encountered in procedural programming has now been solved:

- ➤ If a variable has to be accessed by more than one function, it has to be global
- ➤ The class means that now, a variable can be accessed by a group of (member) functions

❖ Inner classes provide means for further scopes (section 20.10)

18.5 IMPLEMENTING OBJECTS

18.5.1 OBJECTS ON THE RUNTIME STACK

18.5.1.1 Creating Objects Directly

❖ Let us now create the first object: joeBloggs

❖ Since this/he will be of type class Employee, the instantiation is as follows:

```
Employee joeBloggs;
```

18.5.1.1.1 The Dot Operator

❖ The question now is how to access individual members of the class via the object

❖ The dot operator . is used

❖ Thus the following will be required in order to invoke the greet member function of joeBloggs:

```
joeBloggs.greet();
```

❖ While the dot operator can be used for reading and writing to data members, we have implemented our class to prohibit this

- ➤ The access qualifier `private` ensures they cannot be accessed from outside the class
- ➤ Thus the following command

```
joeBloggs.name = "Joseph Bloggs";
```

will cause a compile-time error

❖ Here is the full program so far:

```
#include <iostream>
#include <string>
using namespace std;

// METHOD (1) Declaring and Defining a Class in One
class Employee
```

```
{
      private:
            // Data Member(s)
            string name;
            string nI_Number;
            double salary_Gross;
            double tax_Rate;
            double tax_Amount;
            double salary_Net;

      public:
            // Member Function(s)
            void greet()
            {
                  cout << "Hello, I'm an Employee"<< endl;
            }
};// END...class Employee

int main()
{
      // Creating an OBJECT joeBloggs
      Employee joeBloggs;

      //    Accessing the members of the object...
      //    DIRECTLY via the Object - use .DOT Operator
      joeBloggs.greet(); // OK - greet() is PUBLIC

      // joeBloggs.name = "Joseph Bloggs"; // ILLEGAL - name is PRIVATE

      //»»»»»»»»»»»»»»»»»»»»»»»»»»»»»»»»»»»»»»»»»»»»»»»»»»»»»»
      cout   << endl<< "================"
             << endl<< "END...C++ Class & Object Demo"
             << endl;
      while(true){}
      return 0;

}//END...Function MAIN
```

```
Hello, I'm an Employee

================
END...C++ Class & Object Demo
```

Program Listing/Output 18.2: Class_Object_Client.cpp

18.5.1.1.2 *Equating Objects*

❖ Objects can be set equal to one another

➢ i.e. The following are perfectly acceptable

- Equating two existing objects

```
robertFletcher = joeBloggs;
```

- Defining an object and populating it using another

```
Employee richardSharp = joeBloggs;
```

- There are some subtle but important differences between these two actions, discussed in sections 20.5 and 20.6

➢ Now `robertFletcher` and `richardSharp` will contain the same data member values as `joeBloggs`

- NB. for this example, `joeBloggs` has no values assigned to any of his data members (doing this will be covered in section 18.9 onwards)

18.5.2 ARRAYS OF OBJECTS

❖ Creating a data structure of objects is identical in principle to creating a data structure of a Primitive Type or an ADT

➢ The type must be specified, in this case, the class
➢ A name must be given
➢ In the case of a static array, the size must also be specified
➢ It must then be populated with objects

❖ The following example shows how to create a static array of type Employee

```
#include <iostream>
#include <string>
using namespace std;

// METHOD (1) Declaring and Defining a Class in One
class Employee
{
    private:
        // Data Member(s)
        string name;
        string nI_Number;
        double salary_Gross;
        double tax_Rate;
        double tax_Amount;
        double salary_Net;

    public:
        // Member Function(s)
        void greet()
        {
```

```
                    cout << "Hello, I'm an Employee"<< endl;
                }
};// END...class Employee


int main()
{
        //»»»»»»»»»»»»»»»»»»»»»»»»»»»»»»»»»»»»»»»»»»»»»»»»»»»»
        // Creating OBJECTS
        Employee joeBloggs;
        Employee johnSmith;
        Employee amandaTaylor;

        //»»»»»»»»»»»»»»»»»»»»»»»»»»»»»»»»»»»»»»»»»»»»»»»»»»»»
        // Creating an Array of Objects
        Employee office[] = { joeBloggs, johnSmith, amandaTaylor };
        for( int i = 0; i < 3; ++i )  // Better to use STL Iterators
                                      // (Cybernetics in C++ - Expert)
        {
                office[i].greet();
        }


        //»»»»»»»»»»»»»»»»»»»»»»»»»»»»»»»»»»»»»»»»»»»»»»»»»»»»
        cout  << endl<< "================"
              << endl<< "END...C++ Array of Objects Demo"
              << endl;
        while(true){}
        return 0;

}//END...Function MAIN
```

```
Hello, I'm an Employee
Hello, I'm an Employee
Hello, I'm an Employee

================
END...C++ Array of Objects Demo
```

Program Listing/Output 18.3: Creating an Array of objects of type Employee

❖ Note: While arrays are the most commonly used data structure, they are inherited from C programming and so are not ideal for dealing with objects

➢ It is far better to use a vector or deque for objects - these are covered in Cybernetics in C++ - Expert

18.5.3 POINTERS TO OBJECTS

❖ Pointers to objects are widely used

> ➤ All the advantages of memory management, including high performance programming, can now be fully exploited in object oriented programming
> ➤ The example we have just seen, an array of objects, is in fact of type pointer-to-objects, since an array name is a pointer (section 16)
> ➤ Polymorphism, section 23, draws heavy on pointers to objects

18.5.3.1 Definition and Population

❖ The pointer must be of type Pointer-to-class

❖ It will contain the address of an object

❖ Thus in this case, a pointer to Employee `joeBloggs` will take the following form

```
Employee *ptr_joeBloggs = &joeBloggs;
```

> ➤ The name of the pointer is `ptr_joeBloggs`
> ➤ `Employee` * ensures that it is of type Pointer-to-Employee
> ➤ `&joeBloggs` ensures that it contains the address of the object joeBloggs

18.5.3.2 The "Arrow" Operator

❖ The most powerful and common way to access the individual elements of a class via a pointer is using the "arrow" operator, made up of the characters - and >

❖ Thus in this case, to access the member function greet(), the following will be used:

```
ptr_joeBloggs -> greet();
```

18.5.3.2.1 Preference over the dot operator

❖ It is also possible to use decode using the * operator, along with the dot operator

```
(*ptr_joeBloggs).greet();
```

> ➤ The parentheses are essential

❖ This approach is widely considered more cumbersome and less safe than the arrow operator

❖ **Rule of Thumb:** Use the arrow operator when accessing the elements of any composite type using pointers

18.5.3.3 Using Pointers to Objects

❖ The following example shows how to define, populate and use a pointer-to-object

```cpp
#include <iostream>
#include <string>
using namespace std;

//»»»»»»»»»»»»»»»»»»»»»»»»»»»»»»»»»»»»»»»»»»»»»»»»»»»»»»»»»»»»»»»»»»»»
// METHOD ((1)) Declaring and Defining a Class in One
class Employee
{
     private:
             // Data Member(s)
             string name;
             string nI_Number;
             double salary_Gross;
             double tax_Rate;
             double tax_Amount;
             double salary_Net;

     public:
             // Member Function(s)
             void greet()
             {
                    cout << "Hello, I'm an Employee"<< endl;
             }
};// END...class Employee


int main()
{
        //»»»»»»»»»»»»»»»»»»»»»»»»»»»»»»»»»»»»»»»»»»»»»»»»»»»
        // Creating an OBJECT joeBloggs
        Employee joeBloggs;

        //»»»»»»»»»»»»»»»»»»»»»»»»»»»»»»»»»»»»»»»»»»»»»»»»»»»
        //    Creating & using a POINTER
        Employee *ptr_joeBloggs = &joeBloggs;

        // Access - use ARROW -> Operator
        ptr_joeBloggs -> greet();
        // ptr_joeBloggs -> nI_Number = "GH 78 34 56 V"
        //      ILLEGAL - nI_Number is PRIVATE

        //»»»»»»»»»»»»»»»»»»»»»»»»»»»»»»»»»»»»»»»»»»»»»»»»»»»
        cout << endl<< "================"
             << endl<< "END...C++ Pointer-to-Objects Demo"
             << endl;
        while(true){}
        return 0;
```

```
}//END...Function MAIN
```

```
Hello, I'm an Employee

===============
END...C++ Pointer-to-Objects Demo
```

Program Listing/Output 18.4: Using Pointers to Objects

18.5.4 REFERENCES TO OBJECTS

❖ References to objects are less common than pointers to objects but they are still frequently used

  ➢ They are another means of exploiting the powers of bringing together memory management and object oriented programming

  ➢ The smart pointer (Cybernetics in C++ - Expert) makes use of these

18.5.4.1 Definition and Population

❖ The reference must be of type Reference-to-class

❖ It will contain the address of an object

❖ Thus in this case, a reference to Employee joeBloggs will take the following form

```
Employee& ref_joeBloggs = joeBloggs;
```

  ➢ The name of the reference is `ref_joeBloggs`
  ➢ `Employee&` ensures that it is of type Reference-to-Employee
  ➢ `joeBloggs` ensures that it contains the address of the object joeBloggs
    • Reminder: NOTHING is used to say "the address of" when using a reference

18.5.4.2 Access using the Dot Operator

❖ In order to access the individual elements of a class via a reference, the dot operator . is used

  ➢ The same as when accessing directly using via objects

❖ Thus in this case, to access the member function greet(), the following will be used:

```
ref_joeBloggs.greet();
```

❖ The following example shows how to define, populate and use a reference-to-object:

```cpp
#include <iostream>
#include <string>
using namespace std;

//»»»»»»»»»»»»»»»»»»»»»»»»»»»»»»»»»»»»»»»»»»»»»»»»»»»»»»»»»»»»»»»
// METHOD (1) Declaring and Defining a Class in One
class Employee
{
    private:
        // Data Member(s)
        string name;
        string nI_Number;
        double salary_Gross;
        double tax_Rate;
        double tax_Amount;
        double salary_Net;

    public:
        // Member Function(s)
        void greet()
        {
            cout << "Hello, I'm an Employee"<< endl;
        }
};// END...class Employee


int main()
{
    //»»»»»»»»»»»»»»»»»»»»»»»»»»»»»»»»»»»»»»»»»»»»»»»»»
    // Creating an OBJECT joeBloggs
    Employee joeBloggs;

    //»»»»»»»»»»»»»»»»»»»»»»»»»»»»»»»»»»»»»»»»»»»»»»»»»
    //     Creating & using a REFERENCE
    Employee& ref_joeBloggs = joeBloggs;

    // Access - use - use .DOT Operator
    ref_joeBloggs.greet();
    // ref_joeBloggs.salary_Gross = 17000.0;
    //     ILLEGAL - salary_Gross is PRIVATE


    //»»»»»»»»»»»»»»»»»»»»»»»»»»»»»»»»»»»»»»»»»»»»»»»»»
    cout   << endl<< "==============="
           << endl<< "END...C++ Reference-to-Objects Demo"
           << endl;
    while(true){}
```

```
        return 0;

}//END...Function MAIN
```

```
Hello, I'm an Employee
================
END...C++ Reference-to-Objects Demo
```
Program Listing/Output 18.5: Using References to Objects

18.5.5 OBJECTS ON THE RUNTIME HEAP

18.5.5.1 Definition and Population

❖ When you want to define an object on the runtime heap, the following syntax is used

```
ClassName *pointerName = new ClassName
```

❖ Why must the class name be stated twice?

  ➤ The one on the left hand side of the = is the type of the variable, known at **compile** time

  ➤ The one the right hand side of the = is the type that is bound to at **runtime**

  ➤ In this case, they are the same

  ➤ They can be different - this is the basis of Polymorphism, encountered in section 23.

18.5.5.2 Using Objects on the Runtime Heap

❖ The following program demonstrates the definition and usage of objects in the runtime heap:

```
#include <iostream>
#include <string>
using namespace std;

//»»»»»»»»»»»»»»»»»»»»»»»»»»»»»»»»»»»»»»»»»»»»»»»»»»»»»»»»»»»»»»»»»
// METHOD (1)      Declaring and Defining a Class in One
class Employee
{
      private:
              // Data Member(s)
              string name;
              string nI_Number;
              double salary_Gross;
```

```
            double tax_Rate;
            double tax_Amount;
            double salary_Net;

    public:
            // Member Function(s)
            void greet()
            {
             cout << "Hello, I'm an Employee"<< endl;
  }
};// END...class Employee


int main()
{
        //»»»»»»»»»»»»»»»»»»»»»»»»»»»»»»»»»»»»»»»»»»»»»»»»»»»»»»
        // Creating an OBJECT peterShepherd on the HEAP
        //    Creating & using a POINTER
        //
        Employee *ptr_peterShepherd = new Employee;

        ptr_peterShepherd -> greet();

        //    ILLEGAL - nI_Number is PRIVATE
        // ptr_peterShepherd ->nI_Number = "FG 34 56 93 D"

        //»»»»»»»»»»»»»»»»»»»»»»»»»»»»»»»»»»»»»»»»»»»»»»»»»»»»»»
        cout    << endl<< "================"<< endl
                << "END...C++ Objects on the Runtime Heap Demo"
                << endl;
        while(true){}
        return 0;

}//END...Function MAIN



Hello, I'm an Employee

================

END...C++ Objects on the Runtime Heap Demo
```

Program Listing/Output 18.6: Objects_Heap_Client.cpp

18.5.5.3 *Java & C# - All Objects on the Heap*

❖ A common statement which is heard in both Java and C# is "Primitive Variables are placed on the stack, objects are placed on the heap"

❖ In both languages, the following syntax is required, to create objects:

```
ClassName objectName = new ClassName;
```

> ➤ e.g. In order to create an object joeBloggs, of type Employee, the following line is required

```
Employee joeBloggs = new Employee;
```

❖ Programming in C++ confirms that the new operator is indeed placing the object on the heap

❖ Thus in languages like Java and C#, the runtime heap memory is still there

> ➤ It is just hidden from the programmer
> ➤ This is possible since "Automatic garbage collection" has been engineered for both sorts of runtime memory

18.5.6 OBJECTS AS FUNCTION INPUTS

❖ In order to have a function which takes an object as an input, the following syntax is used:

```
returnType functionName( ClassName objectName )
```

❖ In order to have a function with

> ➤ Name   myFunc
> ➤ Input: Object of type Employee, name Employee_IN
> ➤ Output: None i.e. void

the following prototype is needed:

```
void myFunc( Employee Employee_IN )
```

❖ The following program shows how this can be implemented:

```
#include <iostream>
#include <string>
using namespace std;


// METHOD (1) Declaring and Defining a Class in One
class Employee
{
     private:
             // Data Member(s)
             string name;
             string nI_Number;
             double salary_Gross;
             double tax_Rate;
             double tax_Amount;
             double salary_Net;
```

```
        public:
                // Member Function(s)
                void greet()
                {
                        cout << "Hello, I'm an Employee"<< endl;
                }
};// END...class Employee


// Function Declaration
void state_EmployeeGreeting( Employee );

/*++++++++++++++++++++++++++++++++++++++++++++++++++++++++++++++++++++++++++
*/
int main()
{
        // Creating an OBJECT joeBloggs
        Employee joeBloggs;

        state_EmployeeGreeting( joeBloggs );


        //»»»»»»»»»»»»»»»»»»»»»»»»»»»»»»»»»»»»»»»»»»»»»»»»»»»»»»»»»»»»»»»»»»»»
        cout  << endl<< "==============="
                << endl<< "END...C++ Class & Object Demo"
                << endl;
        while(true){}
        return 0;

}//END...Function MAIN

/*++++++++++++++++++++++++++++++++++++++++++++++++++++++++++++++++++++++++++
        FUNCTION Definition: state_EmployeeGreeting
*/
void state_EmployeeGreeting( Employee Employee_IN )
{
        Employee_IN.greet();

}//END...Function state_EmployeeGreeting
```

```
Hello, I'm an Employee

===============
END...C++ Class & Object Demo
```

Program Listing/Output 18.7: Object_FunctionInput_Client.cpp

18.5.7 OBJECT AS THE FUNCTION OUTPUT

❖ Likewise, in order to implement a function where the output is an object, the prototype is as follows:

```
className functionName( Inputs with Types )
```

❖ In order to have a function with

➢ Name   myFunc
➢ Input: None
➢ Output: Object of type Employee

the following prototype is needed:

```
Employee myFunc( )
```

❖ The following program shows how this can be implemented:

```cpp
#include <iostream>
#include <string>
using namespace std;


 // METHOD (1)     Declaring and Defining a Class in One
 class Employee
 {
      private:
            // Data Member(s)
            string name;
            string nI_Number;
            double salary_Gross;
            double tax_Rate;
            double tax_Amount;
            double salary_Net;

      public:
            // Member Function(s)
            void greet()
            {
                    cout << "Hello, I'm an Employee"<< endl;
            }
};// END...class Employee


// Function Declaration
Employee create_generalEmployee(  );

/*++++++++++++++++++++++++++++++++++++++++++++++++++++++++++++++++++++++++++
*/
int main()
{
```

```
      // Creating an OBJECT joeBloggs
      Employee newEmployee = create_generalEmployee(   );
      newEmployee.greet();



      //»»»»»»»»»»»»»»»»»»»»»»»»»»»»»»»»»»»»»»»»»»»»»»»»»»»»»»»»»»»»»»»»»»
      cout     << endl<< "================"
               << endl<< "END...C++ Class & Object Demo"
               << endl;
      while(true){}
      return 0;

}//END...Function MAIN


/*++++++++++++++++++++++++++++++++++++++++++++++++++++++++++++++++++++
      FUNCTION Definition: state_EmployeeGreeting
*/
Employee create_generalEmployee(   )
{
      Employee local_Employee;
      return local_Employee;

}//END...Function state_EmployeeGreeting
```

```
Hello, I'm an Employee

================
END...C++ Class & Object Demo
```

Program Listing/Output 18.8: Object_FunctionOutput_Client.cpp

18.6 IMPLEMENTING A C++ CLASS - METHOD 2 OF 2

❖ It is possible to write the declaration and definition of a class separately

❖ The declaration presents a list of the class members to the compiler; it contains listings of the

  ➢ Data Members
  ➢ Member Functions

❖ In the definitions, each member function needs to be implemented separately

18.6.1 THE SCOPE RESOLUTION OPERATOR

❖ The scope resolution operator :: is used to indicate which class each member function belongs

❖ The definitions are as follows:

```
returnType ClassName::memberFunctionName( inputs and types )
{

    // Body of Function

}
```

18.6.2 SEPARATING THE DECLARATION & DEFINITIONS LEXICALLY

❖ Thus our Employee class can also be written as follows:

➢ The declaration will be as follows:

```
// Class DECLARATION
class Employee
{
    // Data Members
    private:
        string name;
        string nI_Number;
        double salary_Gross;
        double tax_Rate;
        double tax_Amount;
        double salary_Net;

    // Member Function(s)
    public:
        void greet();

};// END...Class Declaration: Employee
```

➢ So far there is only one member function - it will be defined as follows:

```
void Employee::greet()
{
    cout << "Hello, I'm an Employee"<< endl;

}//END...Member Function greet
```

❖ The program can now be written like this:

```cpp
#include <iostream>
#include <string>
using namespace std;

// METHOD (2)       Declaring and Defining a Class SEPARATELY

// Class DECLARATION
class Employee
{
      // Data Members
      private:
              string name;
              string nI_Number;
              double salary_Gross;
              double tax_Rate;
              double tax_Amount;
              double salary_Net;

      // Member Function(s)
      public:
              void greet();

};// END...Class Declaration: Employee


// Member Function DEFINITION(S):
void Employee::greet()    // :: SCOPE RESOLUTION OPERATOR
{
      cout  << "Hello, I'm an Employee"<< endl;

}//END...Member Function greet

//+++++++++++++++++++++++++++++++++++++++++++++++++++++++++++++++++++++
int main()
{
      // Creating an OBJECT joeBloggs
      Employee joeBloggs;

      //      Accessing the members of the object

      //      (1) DIRECTLY via the Object - use .DOT Operator
      joeBloggs.greet(); //      OK - greet() is PUBLIC

      // joeBloggs.name = "Joseph Bloggs";
              //      ILLEGAL - name is PRIVATE


      // (2)      Via a POINTER - use ARROW -> Operator
      Employee *ptr_joeBloggs = &joeBloggs;
      ptr_joeBloggs -> greet();
```

```
    // ptr_joeBloggs -> nI_Number = "GH 78 34 56 V"
            //        ILLEGAL - nI_Number is PRIVATE

    // (3)        Via a REFERENCE - again  - use .DOT Operator
    Employee & ref_joeBloggs = joeBloggs;
    ref_joeBloggs.greet();

    // ref_joeBloggs.salary_Gross = 17000.0;
            //        ILLEGAL - salary_Gross is PRIVATE


    //»»»»»»»»»»»»»»»»»»»»»»»»»»»»»»»»»»»»»»»»»»»»»»»»»»»»»»»»»»»»»»»»»
    cout   << endl<< "================"<< endl
           << "END...C++ Separating a Class Declaration &
    Definition"
           << endl;
       while(true){}
       return 0;

}//END...Function MAIN
```

```
Hello, I'm an Employee
Hello, I'm an Employee
Hello, I'm an Employee

================
END...C++ Separating a Class Declaration & Definition
```

Program Listing/Output 18.9: Class_LexicalOutside_Client.cpp

❖ An essential point to note is that the declarations and definitions have been separated only lexically

➤ Logically, this is no different from writing them together

❖ The main advantage of this method is it makes it easy to split the class over different files

➤ Common Practice: Put the
  • Declarations in a header file
  • Definitions in a source file

18.6.3 MANAGING A CLASS OVER MULTIPLE FILES

❖ Although our class so far is very small, here's how it can be managed over multiple files

```
#include <iostream>
#include <string>
using namespace std;

#ifndef EMPLOYEE_H
#define EMPLOYEE_H

// Class DECLARATION
class Employee
{
    // Data Members
    private:
        string name;
        string nI_Number;
        double salary_Gross;
        double tax_Rate;
        double tax_Amount;
        double salary_Net;

    public:
        void greet();

};

#endif
```

Program Listing/Output 18.10: Employee.h

```
#include "Employee.h"


// Member Function DEFINITIONS:
void Employee::greet() // :: SCOPE RESOLUTION OPERATOR
{
    cout  <<  "Hello, I'm an Employee"<< endl;

}//END...Member Function greet
```

Program Listing/Output 18.11: Employee.cpp

```
#include <iostream>
#include <string>
using namespace std;

#include "Employee.h"

int main()
 {
    // Creating an OBJECT joeBloggs
    Employee joeBloggs;
```

```
        //       Accessing the members of the object

        //    (1)   DIRECTLY via the Object - use .DOT Operator
        joeBloggs.greet();  //      OK - greet() is PUBLIC

        // joeBloggs.name = "Joseph Bloggs";
             //      ILLEGAL - name is PRIVATE


        //    (2)      Via a POINTER - use ARROW -> Operator
        Employee *ptr_joeBloggs = &joeBloggs;
        ptr_joeBloggs -> greet();

        // ptr_joeBloggs -> nI_Number = "GH 78 34 56 V"
             //      ILLEGAL - nI_Number is PRIVATE

        //    (3)        Via a REFERENCE - again  - use .DOT Operator
        Employee & ref_joeBloggs = joeBloggs;
        ref_joeBloggs.greet();

        // ref_joeBloggs.salary_Gross = 17000.0;
             //      ILLEGAL - salary_Gross is PRIVATE


        //»»»»»»»»»»»»»»»»»»»»»»»»»»»»»»»»»»»»»»»»»»»»»»»»»»»»»»»»»»»»»»»»»
        cout   << endl<< "================"<< endl
             << "END...C++ Managing an OOP Program over Multiple
                             Files"
             << endl;
        while(true){}
        return 0;

}//END...Function MAIN
```

```
Hello, I'm an Employee
Hello, I'm an Employee
Hello, I'm an Employee

================
END...C++ Managing an OOP Program over Multiple Files
```

Program Listing/Output 18.12: Classes_DifferentFiles_Client.cpp

❖ In modern times, as technology becomes increasingly sophisticated, programs are getting longer and more complex

➢ The need to be organised is more important than ever

- ➢ There is no right or wrong way to organise a program, this is down to the programmer (i.e. YOU!:-))
  - This is one of the many ways computer science is an art as well as a science
- ❖ In C++, issues to bear in mind when organising projects include
  - ➢ Splitting a program into Classes
  - ➢ Further dividing the classes using Inner Classes (section 20.10 onwards)
  - ➢ Arranging the classes over different files
- ❖ A common means of organising programs is to
  - ➢ Have two files for each class
    - One Header File for the Declarations
    - One Source File for the Definitions
  - ➢ For much of the rest of this book, and Cybernetics in C++ - Expert, this approach will be taken

18.7 DEFINING A CLASS: DEFINITION & DECLARATIONS, TOGETHER VS SEPARATE

- ❖ There are thus 2 ways to define a class
- ❖ It is down to personal preference which method is preferable
  - ➢ Method 1, where all the member functions are implemented inside the declaration is (almost entirely) used in languages such as Java and C#
  - ➢ Method 2 tends to be a little more popular in C++, since it enables managing a class over different files
    - Common Practice: Have 2 files for every class; put the
      - ○ Declarations in a header file
      - ○ Definitions in a source file
  - ➢ However, in C++, several programs include both styles; several of these will be seen throughout this book and Cybernetics in C++ - Expert

18.8 SPECIAL CLASS MEMBERS

- ❖ Along with the "Normal" data members and member functions encountered so far, there are several special ones which have particular essential features

❖ Special Data Members

> ➤ Static Data Member
> ➤ The "this" Pointer

❖ Special Member Functions

> ➤ Constructors
>   • Default Constructor
>   • Copy Constructor
> ➤ Destructors
> ➤ Mutator Member Function
> ➤ Const Member Functions
> ➤ Static Member Functions
> ➤ Conversion Member Function
> ➤ Virtual Member Function
> ➤ Assignment Operator
> ➤ Printing Member Function
> ➤ Operator Overloading Member Function

18.9 CONSTRUCTORS

18.9.1 WHAT ARE CONSTRUCTORS?

❖ Constructors are special member functions which are invoked automatically when **an object is created**

> ➤ i.e. When an object **comes into scope**

❖ They are invaluable for any initial housekeeping which needs to take place on the object

> ➤ The most common of these tasks is to initialise the values of the data members
>   • In fact this is so pervasive, it is a common practice to define constructors as "member functions for initialising data members"
>   • Our advice: Keep this as a guideline
>   ○ Constructors have several further uses
> ➤ If any resources need to be set, this is a good place to do so, common examples being
>   • Allocating memory on the heap

- Reserving network resources
- Creating database connections
- Initialising file handles
- Initiating communication between internal programs, such as
  - Microsoft's COM references
  - Communications via a DDE, Dynamic Data Exchange, Server
- Groundwork for resources in Concurrent programming (Cybernetics in C++ - Expert) including
  - Threads
    - Mutexes

❖ Note that the name is a slight misnomer:

➢ A constructor does not construct the object, it is called when an object is constructed

➢ This is in conflict with English

- A baker is someone who bakes bread, not someone you call when you bake bread

❖ **Java/C#/Objective-C** Note:

➢ Java and C# also use the term "Constructor"

- The concepts and rules are based on, so identical to, those of C++

➢ Swift and Objective-C use the term "Initialiser"

- The purpose is the same but the concepts and rules are slightly different

18.9.2 RULES OF CONSTRUCTORS

❖ Constructors MUST have the same name as the class

➢ This does indeed mean that the advice, starting the names of (member) functions with a verb cannot be followed on this occasion

❖ Constructors cannot return anything

❖ Constructors can be, and usually are, overloaded

❖ Signature of a Constructor

➢ Two parts of the signature of a "normal" function/member function have been eliminated

- Name
- Output type

➢ We are left with only the types of the input(s)

- The compiler uses these to distinguish between overloaded constructors i.e. different:
  ○ Input types
  ○ Numbers of inputs
  ○ Order of the various types

18.9.3 USING CONSTRUCTORS

18.9.3.1 Defining Constructors

❖ To start with, let us define a constructor which will initialise the `name` data member

➢ It will take a the specified name as a string input

❖ The declaration will be as follows:

```
class Employee
{
    public:
        Employee( string );
};//END...Class Declaration Employee
```

❖ The definition will be as follows:

```
Employee::Employee( string name_IN )
{
    name = name_IN;
}//END... Constructor Employee( string name_IN )
```

18.9.3.2 Invoking Constructors

❖ The constructor gets called automatically as soon as an object is created

➢ There is no further explicit calling for us to do
➢ In order to specify the actual parameters

- On the runtime stack: supply them after the object name, in parentheses
- On the runtime heap: supply them after the name of the runtime type, in parentheses

❖ On the Runtime Stack: In order to specify the `name` data member of the object `joeBloggs` as "Joseph Bloggs", the following is required:

```
Employee joeBloggs("Joseph Bloggs");
```

❖ On the Runtime Heap: In order to specify the `name` data member of the object being pointed to by `ptr_beverleyShepherd` as "Beverley Shepherd", the following is required:

```
Employee *ptr_beverleyShepherd = new Employee("Beverley Shepherd");
```

❖ The full program is as follows:

```
#include <iostream>
#include <string>
using namespace std;

#ifndef EMPLOYEE_H
#define EMPLOYEE_H

// Class DECLARATION
class Employee
{
    // DATA MEMBERS
    private:
            string name;
            string nI_Number;
            double salary_Gross;
            double tax_Rate;
            double tax_Amount;
            double salary_Net;

    // MEMBER FUNCTIONS
    public:
            // ---------- CONSTRUCTOR(S) ------------
            Employee( string );

            // ---------- "NORMAL Member Functions " ------------
            void greet();
            void greet_Name();


};//END...Class Declaration Employee

#endif
```

Program Listing/Output 18.13: Employee.h Declaration of the class Employee

```
#include "Employee.h"

// ---------- CONSTRUCTOR(S) ------------
Employee::Employee( string name_IN )
{
    name = name_IN;
```

```
}//END...Constructor Employee( string name_IN )

// ---------- "NORMAL Member Functions " ------------
 void Employee::greet()
{
       cout << "Hello, I'm an Employee"<< endl;

}//END...Member Function greet

void Employee::greet_Name()
{
       cout    << "Hello, I'm an Employee, "
               << "my name is "<< name
               << endl;

}//END...Member Function greet_Name
```

Program Listing/Output 18.14: Employee.cpp Definitions of the class Employee

```
#include <iostream>
#include <string>
using namespace std;

#include "Employee.h"

int main()
{
       //»»»»»»»»»»»»»»»»»»»»»»»»»»»»»»»»»»»»»»»»»»»»»»»»»»»»»»»»»»
       // Creating an OBJECT joeBloggs
       cout  << endl<< "================"<< endl
             << "Creating an Object on the runtime Stack: joeBloggs
                ..."
             << endl;

       Employee joeBloggs("Joseph Bloggs");
       joeBloggs.greet_Name();

       //»»»»»»»»»»»»»»»»»»»»»»»»»»»»»»»»»»»»»»»»»»»»»»»»»»»»»»»»»»
       cout  << endl<< "================"<< endl
             << "Creating an Object on the runtime Heap: "
             << "accessed via ptr_beverleyShepherd..."
             << endl;

       Employee *ptr_beverleyShepherd = new Employee("Beverley
       Shepherd");
       ptr_beverleyShepherd -> greet_Name();

       //»»»»»»»»»»»»»»»»»»»»»»»»»»»»»»»»»»»»»»»»»»»»»»»»»»»»»»»»»»
       cout  << endl<< "================"<< endl
             << "END...C++ Constructors Demo_01"
             << endl;
```

```
            while(true){}
            return 0;

}//END...Function MAIN
```

```
================
Creating an Object on the runtime Stack: joeBloggs...
Hello, I'm an Employee, my name is Joseph Bloggs

================
Creating an Object on the runtime Heap: accessed via ptr_beverley
Shepherd...
Hello, I'm an Employee, my name is Beverley Shepherd

================
END...C++ Constructors Demo_01
```

Program Listing/Output 18.15: Constructors_01_Client.cpp

18.9.4 OVERLOADING CONSTRUCTORS

❖ It is a common practice to overload constructors

➢ In fact, constructor overloading is required in all but the simplest classes

18.9.4.1 Defining Overloaded Constructors

❖ As discussed in section 18.9.2,

➢ Since the name is predetermined,

➢ There is no output, thus output type,

➢ The only means of distinguishing between Constructors Signatures is the Formal Parameters

• The number of inputs

• The types of the inputs, and the order in which they are implemented

❖ In our class Employee, we now want to create another constructor which will specify the `nI_Number`

➢ The declarations of the constructors will be as follows:

```
Employee( string );
Employee( string, string );
```

➤ The definitions will be as follows:

```
Employee::Employee( string name_IN )
{
        name = name_IN;

}//END...Constructor Employee( string name_IN )

Employee::Employee( string name_IN, string nI_Number_IN )
{
        name = name_IN;
        nI_Number = nI_Number_IN;

}//END...Constructor Employee( string name_IN )
```

18.9.4.2 Invoking Overloaded Constructors

❖ The different constructors need to be invoked individually, with the type of each actual parameter matching the type of each corresponding formal parameter

❖ Thus in our case, the following commands will work:

```
Employee joeBloggs("Joseph Bloggs");
```

➤ Since there is one actual parameter of type string, this object creation invokes the constructor which takes one string input

```
Employee johnSmith("John Smith", "GH 56 34 29 D");
```

➤ Since there are two actual parameters, each of type string, this object creation invokes the constructor which takes two string inputs

❖ Note that the following attempted invocation will cause a compile-time error

```
Employee peterWarner("Peter Warner", "GH 56 34 29 D", "Accounts" );
```

➤ We do not have a constructor which takes 3 strings as the inputs

❖ The full program is as follows:

```
#include <iostream>
#include <string>
using namespace std;

#ifndef EMPLOYEE_H
#define EMPLOYEE_H

// Class DECLARATION
```

```cpp
class Employee
{
        // DATA MEMBERS
        private:
                string name;
                string nI_Number;
                double salary_Gross;
                double tax_Rate;
                double tax_Amount;
                double salary_Net;

        // MEMBER FUNCTIONS
        public:
                // ---------- CONSTRUCTOR(S) ------------
                Employee( string );
                Employee( string, string );

                // ---------- "NORMAL Member Functions " ------------
                void greet();
                void greet_Name();
                void greet_Name_NINumber();

};//END...Class Declaration Employee

#endif
```

Program Listing/Output 18.16: Employee.h Declaration of the class Employee

```cpp
#include "Employee.h"

// ---------- CONSTRUCTOR(S) ------------
Employee::Employee( string name_IN )
{
      name = name_IN;

}//END...Constructor Employee( string name_IN )


Employee::Employee( string name_IN, string nI_Number_IN )
{
        name = name_IN;
        nI_Number = nI_Number_IN;

}//END...Constructor Employee( string name_IN )


// ---------- "NORMAL Member Functions " ------------
void Employee::greet()
{
      cout << "Hello, I'm an Employee"<< endl;

}//END...Member Function greet
```

```
void Employee::greet_Name()
{
        cout << "Hello, I'm an Employee, "
             << "my name is "<< name
             << endl;

}//END...Member Function greet_Name

void Employee::greet_Name_NINumber()
{
        cout << "Hello, I'm an Employee, "
             << "my name is "<< name
             << " and my NI Number is "<< nI_Number
             << endl;

}//END...Member Function greet_Name_NINumber
```

Program Listing/Output 18.17: Employee.cpp Definitions of the class Employee

```
#include <iostream>
#include <string>
using namespace std;

#include "Employee.h"

int main()
 {
      //»»»»»»»»»»»»»»»»»»»»»»»»»»»»»»»»»»»»»»»»»»»»»»»»»»»»»»»»»»»»»»»»»
      // Creating an OBJECT joeBloggs
      cout  << endl<< "================"<< endl
            << "Creating Objects on the runtime Stack..."
            << endl;

      Employee joeBloggs("Joseph Bloggs");
      joeBloggs.greet_Name();

      Employee johnSmith("John Smith", "GH 56 34 29 D");
      johnSmith.greet_Name_NINumber();

//    Employee peterWarner("Peter Warner", "GH 56 34 29 D", "
    Accounts" );

      // ...ERROR: No Constructor which takes 3 string inputs

      //»»»»»»»»»»»»»»»»»»»»»»»»»»»»»»»»»»»»»»»»»»»»»»»»»»»»»»»»»»»»»»»»»
      cout << endl<< "================"<< endl
            << "END...C++ Demo: Overloading Constructors "
            << endl;
      while(true){}
      return 0;
```

```
}//END...Function MAIN
```

```
===============
Creating Objects on the runtime Stack...
Hello, I'm an Employee, my name is Joseph Bloggs
Hello, I'm an Employee, my name is John Smith and my NI
Number is GH 56 34 29 D

===============
END...C++ Demo: Overloading Constructors
```

Program Listing/Output 18.18: Constructors_01_Client.cpp

18.9.5 DEFAULT CONSTRUCTOR

❖ Note that at least one constructor is always there: the Default Constructor

❖ This is a special constructor which:

  ➢ Takes no inputs

  ➢ Performs no functionality

❖ Thus whenever you create a class, the following is "silently" created:

  ➢ Declaration:
  ```
  Classname();
  ```

  ➢ Definition:
  ```
  Classname:: Classname();
  ```

❖ This is invoked every time an object is created and the inputs to the constructor are left blank

  ➢ i.e. We invoked it in sections 18.4, 18.5 and 18.6

18.9.6 ALTERNATIVE MEANS OF INITIALISING DATA MEMBERS

❖ There is a shorthand means of initialising variables in constructors, lexically before the main body,

  ➢ using the colon : operator:
  ```
  ConstructorName( Inputs )
        : variable1( initialValue ), variable1( initialValue ),...
    {
        // Body of Constructor
    }
  ```

❖ In our example, the an alternative way to initialise the variables in the constructors is the following:

```
Employee::Employee( string name_IN )
        : name( name_IN )
{
        // Further Constructor Functionalities

}//END...Constructor Employee( string name_IN )

Employee::Employee( string name_IN, string nI_Number_IN )
        : name( name_IN ), nI_Number( nI_Number_IN )
{
        // Further Constructor Functionalities

}//END...Constructor Employee( string name_IN )
```

❖ There are thus 2 ways of initialising data members:

➢ In the main body of the Constructor

   • Both the = and parentheses operator can be used for the initialisation

➢ Before the main body of the Constructor

   • Only the parentheses operator can be used for the initialisation

❖ Choosing between them is down to personal preference:

➢ Some programmers prefer the latter

   • The view is that initialising data members is the initial part of the housekeeping, not the main part
   • Thus this should be done before the main body
   • The main body should be used only for the further initial housekeeping

➢ Some programmers prefer the former

   • The view is that the whole constructor is for initial housekeeping, initialising data members is a major part of this
   • Thus it should all be done in the main body
   • In all other (member) functions, all the functionalities take place in the main body

18.9.7 FURTHER INITIALISATION WORK IN CONSTRUCTORS

❖ A common piece of advice is if something can be determined at initialisation, ensure all the required work is done within the constructor

➢ Leaving it till later will mean a finite period when it is not done

➤ This could lead to several problems, especially in multithreaded environments (Cybernetics in C++ - Expert)

❖ In our case, when we have a constructor which specifies `salary_Gross`

➤ Immediately, the rest of the numerical data members can be calculated

- `tax_Rate;`
- `tax_Amount;`
- `salary_Net;`

❖ It is strongly advisable to do this work in the constructor

➤ In the forthcoming program, the tax rates have been categorised as follows:

					TAX RATE
• £0	<=	Gross Salary	<	£20000	30%
• £20000	<=	Gross Salary	<	£25000	40%
• £25000		Gross Salary	>=	£25000	50%

➤ The following constructor and data constants facilitate this:

```cpp
const double TAXRATE_THRESHOLD_LOWER = 20000.0;
const double TAXRATE_THRESHOLD_INTERMEDIATE = 25000.0;


const double TAXRATE_RATE_LOWER = 0.3;
const double TAXRATE_RATE_INTERMEDIATE = 0.4;
const double TAXRATE_RATE_HIGHER = 0.5;

Employee::Employee( string name_IN,
                    string nI_Number_IN,
                    double salary_Gross_IN )
{
    name = name_IN;
    nI_Number = nI_Number_IN;
    salary_Gross = salary_Gross_IN;
    //-----------------------------------------------------
    // Calculating the Tax Rate
    //
    if( salary_Gross < TAXRATE_THRESHOLD_LOWER )
    {
        tax_Rate = TAXRATE_RATE_LOWER;
    }
    else
    {
```

```
                if( salary_Gross < TAXRATE_THRESHOLD_INTERMEDIATE )
                {
                        tax_Rate = TAXRATE_RATE_INTERMEDIATE;
                }
                else
                {
                        tax_Rate = TAXRATE_RATE_HIGHER;
                }
        }

        //------------------------------------------------------
        //    Calculating the Tax Amount & Net Salary
        //
        tax_Amount = tax_Rate * salary_Gross;
        salary_Net = salary_Gross - tax_Amount;


        // Incrementing the Number of Objects/Employees

}/* END...Constructor Employee( string name_IN,
                               string nI_Num_IN,
                               double salary_Gross_IN )
*/
```

Program Listing/Output 18.19: Constructor and constant data for initialising Gross Salary and remaining financial data

❖ The new full program is as follows:

```
#include <iostream>
#include <string>
using namespace std;

#ifndef EMPLOYEE_H
#define EMPLOYEE_H


const double TAXRATE_THRESHOLD_LOWER = 20000.0;
const double TAXRATE_THRESHOLD_INTERMEDIATE = 25000.0;


const double TAXRATE_RATE_LOWER = 0.3;
const double TAXRATE_RATE_INTERMEDIATE = 0.4;
const double TAXRATE_RATE_HIGHER = 0.5;


// Class DECLARATION
class Employee
{
        // DATA MEMBERS
        private:
                string name;
```

```
            string nI_Number;
            double salary_Gross;
            double tax_Rate;
            double tax_Amount;
            double salary_Net;

    // MEMBER FUNCTIONS
    public:
            // ---------- CONSTRUCTOR(S) ------------
            Employee( string );
            Employee( string, string );
            Employee( string name_IN,
                    string nI_Number,
                    double salary_Gross_IN );


            // ---------- "NORMAL Member Functions " ------------
            void greet();
            void greet_Name();
            void greet_Name_NINumber();
            void greet_Name_SalaryGross();
            void greet_AllAttributes();


};//END...Class Declaration Employee

#endif
```

Program Listing/Output 18.20: Employee.h Declaration of the class Employee

```
#include "Employee.h"

// ---------- CONSTRUCTOR(S) ------------
Employee::Employee( string name_IN )
{
      name = name_IN;

}//END...Constructor Employee( string name_IN )


Employee::Employee( string name_IN, string nI_Number_IN )
{
      name = name_IN;
      nI_Number = nI_Number_IN;

}//END...Constructor Employee( string name_IN )

Employee::Employee( string name_IN,
                    string nI_Number_IN,
                    double salary_Gross_IN )
{
      name = name_IN;
```

```
        nI_Number = nI_Number_IN;
        salary_Gross = salary_Gross_IN;
        //------------------------------------------------------
        //      Calculating the Tax Rate
        //
        if( salary_Gross < TAXRATE_THRESHOLD_LOWER )
        {
                tax_Rate = TAXRATE_RATE_LOWER;
        }
        else
        {
                if( salary_Gross < TAXRATE_THRESHOLD_INTERMEDIATE )
                {
                        tax_Rate = TAXRATE_RATE_INTERMEDIATE;
                }
                else
                {
                        tax_Rate = TAXRATE_RATE_HIGHER;
                }
        }


        //------------------------------------------------------
        //      Calculating the Tax Amount & Net Salary
        //
        tax_Amount = tax_Rate * salary_Gross;
        salary_Net = salary_Gross - tax_Amount;


        // Incrementing the Number of Objects/Employees

}/* END...Constructor Employee( string name_IN,
                                string nI_Num_IN,
                                double salary_Gross_IN )
*/
// ---------- "NORMAL Member Functions " ------------
void Employee::greet()
{
      cout << "Hello, I'm an Employee"<< endl;

}//END...Member Function greet

void Employee::greet_Name()
{
        cout  << "Hello, I'm an Employee, "
              << "my name is "<< name
              << endl;

}//END...Member Function greet_Name

void Employee::greet_Name_NINumber()
{
        cout  << "Hello, I'm an Employee, "
              << "my name is "<< name
```

```
                   << " and my NI Number is "<< nI_Number
                   << endl;

}//END...Member Function greet_Name_NINumber


void Employee::greet_Name_SalaryGross()
{
       cout   << "Hello, I'm an Employee, my name is "<< name
              << ", my Gross Salary is "<< salary_Gross
              << endl;

}//END...Member Function greet_Name_SalaryGross()


void Employee::greet_AllAttributes()
{
       cout   << endl
              << "====================="
              << endl
              << "Hello, I am an employee, my name is "<< name<< ", "
              << endl
              << "my NI Number is "<< nI_Number<< ","
              << endl
              << "my Gross Salary is "<< salary_Gross
              << endl
              << "my tax Rate is "<< tax_Rate<< "%"
              << endl
              << "so the amount I have to pay in tax is "<<
                  tax_Amount
              << endl
              << "thus my Net Salary is "<< salary_Net
              << endl
              << "====================="<< endl;
}// END...MF greet_AllAttributes()
```

Program Listing/Output 18.21: Employee.cpp Definitions of the class Employee

```
/*      ************************************************************
        CONSTRUCTORS DEMO
        ************************************************************
        -       This program demonstrates initialising data members
                using  Constructors
*/
#include <iostream>
#include <string>
using namespace std;

#include "Employee.h"

int main()
{
```

```
//»»»»»»»»»»»»»»»»»»»»»»»»»»»»»»»»»»»»»»»»»»»»»»»»»»»»»»»»»»»»»»»
// Creating OBJECTS
cout    << endl<< "==============="<< endl
        << "Creating Objects on the runtime Stack..."
        << endl;

Employee joeBloggs("Joseph Bloggs", "PH 24 25 67 G",
                          18000.0 );
joeBloggs.greet_AllAttributes();


Employee johnSmith("John Smith", "GH 56 34 29 D",
                          23000.0 );
johnSmith.greet_AllAttributes();

Employee amandaTaylor("Amanda Taylor", "RG 35 23 56 T",
                          25500.0);
amandaTaylor.greet_AllAttributes();


//»»»»»»»»»»»»»»»»»»»»»»»»»»»»»»»»»»»»»»»»»»»»»»»»»»»»»»»»»»»»»»»
cout    << endl<< "==============="<< endl
        << "END...C++ Demo: Overloading & Using Constructors "
        << endl;
while(true){}
return 0;

}//END...Function MAIN
```

```
===============
Creating Objects on the runtime Stack...

====================
Hello, I am an employee, my name is Joseph Bloggs,
my NI Number is PH 24 25 67 G,
my Gross Salary is 18000
my tax Rate is 0.3%
so the amount I have to pay in tax is 5400
thus my Net Salary is 12600
====================

====================
Hello, I am an employee, my name is John Smith,
my NI Number is GH 56 34 29 D,
my Gross Salary is 23000
my tax Rate is 0.4%
so the amount I have to pay in tax is 9200
thus my Net Salary is 13800
====================

====================
```

```
Hello, I am an employee, my name is Amanda Taylor,
my NI Number is RG 35 23 56 T,
my Gross Salary is 25500
my tax Rate is 0.5%
so the amount I have to pay in tax is 12750
thus my Net Salary is 12750
=====================


===============
END...C++ Demo: Overloading & Using Constructors
```

Program Listing/Output 18.22: Constructors_03_Client.cpp

18.10 DESTRUCTORS

18.10.1 WHAT ARE DESTRUCTORS?

❖ Destructors are special member functions which are invoked automatically when **an object goes out of scope**

➢ i.e. an object is destroyed

➢ It is thus the opposite of a constructor

❖ They are invaluable for any final housekeeping work which needs to take place regarding the object

➢ Common examples comprise freeing up resources reserved by the object, including:
  • Runtime Heap memory
  • Network resources
  • Database connections
  • Shared resources between applications, via a DDE Dynamic Data Exchange, Server

❖ Note that like Constructors, the name is a slight misnomer:

➢ A destructor does not destroy the object, it is called when an object is destroyed

❖ Most OOP languages hide destructors from the programmer

➢ This includes Java, Perl, Python and Web Programming Languages

➢ The main reason is because they also hide most, if not all, resource management
  • A major one of these is the runtime heap memory

➢ Part of the power of C++ is that the programmer (i.e. You! :-)) are/is able to program all of these explicitly

- Thus in C++, destructors are essential

18.10.2 RULES OF DESTRUCTORS

❖ Destructors MUST have the same name as the class, preceded by the tilde ~ character

➢ This does indeed mean that again, the advice of starting the names of (member) functions with a verb cannot be followed

❖ Destructors cannot take any inputs

➢ This makes logical sense: You do not want to give data to an object when it is being destroyed

❖ Destructors cannot return anything

➢ This also makes logical sense: You do not want to receive data from an object when it is being destroyed

❖ Signature of a Destructor

➢ Only the name is present
➢ Thus a declaration will look like this:

```
~className();
```

➢ A definition will look like this:

```
className::~className()
{
    // Destructor functionalities
}
```

18.10.3 USING DESTRUCTORS

❖ For demonstration purposes, for this program, a destructor will be created when a message will be output, reporting which object is going out of scope:

➢ Declaration

```
~Employee( );
```

➢ Definition

```
Employee::~Employee( )
{
      cout  << "Goodbye, I, "<< name
            << " am  going out of scope now."
            << endl;
}// END...Destructor ~Employee
```

❖ The full program is as follows:

```
#include <iostream>
#include <string>
using namespace std;

#ifndef EMPLOYEE_H
#define EMPLOYEE_H


const double TAXRATE_THRESHOLD_LOWER = 20000.0;
const double TAXRATE_THRESHOLD_INTERMEDIATE = 25000.0;


const double TAXRATE_RATE_LOWER = 0.3;
const double TAXRATE_RATE_INTERMEDIATE = 0.4;
const double TAXRATE_RATE_HIGHER = 0.5;


// Class DECLARATION
class Employee
{
      // DATA MEMBERS
      private:
              string name;
              string nI_Number;
              double salary_Gross;
              double tax_Rate;
              double tax_Amount;
              double salary_Net;

      // MEMBER FUNCTIONS
      public:
              // ---------- CONSTRUCTOR(S) ------------
              Employee( string );
              Employee( string, string );
              Employee(    string name_IN,
                              string nI_Number,
                              double salary_Gross_IN );


              // ---------- DESTRUCTOR ------------
              ~Employee(  );
```

```
              // ---------- "NORMAL Member Functions " ------------

          void greet();
          void greet_Name();
          void greet_Name_NINumber();
          void greet_Name_SalaryGross();
          void greet_AllAttributes();


};//END...Class Declaration Employee

#endif
```

Program Listing/Output 18.23: Employee.h Declaration of the class Employee

```
#include "Employee.h"

// ---------- CONSTRUCTOR(S) ------------
Employee::Employee( string name_IN )
{
      name = name_IN;

}//END...Constructor Employee( string name_IN )


Employee::Employee( string name_IN, string nI_Number_IN )
{
      name = name_IN;
      nI_Number = nI_Number_IN;

}//END...Constructor Employee( string name_IN )

Employee::Employee( string name_IN,
                  string nI_Number_IN,
                  double salary_Gross_IN )
{
    name = name_IN;
    nI_Number = nI_Number_IN;
    salary_Gross = salary_Gross_IN;
    //----------------------------------------------------
    //     Calculating the Tax Rate
    //
    if( salary_Gross < TAXRATE_THRESHOLD_LOWER )
    {
          tax_Rate = TAXRATE_RATE_LOWER;
    }
    else
    {
          if( salary_Gross < TAXRATE_THRESHOLD_INTERMEDIATE )
          {
                tax_Rate = TAXRATE_RATE_INTERMEDIATE;
          }
          else
```

```
            {
                tax_Rate = TAXRATE_RATE_HIGHER;
            }
        }

    //-------------------------------------------------------
    //     Calculating the Tax Amount & Net Salary
    //
    tax_Amount = tax_Rate * salary_Gross;
    salary_Net = salary_Gross - tax_Amount;


    // Incrementing the Number of Objects/Employees

}/* END...Constructor Employee( string name_IN,
                               string nI_Num_IN,
                               double salary_Gross_IN )
*/

// ---------- DESTRUCTOR ------------
Employee::~Employee(  )
{
    cout  << "Goodbye, I, "<< name
          << " am  going out of scope now."
          << endl;

}// END...Destructor ~Employee



// ---------- "NORMAL Member Functions " ------------
void Employee::greet()
{
    cout << "Hello, I'm an Employee"<< endl;

}//END...Member Function greet

void Employee::greet_Name()
{
    cout << "Hello, I'm an Employee, "
         << "my name is "<< name
         << endl;

}//END...Member Function greet_Name

void Employee::greet_Name_NINumber()
{
    cout << "Hello, I'm an Employee, "
         << "my name is "<< name
         << " and my NI Number is "<< nI_Number
         << endl;

}//END...Member Function greet_Name_NINumber
```

```
void Employee::greet_Name_SalaryGross()
{
      cout << "Hello, I'm an Employee, my name is "<< name
           << ", my Gross Salary is "<< salary_Gross
           << endl;

}//END...Member Function greet_Name_SalaryGross()


void Employee::greet_AllAttributes()
{
      cout << endl
           << "====================="
           << endl
           << "Hello, I am an employee, my name is "<< name<< ", "
           << endl
           << "my NI Number is "<< nI_Number<< ","
           << endl
           << "my Gross Salary is "<< salary_Gross
           << endl
           << "my tax Rate is "<< tax_Rate<< "%"
           << endl
           << "so the amount I have to pay in tax is "<< tax_Amount
           << endl
           << "thus my Net Salary is "<< salary_Net
           << endl
           << "====================="<< endl;
}// END...MF greet_AllAttributes()
```

Program Listing/Output 18.24: Employee.cpp Definitions of the class Employee

```
/*      ************************************************************
        DEMO: OBJECTS ON THE RUNTIME HEAP
        ************************************************************
*/
#include <iostream>
#include <string>
using namespace std;

#include "Employee.h"

void createOffice_01();

int main()
{
      //»»»»»»»»»»»»»»»»»»»»»»»»»»»»»»»»»»»»»»»»»»»»»»»»»»»»»»»»»»»»»»»»»»»»»»
      // Creating OBJECTS

      createOffice_01();
```

```
        //»»»»»»»»»»»»»»»»»»»»»»»»»»»»»»»»»»»»»»»»»»»»»»»»»»»»»»»»»»»»»»»»»»
        cout   << endl<< "================"<< endl
               << "END...C++ Demo: Using Destructors "
               << endl;
        while(true){}
        return 0;

}//END...Function MAIN


// Declaration: Function which will create Objects on the runtime
    stack
// They will go out of scope when the function exits
//
void createOffice_01()
{
        Employee joeBloggs("Joseph Bloggs", "PH 24 25 67 G",
                                18000.0 );
        joeBloggs.greet_Name();

        Employee johnSmith("John Smith", "GH 56 34 29 D",
                                23000.0 );
        johnSmith.greet_Name();

        Employee amandaTaylor("Amanda Taylor", "RG 35 23 56 T",
                                25500.0);
        amandaTaylor.greet_Name();

}//END...Function createOffice_01
```

```
Hello, I'm an Employee, my name is Joseph Bloggs
Hello, I'm an Employee, my name is John Smith
Hello, I'm  an Employee, my name is Amanda Taylor
Goodbye, I, Amanda Taylor am  going out of scope now.
Goodbye, I, John Smith am  going out of scope now.
Goodbye, I, Joseph Bloggs am  going out of scope now.

================
END...C++ Demo: Using Destructors
```

Program Listing/Output 18.25: Destructors_01_Client.cpp

18.10.4 DEFAULT DESTRUCTOR

❖ Like the default constructor, the default destructor is always there

  ➢ It simply performs no functionality

❖ Thus whenever you create a class, the following is also "silently" created:

➢ Declaration:

```
~Classname();
```

➢ Definition:

```
Classname::~Classname();
```

❖ This is invoked every time an object goes out of scope, where no destructor has been coded

➢ i.e. We invoked it in the sections prior to 18.10.3

18.11 CONSTRUCTORS AND DESTRUCTORS FOR OBJECTS ON THE RUNTIME HEAP

❖ We are now in a position to fully analyse what was mentioned in section 14

➢ When using the runtime heap in C++

- `new` and `delete` should be given preference
- over `malloc()` and `free()`

❖ `new` and `delete` can deal with objects, `malloc()` and `free()` cannot

➢ `new`   Will call the Constructor of the Object
➢ `delete`   Will call the Destructor of the Object

❖ The following program demonstrates this:

➢ Note that this time, the destructors are called in the order we "destroyed" the heap objects

- This confirms that we are controlling their scope
- The good practice of setting each of the pointers to these heap objects to NULL after they have been destroyed, has been followed here:

```cpp
#include <iostream>
#include <string>
using namespace std;

#ifndef EMPLOYEE_H
#define EMPLOYEE_H


const double TAXRATE_THRESHOLD_LOWER = 20000.0;
const double TAXRATE_THRESHOLD_INTERMEDIATE = 25000.0;
```

```
const double TAXRATE_RATE_LOWER = 0.3;
const double TAXRATE_RATE_INTERMEDIATE = 0.4;
const double TAXRATE_RATE_HIGHER = 0.5;


// Class DECLARATION
class Employee
{
        // DATA MEMBERS
        private:
                string name;
                string nI_Number;
                double salary_Gross;
                double tax_Rate;
                double tax_Amount;
                double salary_Net;

        // MEMBER FUNCTIONS
        public:
                // ---------- CONSTRUCTOR(S) ------------
                Employee( string );
                Employee( string, string );
                Employee( string name_IN,
                          string nI_Number,
                          double salary_Gross_IN );


                // ---------- DESTRUCTOR ------------
                ~Employee(   );


                // ---------- "NORMAL Member Functions " ------------
                void greet();
                void greet_Name();
                void greet_Name_NINumber();
                void greet_Name_SalaryGross();
                void greet_AllAttributes();


};//END...Class Declaration Employee

#endif
```

Program Listing/Output 18.26: Employee.h

```
#include "Employee.h"

// ---------- CONSTRUCTOR(S) ------------
Employee::Employee( string name_IN )
{
        name = name_IN;
```

```
}//END...Constructor Employee( string name_IN )


Employee::Employee( string name_IN, string nI_Number_IN )
{
      name = name_IN;
      nI_Number = nI_Number_IN;

}//END...Constructor Employee( string name_IN )

Employee::Employee( string name_IN,
                                string nI_Number_IN,
                                double salary_Gross_IN )
{
      name = name_IN;
      nI_Number = nI_Number_IN;
      salary_Gross = salary_Gross_IN;
      //----------------------------------------------------
      //      Calculating the Tax Rate
      //
      if( salary_Gross < TAXRATE_THRESHOLD_LOWER )
      {
            tax_Rate = TAXRATE_RATE_LOWER;
      }
      else
      {
            if( salary_Gross < TAXRATE_THRESHOLD_INTERMEDIATE )
            {
                  tax_Rate = TAXRATE_RATE_INTERMEDIATE;
            }
            else
            {
                  tax_Rate = TAXRATE_RATE_HIGHER;
            }
      }

      //----------------------------------------------------
      //      Calculating the Tax Amount & Net Salary
      //
      tax_Amount = tax_Rate * salary_Gross;
      salary_Net = salary_Gross - tax_Amount;


      // Incrementing the Number of Objects/Employees

}/* END...Constructor Employee( string name_IN,
                                        string nI_Num_IN,
                                        double
                                        salary_Gross_IN )
*/

// ---------- DESTRUCTOR ------------
```

```cpp
Employee::~Employee(  )
{
      cout   << "Goodbye, I, "<< name
             << " am  going out of scope now."
             << endl;

}// END...Destructor ~Employee




// ---------- "NORMAL Member Functions " ------------
void Employee::greet()
{
      cout   << "Hello, I'm an Employee"<< endl;

}//END...Member Function greet

void Employee::greet_Name()
{
      cout   << "Hello, I'm an Employee, "
             << "my name is "<< name
             << endl;

}//END...Member Function greet_Name

void Employee::greet_Name_NINumber()
{
      cout   << "Hello, I'm an Employee, "
             << "my name is "<< name
             << " and my NI Number is "<< nI_Number
             << endl;

}//END...Member Function greet_Name_NINumber


void Employee::greet_Name_SalaryGross()
{
      cout   << "Hello, I'm an Employee, my name is "<< name
             << ", my Gross Salary is "<< salary_Gross
             << endl;

}//END...Member Function greet_Name_SalaryGross()


void Employee::greet_AllAttributes()
{
        cout << endl
             << "====================="
             << endl
             << "Hello, I am an employee, my name is "<< name<< ", "
             << endl
             << "my NI Number is "<< nI_Number<< ","
             << endl
```

```
                 << "my Gross Salary is "<< salary_Gross
                 << endl
                 << "my tax Rate is "<< tax_Rate<< "%"
                 << endl
                 << "so the amount I have to pay in tax is "<< tax_Amount
                 << endl
                 << "thus my Net Salary is "<< salary_Net
                 << endl
                 << "====================="<< endl;
}// END...MF greet_AllAttributes()
```

Program Listing/Output 18.27: Employee.cpp

```cpp
/*      ***********************************************************
        DESTRUCTORS DEMO
        ***********************************************************
        -       This program demonstrates using Destructors
*/
#include <iostream>
#include <string>
using namespace std;

#include "Employee.h"


int main()
{
    //»»»»»»»»»»»»»»»»»»»»»»»»»»»»»»»»»»»»»»»»»»»»»»»»»»»»»»»»»»»»»»
    // Creating OBJECTS on the runtime HEAP
    //
    Employee *ptr_scottHamilton = new Employee( "Scott Hamilton",
                                                "FG 45 90 87 D",
                                                19750.0 );
    ptr_scottHamilton -> greet_AllAttributes();

    Employee *ptr_susanFletcher = new Employee( "Susan Fletcher",
                                                "EG 34 98 67 C",
                                                21750.0 );
    ptr_susanFletcher ->greet_AllAttributes();

    Employee *ptr_patriciaCooper = new Employee("Patricia Cooper",
                                                "SD 45 78 63 P",
                                                23500.0 );
    ptr_patriciaCooper ->greet_AllAttributes();


    //»»»»»»»»»»»»»»»»»»»»»»»»»»»»»»»»»»»»»»»»»»»»»»»»»»»»»»»»»»»»»»
    // Deleting OBJECTS from the runtime HEAP
    //
    delete ptr_susanFletcher;
    ptr_susanFletcher = NULL;
```

```
      delete ptr_scottHamilton;
      ptr_susanFletcher = NULL;

      delete ptr_patriciaCooper;
      ptr_susanFletcher = NULL;


      //v»»»»»»»»»»»»»»»»»»»»»»»»»»»»»»»»»»»»»»»»»»»»»»»»»»»»»»»»»»»»»
      cout  << endl<< "==============="<< endl
            << "END...C++ Demo: Constructor & Destructors invocations "
            << "for Objects on the Runtime Heap"
            << endl;
      while(true){}
      return 0;

}//END...Function MAIN
```

```
====================
Hello, I am an employee, my name is Scott Hamilton,
my NI Number is FG 45 90 87 D,
my Gross Salary is 19750
my tax Rate is 0.3%
so the amount I have to pay in tax is 5925
thus my Net Salary is 13825
====================


====================
Hello, I am an employee, my name is Susan Fletcher,
my NI Number is EG 34 98 67 C,
my Gross Salary is 21750
my tax Rate is 0.4%
so the amount I have to pay in tax is 8700
thus my Net Salary is 13050
====================


====================
Hello, I am an employee, my name is Patricia Cooper,
my NI Number is SD 45 78 63 P,
my Gross Salary is 23500
my tax Rate is 0.4%
so the amount I have to pay in tax is 9400
thus my Net Salary is 14100
====================
Goodbye, I, Susan Fletcher am  going out of scope now.
Goodbye, I, Scott Hamilton am  going out of scope now.
Goodbye, I, Patricia Cooper am  going out of scope now.

===============
```

```
END...C++ Demo: Constructor & Destructors invocations for Objects on
    the Runtime
  Heap
```

Program Listing/Output 18.28: Objects_Heap_Client.cpp

19. OBJECT ORIENTED DESIGN USING UML

19.1 WHAT IS UML?

❖ The Unified Modelling Language is a formal language devised to communicate programming design to humans, especially non-programmers

➢ Programming languages are unsuitable for modelling since they
- Contain all the details, never any design overviews
- Cannot be understood by people unfamiliar with the language
➢ Natural human languages
- Suffer from verbosity
- Lack the precision required to avoid ambiguities

❖ UML strikes a balance between the two, and is widely used in software design, IT, and even project management

19.2 MODELLING IN UML

❖ UML was introduced as a means of representing object oriented designs

➢ The idea was that these could be discussed and modified by people, often including non-programmers
➢ They would be kept separate from the final OOP language itself

❖ It quickly progressed, with further diagrams being added, for things like how the end human user will interact with the system

19.2.1 UML DIAGRAMS

❖ Structure

➢ Class diagram
➢ Component diagram
➢ Composite structure diagram
➢ Deployment diagram
➢ Object diagram

➢ Package diagram

➢ Profile diagram

❖ Behaviour

➢ Activity diagram

➢ State machine diagram

➢ Use case diagram

❖ Interaction

➢ Communication diagram

➢ Interaction overview diagram

➢ Sequence diagram

➢ Timing diagrams

19.2.2 CLASS DIAGRAMS

❖ Class diagrams are essentially a list of the data members and member functions

❖ Our class Employee, can be represented diagrammatically IN UML as follows

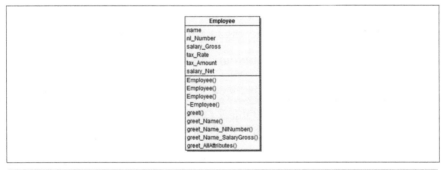

Figure 19.1: UML Class Diagram: Employee

❖ The ideas are

➢ All details, such as how member functions will achieve their objectives, are not included

➢ No final OOP language has been mentioned so far

• It could be passed on to a C++ team, a Java team...

19.2.3 OBJECT DIAGRAMS

❖ Our 3 objects: `joeBloggs`, `johnSmth` and `amandaTaylor`, each with their own data members/attributes, can be represented in an object diagram as follows:

Figure 19.2: UML Object Diagram: Instances of Class Employee

❖ As this shows, a large complex project, with several classes and objects, can be represented in a compact organised way using UML

➤ It will be intelligible to non-programmers

19.3 BEAUTIES OF OBJECT ORIENTED PROGRAMMING

❖ There are many reasons why Object Oriented Programming is making a significant contribution to the modern world, as well as being a joy to work with:

➤ Dealing with complexities
  • An inevitable consequence of modern day demands is that programs get very long, and there is a lot of data to deal with
  • Object Oriented Programming enables us to organise our code in classes, and so manage large amounts of complexities
  • Since a hierarchy can be built, there is no longer the compulsion to remember the name and functionality of each individual function, as is the case in procedural programming

➤ New level of thinking and software architecture

- The Abstraction means we can now think at a different higher level
- Instead of thinking about individual functions and data items, we can think of them in groups, and interactions between these groups
- This will be extended when looking at
  - Inner Classes
  - Inheritance

➤ Encapsulation

- In procedural programming, there are only two scopes: as soon as a variable is required by more than one function, it has to become global
- In object oriented programming, variables can have varying scope: a data member is visible by a "group of operations"

➤ Representation of the real world

- As we have been seeing, since the Abstraction and Encapsulation are powerful ways of thinking outside the computer world, most real world problems can be implemented directly into an object oriented program
  - The "transitional" phase is fairly minimal
  - It is larger in C and much larger in Assembler
- It is largely for these reasons, it has been realised that Object Oriented thinking has advantages which can be realised in far more than just software engineering
  - Object Oriented Systems are now widely used in
    - IT
    - Company management

➤ Reusability

- Once a class has been written, objects can be created from it indefinitely

20. ABSTRACTION & ENCAPSULATION USING CLASSES & OBJECTS (2)

❖ Having looked at what are classes and objects, this section now looks at further features, including

➤ Further special members
➤ Members which perform functionalities "behind the scenes"

20.1 MUTATOR MEMBER FUNCTION

20.1.1 WHAT ARE MUTATORS?

❖ Mutators are special member functions which are used to control changes to data members

> Other names include "Properties," used in Microsoft environments such as .NET

❖ Common Practice: Have 2 mutator member functions for each data member for which changes need to be controlled:

> `get_DataMemberName`:

• reads (and returns) the data member value

> `set_DataMemberName`:

• Writes a value to the data member

❖ Rough Guides:

> Constructors are for initialising data members

> Mutators are for controlling data members thereafter

• Have 2 mutator member functions for each important (i.e. all but the most trivial) data members

❖ They are usually essential:

> Validity checks

• e.g. Your program is taking in dates, check here that 31 June, 30 February cannot be entered

> Data representation: Often your class will represent data differently from other parts of the program, let alone the outside world

• e.g. Your program represents temperature in Kelvin,

    ○ Other parts of the program use Celsius or Fahrenheit

• Mutators are ideal for conversions

> Enables access from outside the program

• Can change data members after initialisation

    ○ This does not violate privacy (see below)

> Can often aid debugging

• If all changes to a data member are via a mutator, tracking down bugs is much easier

20.1.1.1 Mutators do not violate Encapsulation

❖ A common concern is "If we are giving access to private data members, are we not breaching the encapsulation, so carefully engineered into object oriented programming?!"

➢ No, we are not undoing the privacy, we are giving access which is
- Restricted
- Something we, as the programmer of the class, have given authority to

❖ Analogy: Endoscopy, allows a doctor restricted access to your inner organs

❖ Mutators allow restricted access to your private data members, one at a time

❖ In a similar manner, Constructors also allow access to Data Members

20.1.2 DEFINING AND USING MUTATORS

❖ There are no further features engineered into C++ for facilitating mutators

➢ As far as the compiler and linker are concerned, they are normal member functions
➢ We thus need to be extra careful with the mutator naming, to ensure no programmer gets mixed up

❖ In our Employee example, we want to control changes to the data member `salary_Gross`

➢ The following two mutators need to be created
- `get_salary_Gross`
  - This will read and return the value of `salary_Gross`
- `set_salary_Gross`
  - This will write a value to `salary_Gross`
  - Like the constructor(s) with which initialise `salary_Gross`, it will also calculate the new values of `tax_Rate`, `tax_Amount` and `salary_Net`,

❖ The full modified program is now as follows:

```
#include <iostream>
#include <string>
using namespace std;
```

```cpp
#ifndef EMPLOYEE_H
#define EMPLOYEE_H



const double TAXRATE_THRESHOLD_LOWER = 20000.0;
const double TAXRATE_THRESHOLD_INTERMEDIATE = 25000.0;


const double TAXRATE_RATE_LOWER = 0.3;
const double TAXRATE_RATE_INTERMEDIATE = 0.4;
const double TAXRATE_RATE_HIGHER = 0.5;


// Class DECLARATION
class Employee
{
        // Data Members
        private:
                string name;
                string nI_Number;
                double salary_Gross;
                double tax_Rate;
                double tax_Amount;
                double salary_Net;

        public:
                // ---------- CONSTRUCTORS ------------
                Employee(    string name_IN );

                Employee(    string name_IN,
                             string nI_Number );

                Employee(    string name_IN,
                             string nI_Number,
                             double salary_Gross_IN );

                // ---------- DESTRUCTOR ------------
                ~Employee(  );

                // ---------- MUTATORS ------------
                double get_salary_Gross(  );
                void set_salary_Gross( double salary_Gross_IN );

                //------- "Normal" Member Functions -------
                void greet();
                void greet_Name();
                void greet_Name_SalaryGross();
                void greet_AllAttributes();
```

```
};//END...Class Declaration Employee

#endif
```

```cpp
#include "Employee.h"

// ---------- CONSTRUCTORS ------------
Employee::Employee( string name_IN )
{
      name = name_IN;

}/* END...Constructor Employee( string name_IN )
 */


Employee::Employee( string name_IN,
                              string nI_Number_IN )
{
      name = name_IN;
      nI_Number = nI_Number_IN;

}/* END...Constructor Employee( string name_IN,
                                    string nI_Num_IN )
  */


Employee::Employee( string name_IN,
                           string nI_Number_IN,
                           double salary_Gross_IN )
{
      name = name_IN;
      nI_Number = nI_Number_IN;
      salary_Gross = salary_Gross_IN;
      //-----------------------------------------------------
      // Calculating the Tax Rate
      //
      if( salary_Gross < TAXRATE_THRESHOLD_LOWER )
      {
            tax_Rate = TAXRATE_RATE_LOWER;
      }
      else
      {
            if( salary_Gross < TAXRATE_THRESHOLD_INTERMEDIATE )
            {
                tax_Rate = TAXRATE_RATE_INTERMEDIATE;
            }
            else
            {
                tax_Rate = TAXRATE_RATE_HIGHER;
            }
```

```
      }

      //-----------------------------------------------------
      //    Calculating the Tax Amount & Net Salary
      //
      tax_Amount = tax_Rate * salary_Gross;
      salary_Net = salary_Gross - tax_Amount;


      // Incrementing the Number of Objects/Employees

}/* END...Constructor Employee( string name_IN,
                                string nI_Num_IN,
                                double salary_Gross_IN )
 */

// ---------- DESTRUCTOR ------------
Employee::~Employee(  )
{
      cout << "Goodbye, I, "<< name<<
                " am  going out of scope now."
           << endl;

}//END...Destructor ~Employee


// ---------- MUTATORS ------------
double Employee::get_salary_Gross(  )
{
      return salary_Gross;

}//END...Mutator get_salary_Gross


void Employee::set_salary_Gross( double salary_Gross_IN )
{
      salary_Gross = salary_Gross_IN;

      //-----------------------------------------------------
      // Calculating the NEW Tax Rate
      //
      if( salary_Gross < TAXRATE_THRESHOLD_LOWER )
      {
            tax_Rate = TAXRATE_RATE_LOWER;
      }
      else
      {
            if( salary_Gross < TAXRATE_THRESHOLD_INTERMEDIATE )
            {
                  tax_Rate = TAXRATE_RATE_INTERMEDIATE;
            }
            else
            {
```

```
                tax_Rate = TAXRATE_RATE_HIGHER;
                }
        }

        //------------------------------------------------------
        //     Calculating the NEW Tax Amount & Net Salary
        //
        tax_Amount = tax_Rate * salary_Gross;
        salary_Net = salary_Gross - tax_Amount;

}//END...Mutator set_salary_Gross



// Member Function DEFINITIONS:
void Employee::greet()
{
        cout  << "Hello, I'm an Employee"<< endl;

}//END...Member Function greet()


void Employee::greet_Name()
{
        cout  << "Hello, I'm an Employee, my name is "<< name
              << endl;

}//END...Member Function greet_Name()

void Employee::greet_Name_SalaryGross()
{
        cout  << "Hello, I'm an Employee, my name is "<< name
              << ", my Gross Salary is "<< get_salary_Gross()
              << endl;

}//END...Member Function greet_Name_SalaryGross()


void Employee::greet_AllAttributes()
{
        //v»»»»»»»»»»»»»»»»»»»»»»»»»»»»»»»»»»»»»»»»»»»»»»»»»»»»»»»»»»»ż
        cout << endl
             << "====================="
             << endl
             << "Hello, I am an employee, my name is "<< name<< ", "
             << endl
             << "my NI Number is "<< nI_Number<< ","
             << endl
             << "my Gross Salary is "<< salary_Gross
             << endl
             << "my tax Rate is "<< tax_Rate<< "%"
             << endl
             << "so the amount I have to pay in tax is "<< tax_Amount
```

```
            << endl
            << "thus my Net Salary is "<< salary_Net
            << endl
            << "====================="<< endl;
}// END...MF greet_AllAttributes()
```

```cpp
/*      ***********************************************************
        Const Mutators Demo
        ***********************************************************
        -       This program demonstrates controlling data members using
                Mutator Member Functions
*/

#include <iostream>
#include <string>
using namespace std;

#include "Employee.h"

int main()
{
    //∨»»»»»»»»»»»»»»»»»»»»»»»»»»»»»»»»»»»»»»»»»»»»»»»»»»»»»»»ż
    Employee joeBloggs( "Joseph Bloggs", "PH 45 67 34 D", 17000.0 );
    joeBloggs.greet_AllAttributes();

    joeBloggs.set_salary_Gross( 17500.0 );
    joeBloggs.greet_AllAttributes();


    //∨»»»»»»»»»»»»»»»»»»»»»»»»»»»»»»»»»»»»»»»»»»»»»»»»»»»»»»»ż
    cout    << endl<< "================"<< endl
            << "END...C++ Class Mutator Member Function Demo"
            << endl;
    while(true){}
    return 0;

}//END...Function MAIN
```

```
====================
Hello, I am an employee, my name is Joseph Bloggs,
my NI Number is PH 45 67 34 D,
my Gross Salary is 17000
my tax Rate is 0.3%
so the amount I have to pay in tax is 5100
thus my Net Salary is 11900
====================

====================
Hello, I am an employee, my name is Joseph Bloggs,
my NI Number is PH 45 67 34 D,
```

```
my Gross Salary is 17500
my tax Rate is 0.3%
so the amount I have to pay in tax is 5250
thus my Net Salary is 12250
======================

===============
END...C++ Class Mutator Member Function Demo
```

Program Listing/Output 20.3: MutatorsDemo_Client.cpp - using Mutator Member Functions

EXERCISE

❖ *Write a mutator member function which will check for the validity of UK National Insurance numbers. i.e. it should be*

  ➢ *2 characters, followed by a space*
  ➢ *3 sets of 2 digits, each followed by a space*
  ➢ *One final character*

❖ Examples:

  ➢ FG 34 89 34 F   Valid
  ➢ FGX 23 45 69 P   Invalid
  ➢ DF 23 98 23 S   Valid
  ➢ SH 234 89 23 C   Invalid
  ➢ DF 56 34 28 CD   Invalid

20.2 CONST MEMBER FUNCTIONS

20.2.1 WHAT IS A CONST MEMBER FUNCTION?

❖ Revision: `const` forces an entity to be read-only

❖ In Object Oriented Programming, a `const` member function is one that cannot write any value to any data member

❖ The main purpose is for future proofing your programs

  ➢ Preventing unintentional changes
    • If you know a member function is not write to any data member, you can be sure that nobody who maintains your program in future will erroneously add code to write to the data members
  ➢ Common Practice: Make the mutator member functions which read and return the values of the data members const

20.2.2 MAKING A MEMBER FUNCTION CONST

❖ In order to make a member function `const`, simply add the `const` keyword after the prototype in both the

➢ Declaration

➢ Definition

20.2.3 USING CONST MEMBER FUNCTION

❖ Let us now modify the "get" mutator member function by making it const

➢ The declaration, in the class Employee, will now have the keyword const

```
double get_salary_Gross( ) const;
```

➢ The definition is as follows:

```
double Employee::get_salary_Gross( ) const
{
      // tax_Rate = 0.6;
            // ILLEGAL - const member function cannot
      //           modify any data members

      return salary_Gross;

}//END...Mutator get_salary_Gross
```

Program Listing/Output 20.4: const Mutator member function

❖ The full program is as follows:

```
#include <iostream>
#include <string>
using namespace std;

#ifndef EMPLOYEE_H
#define EMPLOYEE_H


const double TAXRATE_THRESHOLD_LOWER = 20000.0;
const double TAXRATE_THRESHOLD_INTERMEDIATE = 25000.0;


const double TAXRATE_RATE_LOWER = 0.3;
const double TAXRATE_RATE_INTERMEDIATE = 0.4;
const double TAXRATE_RATE_HIGHER = 0.5;


// Class DECLARATION
```

```
class Employee
{
      // DATA MEMBERS
      private:
            string name;
            string nI_Number;
            double salary_Gross;
            double tax_Rate;
            double tax_Amount;
            double salary_Net;

      // MEMBER FUNCTIONS
      public:
            // ---------- CONSTRUCTOR(S) ------------
            Employee( string );
      Employee( string, string );
      Employee(    string name_IN,
                                   string nI_Number,
                                   double salary_Gross_IN );

            // ---------- DESTRUCTOR ------------
            ~Employee(  );

            // ---------- MUTATORS ------------
            double get_salary_Gross(  ) const;
            void set_salary_Gross( double salary_Gross_IN );

            // ---------- "NORMAL Member Functions " ------------
            void greet();
            void greet_Name();
            void greet_Name_NINumber();
            void greet_Name_SalaryGross();
            void greet_AllAttributes();


};//END...Class Declaration Employee

#endif
```

Program Listing/Output 20.5: Employee.h

```
#include "Employee.h"

// ---------- CONSTRUCTOR(S) ------------
Employee::Employee( string name_IN )
{
      name = name_IN;

}//END...Constructor Employee( string name_IN )


Employee::Employee( string name_IN, string nI_Number_IN )
```

```
{
      name = name_IN;
      nI_Number = nI_Number_IN;

}//END...Constructor Employee( string name_IN )

Employee::Employee( string name_IN,
                                 string nI_Number_IN,
                                 double salary_Gross_IN )
{
      name = name_IN;
      nI_Number = nI_Number_IN;
      salary_Gross = salary_Gross_IN;
      //----------------------------------------------------
      // Calculating the Tax Rate
      //
      if( salary_Gross < TAXRATE_THRESHOLD_LOWER )
      {
            tax_Rate = TAXRATE_RATE_LOWER;
      }
      else
      {
            if( salary_Gross < TAXRATE_THRESHOLD_INTERMEDIATE )
            {
                  tax_Rate = TAXRATE_RATE_INTERMEDIATE;
            }
            else
            {
                  tax_Rate = TAXRATE_RATE_HIGHER;
            }
      }

      //----------------------------------------------------
      //     Calculating the Tax Amount & Net Salary
      //
      tax_Amount = tax_Rate * salary_Gross;
      salary_Net = salary_Gross - tax_Amount;


      // Incrementing the Number of  Objects/Employees

}/* END...Constructor Employee( string name_IN,
                                       string nI_Num_IN,
                                       double salary_Gross_IN )
 */

// ---------- DESTRUCTOR ------------
Employee::~Employee(  )
{
      cout  << "Goodbye, I, "<< name
            << " am  going out of scope now."
            << endl;
```

```
}// END...Destructor ~Employee

// ---------- MUTATORS ------------
// const keyword is required here as well
double Employee::get_salary_Gross( ) const
{
       // tax_Rate = 0.6;
              // ILLEGAL - const member function cannot
              //                   modify any data members

       return salary_Gross;

}//END...Mutator get_salary_Gross


void Employee::set_salary_Gross( double salary_Gross_IN )
{
       salary_Gross = salary_Gross_IN;

       //----------------------------------------------------
       //     Calculating the NEW Tax Rate
       //
       if( salary_Gross < TAXRATE_THRESHOLD_LOWER )
       {
              tax_Rate = TAXRATE_RATE_LOWER;
       }
       else
       {
              if( salary_Gross < TAXRATE_THRESHOLD_INTERMEDIATE )
              {
                     tax_Rate = TAXRATE_RATE_INTERMEDIATE;
              }
              else
              {
                     tax_Rate = TAXRATE_RATE_HIGHER;
              }
       }

       //----------------------------------------------------
       //     Calculating the NEW Tax Amount & Net Salary
       //
       tax_Amount = tax_Rate * salary_Gross;
       salary_Net = salary_Gross - tax_Amount;

}//END...Mutator set_salary_Gross


// ---------- "NORMAL Member Functions " ------------
void Employee::greet()
{
       cout  << "Hello, I'm an Employee"<< endl;

}//END...Member Function greet
```

```
void Employee::greet_Name()
{
      cout   << "Hello, I'm an Employee, "
             << "my name is "<< name
             << endl;

}//END...Member Function greet_Name

void Employee::greet_Name_NINumber()
{
      cout   << "Hello, I'm an Employee, "
             << "my name is "<< name
             << " and my NI Number is "<< nI_Number
             << endl;

}//END...Member Function greet_Name_NINumber


void Employee::greet_Name_SalaryGross()
{
      cout   << "Hello, I'm an Employee, my name is "<< name
             << ", my Gross Salary is "<< salary_Gross
             << endl;

}//END...Member Function greet_Name_SalaryGross()


void Employee::greet_AllAttributes()
{
      cout   << endl
             << "====================="
             << endl
             << "Hello, I am an employee, my name is "<< name<< ", "
             << endl
             << "my NI Number is "<< nI_Number<< ","
             << endl
             << "my Gross Salary is "<< salary_Gross
             << endl
             << "my tax Rate is "<< tax_Rate<< "%"
             << endl
             << "so the amount I have to pay in tax is "<< tax_Amount
             << endl
             << "thus my Net Salary is "<< salary_Net
             << endl
             << "====================="<< endl;
}// END...MF greet_AllAttributes()
```

Program Listing/Output 20.6: Employee.cpp

```
/*    ***************************************************************
      Const Mutators Demo
```

```
    *************************************************************
    -       This program demonstrates controlling data members using
            Mutator Member Functions, one of them being const
*/

#include <iostream>
#include <string>
using namespace std;

#include "Employee.h"

int main()
{
    //»»»»»»»»»»»»»»»»»»»»»»»»»»»»»»»»»»»»»»»»»»»»»»»»»»»»»»»»»»»»»»»»»»
    Employee joeBloggs( "Joseph Bloggs", "PH 45 67 34 D", 17000.0 );
    joeBloggs.greet_AllAttributes();

    joeBloggs.set_salary_Gross( 17500.0 );
    joeBloggs.greet_AllAttributes();


    //»»»»»»»»»»»»»»»»»»»»»»»»»»»»»»»»»»»»»»»»»»»»»»»»»»»»»»»»»»»»»»»»»»
     cout << endl<< "==============="<< endl
          << "END...C++ Class Mutator Member Function Demo"
          << endl;
  while(true){}
  return 0;

}//END...Function MAIN
```

```
====================
Hello, I am an employee, my name is Joseph Bloggs,
my NI Number is PH 45 67 34 D,
my Gross Salary is 17000
my tax Rate is 0.3%
so the amount I have to pay in tax is 5100
thus my Net Salary is 11900
====================


====================
Hello, I am an employee, my name is Joseph Bloggs,
my NI Number is PH 45 67 34 D,
my Gross Salary is 17500
my tax Rate is 0.3%
so the amount I have to pay in tax is 5250
thus my Net Salary is 12250
====================


===============
END...C++ Class Mutator Member Function Demo
```

Program Listing/Output 20.7: ConstMutatorsDemo_Client.cpp

20.3 STATIC DATA MEMBERS

20.3.1 WHAT ARE STATIC DATA MEMBERS?

❖ A `static` data member is one where every object of that class shares the same value

➢ Such a data member is CLASS-specific

➢ The majority of data members, which are "non-static" are object specific e.g For our employee class, each object will have its own data

- `joeBloggs` will have his own `salary_Gross`
- `amandaTaylor` will have her own `salary_Gross`

❖ Common uses include

➢ A variable to count the number of objects created

➢ An object which all objects use,

- e.g. All Employees might use the same canteen
- Data shared by all objects e.g. All footballers should have a blood pressure between 90 and 120 bars

20.3.2 INITIALISING STATIC DATA MEMBERS

❖ Static data members must be initialised in the class

➢ This is essential since they are class specific

➢ Initialisation cannot be carried out using constructors, since these are used for "non-static" i.e. object specific data members

❖ They must be initialised outside the class lexically, using the scope resolution operator (see below for the example)

20.3.3 USING STATIC DATA MEMBERS

❖ In our Employee program, a static data member `num_Employees` will be implemented to monitor the number of Employee objects created

❖ Its declaration, in the class Employee, will be as follows:

```
static int num_Employees;
```

❖ Its initialisation, must be as follows:

```
int Employee::num_Employees = 0;
```

❖ **Common Interview Question:**

(Q)   How can a class be implemented so that the static data member num_Employees is
   ❖ Incremented every time a new object is created
   ❖ Decremented every time an new object is destroyed

(A)   ❖ Increment it in every CONSTRUCTOR
   ❖ Decrement it in every DESTRUCTOR

❖ In the full Employee program which follows

➢ An additional mutator member function, which returns the current number of objects, has been included:

```cpp
#include <iostream>
#include <string>
using namespace std;

#ifndef EMPLOYEE_H
#define EMPLOYEE_H


 const double TAXRATE_THRESHOLD_LOWER = 20000.0;
 const double TAXRATE_THRESHOLD_INTERMEDIATE = 25000.0;


 const double TAXRATE_RATE_LOWER = 0.3;
 const double TAXRATE_RATE_INTERMEDIATE = 0.4;
 const double TAXRATE_RATE_HIGHER = 0.5;


 // Class DECLARATION
 class Employee
 {
      // DATA MEMBERS
      private:
            string name;
            string nI_Number;
            double salary_Gross;
            double tax_Rate;
            double tax_Amount;
            double salary_Net;

            // ---------- STATIC Data Member(s) ------------
            static int num_Employees;

  // MEMBER FUNCTIONS
  public:
        // ---------- CONSTRUCTOR(S) ------------
  Employee( string );
  Employee( string, string );
```

```
  Employee( string name_IN,
                        string nI_Number,
                        double salary_Gross_IN );

        // ---------- DESTRUCTOR ------------
         ~Employee(   );

        // ---------- MUTATORS ------------
        double get_salary_Gross(   ) const;
        void set_salary_Gross( double salary_Gross_IN );

        int get_num_Employees();

        // ---------- "NORMAL Member Functions " ------------
        void greet();
        void greet_Name();
        void greet_Name_NINumber();
        void greet_Name_SalaryGross();
        void greet_AllAttributes();


};//END...Class Declaration Employee

#endif
```

Program Listing/Output 20.8: Employee.h

```
  #include "Employee.h"

// ---------- STATIC Data Member(s) ------------
int Employee::num_Employees = 0;


// ---------- CONSTRUCTOR(S) ------------
Employee::Employee( string name_IN )
{
        name = name_IN;

      // Keeping track of the number of objects
      ++num_Employees;

}//END...Constructor Employee( string name_IN )


Employee::Employee( string name_IN, string nI_Number_IN )
{
      name = name_IN;
      nI_Number = nI_Number_IN;

      // Keeping track of the number of objects
      ++num_Employees;
```

```
}//END...Constructor Employee( string name_IN )

Employee::Employee( string name_IN,
                                  string nI_Number_IN,
                                  double salary_Gross_IN )
{
      name = name_IN;
      nI_Number = nI_Number_IN;
      salary_Gross = salary_Gross_IN;
      //--------------------------------------------------
      // Calculating the Tax Rate
      //
      if( salary_Gross < TAXRATE_THRESHOLD_LOWER )
      {
           tax_Rate = TAXRATE_RATE_LOWER;
 }
 else
 {
          if( salary_Gross < TAXRATE_THRESHOLD_INTERMEDIATE )
          {
              tax_Rate = TAXRATE_RATE_INTERMEDIATE;
          }
          else
          {
              tax_Rate = TAXRATE_RATE_HIGHER;
          }
 }

 //----------------------------------------------------
 //     Calculating the Tax Amount & Net Salary
 //
 tax_Amount = tax_Rate * salary_Gross;
 salary_Net = salary_Gross - tax_Amount;


 // Incrementing the Number of Objects/Employees

 // Keeping track of the number of objects
 ++num_Employees;

}/* END...Constructor Employee( string name_IN,
                                 string nI_Num_IN,
                                 double salary_Gross_IN )
 */

// ---------- DESTRUCTOR ------------
Employee::~Employee(  )
{
      cout << "Goodbye, I, "<< name
           << " am  going out of scope now."
           << endl;
```

```
        // Keeping track of the number of objects
        --num_Employees;

}// END...Destructor ~Employee

// ---------- MUTATORS ------------
// const keyword is required here as well
double Employee::get_salary_Gross(  ) const
{
        // tax_Rate = 0.6;
                // ILLEGAL - const member function cannot
                //           modify any data members

        return salary_Gross;

}//END...Mutator get_salary_Gross


void Employee::set_salary_Gross( double salary_Gross_IN )
{
        salary_Gross = salary_Gross_IN;

        //-----------------------------------------------------
        // Calculating the NEW Tax Rate
        //
        if( salary_Gross < TAXRATE_THRESHOLD_LOWER )
        {
            tax_Rate = TAXRATE_RATE_LOWER;
        }
        else
        {
            if( salary_Gross < TAXRATE_THRESHOLD_INTERMEDIATE )
            {
                tax_Rate = TAXRATE_RATE_INTERMEDIATE;
            }
            else
            {
                tax_Rate = TAXRATE_RATE_HIGHER;
            }
    }
    //-----------------------------------------------------
    //     Calculating the NEW Tax Amount & Net Salary
    //
    tax_Amount = tax_Rate * salary_Gross;
    salary_Net = salary_Gross - tax_Amount;

}//END...Mutator set_salary_Gross

int Employee::get_num_Employees()
{
        return num_Employees;

}//END...Mutator get_num_Employees
```

```cpp
// ---------- "NORMAL Member Functions " ------------
void Employee::greet()
{
        cout << "Hello, I'm an Employee"<< endl;

}//END...Member Function greet

void Employee::greet_Name()
{
      cout << "Hello, I'm an Employee, "
            << "my name is "<< name
            << endl;

}//END...Member Function greet_Name

void Employee::greet_Name_NINumber()
{
        cout << "Hello, I'm an Employee, "
            << "my name is "<< name
            << " and my NI Number is "<< nI_Number
            << endl;

}//END...Member Function greet_Name_NINumber


void Employee::greet_Name_SalaryGross()
{
      cout << "Hello, I'm an Employee, my name is "<< name
            << ", my Gross Salary is "<< salary_Gross
            << endl;

}//END...Member Function greet_Name_SalaryGross()


void Employee::greet_AllAttributes()
{
      cout << endl
      << "====================="
      << endl
      << "Hello, I am an employee, my name is "<< name<< ", "
      << endl
      << "my NI Number is "<< nI_Number<< ","
      << endl
      << "my Gross Salary is "<< salary_Gross
      << endl
      << "my tax Rate is "<< tax_Rate<< "%"
      << endl
      << "so the amount I have to pay in tax is "<< tax_Amount
      << endl
      << "thus my Net Salary is "<< salary_Net
      << endl
```

```
         << "===================="<< endl;
 }// END...MF greet_AllAttributes()
```

Program Listing/Output 20.9: Employee.cpp

```cpp
#include <iostream>
#include <string>
using namespace std;

#include "Employee.h"

 int main()
 {

     cout << endl
         << "============"<< endl
         << "Object joeBloggs being created"
         << endl;
  Employee joeBloggs( "Joseph Bloggs", "PH 45 67 34 D", 17000.0 );
  joeBloggs.greet_Name_SalaryGross();
  cout  << "The current number of employees is, "
        << "accessing from joeBloggs: "
        << joeBloggs.get_num_Employees()
        << endl;

  cout << endl
       << "============"<< endl
       << "Object johnSmith being created"
       << endl;
  Employee johnSmith( "John Smith", "GH 67 34 67 D", 18000.0 );
  johnSmith.greet_Name_SalaryGross();

  cout << "The current number of employees is, "
       << "accessing from joeBloggs:   "
       << joeBloggs.get_num_Employees()
       << endl;
  cout << "The current number of employees is, "
       << "accessing from johnSmith: "
       << johnSmith.get_num_Employees()
       << endl;

  cout << endl
       << "============"<< endl
       << "Object amandaTaylor being created"
       << endl;
  Employee amandaTaylor("Amanda Taylor", "SG 36 78 24 F", 22000.0 );
  amandaTaylor.greet_Name_SalaryGross();

  cout << "The current number of employees is, "
       << "accessing from joeBloggs:   "
       << joeBloggs.get_num_Employees()
       << endl;
```

```
  cout << "The current number of employees is, "
       << "accessing from johnSmith: "
       << johnSmith.get_num_Employees()
       << endl;
  cout << "The current number of employees is, "
       << "accessing from amandaTaylor: "
       << amandaTaylor.get_num_Employees()
       << endl;


  //»»»»»»»»»»»»»»»»»»»»»»»»»»»»»»»»»»»»»»»»»»»»»»»»»»»»»»»»»»»»ż
  cout << endl<< "================"<< endl
       << "END...C++ Demo: Static Data Member"
       << endl;
  while(true){}
  return 0;

}//END...Function MAIN
```

```
============
Object joeBloggs being created
Hello, I'm an Employee, my name is Joseph Bloggs, my Gross
Salary is 17000
The current number of employees is, accessing from joeBloggs: 1

============
Object johnSmith being created
Hello,  I'm an Employee, my name is John Smith, my Gross
Salary is 18000
The current number of employees is, accessing from joeBloggs:  2
The current number of employees is, accessing from johnSmith: 2

============
Object amandaTaylor being created
Hello,  I'm an Employee, my name is Amanda Taylor, my Gross
Salary is 22000
The current number of employees is, accessing from joeBloggs:  3
The current number of employees is, accessing from johnSmith: 3
The current number of employees is, accessing from amandaTaylor: 3

================
END...C++ Demo: Static Data Member
```

Program Listing/Output 20.10: StaticDataMember_Demo_Client.cpp

20.4 STATIC MEMBER FUNCTIONS

20.4.1 WHAT ARE STATIC MEMBER FUNCTIONS?

❖ A `static` Member Function is a special member function which can be called via the class name as well as via any objects

➢ The bulk of member functions are "non-static" i.e. they can only ever be called via an object

❖ While reading the static data member using an object is very useful, there is one essential special case it cannot deal with

➢ When the number of objects is ZERO

❖ We need a member function which does not require an object

❖ In our case, we want to be able to say

```
Employee::get_num_Employees()
```

➢ so the number can be read whether there are zero objects or not

20.4.1.1 Rules of Static Member Functions?

❖ The keyword `static` must be used in the declaration

❖ It must not be duplicated in the definition

❖ Operators for accessing it:

➢ SCOPE RESOLUTION OPERATOR when accessing via the Class
➢ DOT OPERATOR when accessing via an object
  • NB. Most OOP languages, including Java, use the dot operator for both

❖ Static Member Functions can only ever access static data members

➢ Makes sense:
  • Take the hypothetical case: a static member function, `get_salary_ Gross` can access any data member
  • We have many Employees inc `joeBloggs` and `amandaTaylor`
    ○ each has their OWN `salary_Gross`
  • We call the static member function via the class
    ```
    Employee::get_ salary_Gross
    ```
  • Whose `salary_Gross` will it access?...`joeBloggs`' s, `amandaTaylor`' s, someone else's?!
➢ Static Data members have no problem: if all objects share the same value, no problem accessing via the class

20.4.2 USING THE STATIC MEMBER FUNCTIONS

❖ In our Employee example, we will make the mutator member function `get_num_Employees()` static

❖ Its declaration will become:

```
static int get_num_Employees();
```

❖ Its definition will be as before (since the keyword `static` must not be duplicated)

❖ The full program is as follows:

```cpp
#include <iostream>
#include <string>
using namespace std;

#ifndef EMPLOYEE_H
#define EMPLOYEE_H


const double TAXRATE_THRESHOLD_LOWER = 20000.0;
const double TAXRATE_THRESHOLD_INTERMEDIATE = 25000.0;


const double TAXRATE_RATE_LOWER = 0.3;
const double TAXRATE_RATE_INTERMEDIATE = 0.4;
const double TAXRATE_RATE_HIGHER = 0.5;


// Class DECLARATION
class Employee
{
    // DATA MEMBERS
    private:
        string name;
        string nI_Number;
        double salary_Gross;
        double tax_Rate;
        double tax_Amount;
        double salary_Net;

        // ---------- STATIC Data Member(s) ------------
        static int num_Employees;

    // MEMBER FUNCTIONS
    public:
        // ---------- CONSTRUCTOR(S) ------------
            Employee( string );
            Employee( string, string );
            Employee( string name_IN,
            string nI_Number,
```

```
                    double salary_Gross_IN );

        // ---------- DESTRUCTOR ------------
        ~Employee( );

        // ---------- MUTATORS ------------
        double get_salary_Gross( ) const;
        void set_salary_Gross( double salary_Gross_IN );

        // ---------- STATIC MEMBER FUNCTION(S) ------------
        // ...also a Mutator
        static int get_num_Employees();

        // ---------- "NORMAL Member Functions " ------------
        void greet();
        void greet_Name();
        void greet_Name_NINumber();
        void greet_Name_SalaryGross();
        void greet_AllAttributes();


};//END...Class Declaration Employee

#endif
```

Program Listing/Output 20.11: Employee.h

```
#include "Employee.h"

// ---------- STATIC Data Member(s) ------------
int Employee::num_Employees = 0;


// ---------- CONSTRUCTOR(S) ------------
Employee::Employee( string name_IN )
{
        name = name_IN;

      // Keeping track of the number of objects
      ++num_Employees;

}//END...Constructor Employee( string name_IN )


Employee::Employee( string name_IN, string nI_Number_IN )
{
      name = name_IN;
      nI_Number = nI_Number_IN;

    // Keeping track of the number of objects
    ++num_Employees;
```

```
}//END...Constructor Employee( string name_IN )

Employee::Employee( string name_IN,
                     string nI_Number_IN,
                     double salary_Gross_IN )
{
      name = name_IN;
      nI_Number = nI_Number_IN;
      salary_Gross = salary_Gross_IN;
      //--------------------------------------------------
      //     Calculating the Tax Rate
      //
      if( salary_Gross < TAXRATE_THRESHOLD_LOWER )
      {
            tax_Rate = TAXRATE_RATE_LOWER;
      }
      else
      {
            if( salary_Gross < TAXRATE_THRESHOLD_INTERMEDIATE )
            {
                  tax_Rate = TAXRATE_RATE_INTERMEDIATE;
            }
            else
            {
                  tax_Rate = TAXRATE_RATE_HIGHER;
            }
      }

  //--------------------------------------------------------
  //     Calculating the Tax Amount \& Net Salary
  //
  tax_Amount = tax_Rate * salary_Gross;
  salary_Net = salary_Gross - tax_Amount;


  // Incrementing the Number of Objects/Employees

  // Keeping track of the number of objects
  ++num_Employees;

}/* END...Constructor Employee( string name_IN,
                                string nI_Num_IN,
                                double salary_Gross_IN )
  */

// ---------- DESTRUCTOR ------------
Employee::~Employee(  )
{
        cout << "Goodbye, I, "<< name
             << " am  going out of scope now."
             << endl;
```

```
        // Keeping track of the number of objects
        --num_Employees;

}// END...Destructor ~Employee

// ---------- MUTATORS ------------
// const keyword is required here as well
double Employee::get_salary_Gross( ) const
{
        // tax_Rate = 0.6;
                // ILLEGAL - const member function cannot
                //                   modify any data members

        return salary_Gross;

}//END...Mutator get_salary_Gross


void Employee::set_salary_Gross( double salary_Gross_IN )
{
        salary_Gross = salary_Gross_IN;

        //----------------------------------------------------
        //     Calculating the NEW Tax Rate
        //
        if( salary_Gross < TAXRATE_THRESHOLD_LOWER )
        {
                tax_Rate = TAXRATE_RATE_LOWER;
        }
        else
        {
                if( salary_Gross < TAXRATE_THRESHOLD_INTERMEDIATE )
                {
                        tax_Rate = TAXRATE_RATE_INTERMEDIATE;
                }
                else
                {
                        tax_Rate = TAXRATE_RATE_HIGHER;
                }
   }

  //----------------------------------------------------
  //     Calculating the NEW Tax Amount \& Net Salary
  //
  tax_Amount = tax_Rate * salary_Gross;
  salary_Net = salary_Gross - tax_Amount;

}//END...Mutator set_salary_Gross


// ---------- STATIC MEMBER FUNCTION(S) ------------
// ...also a Mutator
int Employee::get_num_Employees()
```

```
{
        return num_Employees;

}//END...Mutator get_num_Employees


// ---------- "NORMAL Member Functions " ------------
void Employee::greet()
{
        cout << "Hello, I'm an Employee"<< endl;

}//END...Member Function greet

void Employee::greet_Name()
{
         cout << "Hello, I'm an Employee, "
              << "my name is "<< name
              << endl;

}//END...Member Function greet_Name

void Employee::greet_Name_NINumber()
{
        cout << "Hello, I'm an Employee, "
             << "my name is "<< name
             << " and my NI Number is "<< nI_Number
             << endl;

}//END...Member Function greet_Name_NINumber


void Employee::greet_Name_SalaryGross()
{
        cout << "Hello, I'm an Employee, my name is "<< name
             << ", my Gross Salary is "<< salary_Gross
             << endl;

}//END...Member Function greet_Name_SalaryGross()


void Employee::greet_AllAttributes()
{
        cout << endl
        << "====================="
        << endl
        << "Hello, I am an employee, my name is "<< name<< ", "
        << endl
        << "my NI Number is "<< nI_Number<< ","
        << endl
        << "my Gross Salary is "<< salary_Gross
        << endl
        << "my tax Rate is "<< tax_Rate<< "%"
        << endl
```

```
        << "so the amount I have to pay in tax is "<< tax_Amount
        << endl
        << "thus my Net Salary is "<< salary_Net
        << endl
        << "===================="<< endl;
}// END... Member Function greet_AllAttributes()
```

Program Listing/Output 20.12: Employee.cpp

```cpp
#include <iostream>
#include <string>
using namespace std;

#include "Employee.h"

int main()
{
    //»»»»»»»»»»»»»»»»»»»»»»»»»»»»»»»»»»»»»»»»»»»»»»»»»»»»»»»»»»»»»»»»»»»»
    // Calling a static data member via the CLASS, not an object
    // Need the SCOPE RESOLUTION OPERATOR, not the Dot operator
    // (Most languages, inc Java, use the dot operator)
    //
    cout << "The current number of employees is, "
         << "accessing via the class Employee: "
         << Employee::get_num_Employees()
         << endl;


    //»»»»»»»»»»»»»»»»»»»»»»»»»»»»»»»»»»»»»»»»»»»»»»»»»»»»»»»»»»»»»»»»»»»»
    cout << endl
         << "============="<< endl
         << "Object joeBloggs being created"
         << endl;
    Employee joeBloggs( "Joseph Bloggs", "PH 45 67 34 D", 17000.0 );
    joeBloggs.greet_Name_SalaryGross();

    cout << "The current number of employees is, "
         << "accessing via the class Employee: "
         << Employee::get_num_Employees()
         << endl;

    cout << "The current number of employees is, "
         << "accessing from joeBloggs: "
         << joeBloggs.get_num_Employees()
         << endl;

    cout << endl
         << "============="<< endl
         << "Object johnSmith being created"
         << endl;
    Employee johnSmith( "John Smith", "GH 67 34 67 D", 18000.0 );
    johnSmith.greet_Name_SalaryGross();
```

```
        cout << "The current number of employees is, "
             << "accessing via the class Employee: "
             << Employee::get_num_Employees()<< endl;

        cout << "The current number of employees is, "
             << "accessing from joeBloggs:   "
             << joeBloggs.get_num_Employees()<< endl;

        cout << "The current number of employees is, "
             << "accessing from johnSmith: "
             << johnSmith.get_num_Employees()<< endl;


        cout << endl
             << "============"<< endl
             << "Object amandaTaylor being created"
             << endl;
        Employee amandaTaylor( "Amanda Taylor", "SG 36 78 24 F",
                               22000.0 );
        amandaTaylor.greet_Name_SalaryGross();

        cout << "The current number of employees is, "
             << "accessing via the class Employee: "
             << Employee::get_num_Employees()
             << endl;

        cout << "The current number of employees is, "
             << "accessing from joeBloggs:   "
             << joeBloggs.get_num_Employees()
             << endl;

        cout << "The current number of employees is, "
             << "accessing from johnSmith: "
             << johnSmith.get_num_Employees()
             << endl;

        cout << "The current number of employees is, "
             << "accessing from amandaTaylor: "
             << amandaTaylor.get_num_Employees()
             << endl;

    //»»»»»»»»»»»»»»»»»»»»»»»»»»»»»»»»»»»»»»»»»»»»»»»»»»»»»»»»»»»»»»»»»»»ż
        cout << endl<< "================"
             << endl<< "END...C++ Demo: Static Member Function"
             << endl;
    while(true){}
    return 0;

}//END...Function MAIN
```

```
The current number of employees is, accessing via the
class  Employee: 0


============
Object joeBloggs being created
Hello, I'm an Employee, my name is Joseph Bloggs, my Gross
Salary is 17000
The current number of employees is, accessing via the
class Employee: 1
The current number of employees is, accessing from joeBloggs: 1


============
Object johnSmith being created
Hello, I'm an Employee, my name is John Smith, my Gross
Salary is 22000
The current number of employees is, accessing via the
class Employee: 3
The current number of employees is, accessing from joeBloggs:  3
The current number of employees is, accessing from johnSmith: 3
The current number of employees is, accessing from amandaTaylor: 3


===============
END...C++ Demo: Static Member Function
```

Program Listing/Output 20.13: StaticMemberFunction_Demo_Client.cpp

20.5 ASSIGNMENT OPERATOR

20.5.1 WHAT IS THE ASSIGNMENT OPERATOR?

❖ The assignment operator, silently built into every class, is a special operator used to **assign** values to one **existing** object by copying values from another **existing** object

➤ The usage is subtle as the following shows:

```
Employee robertFletcher( /* Constructor Inputs */ );
Employee joeBloggs( /* Constructor Inputs */ );

robertFletcher = joeBloggs; // ASSIGNMENT OPERATOR called
```

❖ **Common Interview Question:**

(Q) If it is illegal to try setting two arrays equal to one another, why is it acceptable to set two objects equal to one another, even when one of more of the data members is/are (an) array(s)?

(A) Although the same symbol is used, =, it is actually performing 2 different actions in each case

➢ For the array, it is attempting to perform the equality operation (section 2.2.2.1)

• i.e. take the value of the right hand side, put it into the address of the left

➢ For the objects, it is calling the assignment operator

❖ It is thus important to note that the = symbols mean different operations according to the data items it is operating on

❖ `primitiveVariable = primitiveVariable`

➢ This will call the standard assignment operator

❖ `object = object`

➢ This will call the object oriented assignment operator, built into the class

20.5.2 USING THE ASSIGNMENT OPERATOR

❖ When we have a statement as follows:

`peterJohnson = jonathanSharp;`

➢ ...where `jonathanSharp` and `peterJohnson` are both existing objects of type `Employee`

➢ This goes through all the members of the source object and systematically copies them into the destination object

❖ we could equally have written:

`peterJohnson.operator = jonathanSharp;`

➢ This latter form is far messier and so hardly ever used

20.5.3 DEFAULT ASSIGNMENT OPERATOR

❖ If you do not explicitly program the assignment operator, the default one will be used, which is always there

➢ This is the one we were using

❖ It is possible to give our own meaning to this

> This is part of Operator Overloading, covered in section 21.

20.6 COPY CONSTRUCTOR

20.6.1 WHAT IS THE COPY CONSTRUCTOR?

❖ Revision: A constructor is called whenever an object is created

❖ The **Copy** Constructor is called whenever a **copy** of an object is created

> i.e. whenever an object is passed by **value**

❖ This is very common, situations include:

> Passing the object to a function

> Returning the object from a function

> Initialising an object using another object of the same class type

20.6.1.1 Assignment vs. Initialisation

❖ In C programming, there is rarely any difference between initialisation and assignment

❖ In C++, since we are dealing with objects, the distinction is very important

❖ Consider the following case:

```
Employee patriciaMcKenzie( /* Constructor Inputs */ );


Employee rachelWorthington = patriciaMcKenzie;
```

> The second line is in fact an *initialisation*

• It looks like, but is not an assignment

❖ In object oriented programming

> Assignment is the practice of changing the value/state of an **existing** object

> Initialisation is the practice of creating a **new** object and giving it a value/state at the same time

❖ The Copy Constructor is used to optimise the process of initialisation

> A default constructor could be used to create the new object

> An assignment operator could then be used to give it values

> However, it is wasteful, especially for a large object, to give an object an initial state, only to override it immediately

❖ Summarising:

➢ Initial Definitions	`Employee peterJohnson( /*Constructor Inputs*/ );`
	`Employee jonathanSharp( /*Constructor Inputs*/ );`
	`Employee patriciaMcKenzie( /*Constructor Inputs*/ );`

➢ Assignment • 2 EXISTING Objects	`peterJohnson = jonathanSharp;`
➢ Initialisation • Create a new Object from an Existing one	`Employee rachelWorthington = patriciaMcKenzie;`

20.6.1.2 Essential Nature of the Copy Constructor

❖ Without such definitions, the default behaviour is used, which may well lead to problems:

❖ When passing an object to a function by value, a copy of the object is made

> If the object contains a pointer to dynamic heap memory, then both It and the original object will be pointing to the allocated memory

> The problem of 2 or more pointers to the heap (section 14.5) will now be manifest:

   • The copy may well manipulate the heap memory
   • When the function call is complete and the copy is destroyed, its destructor will be called
      ○ If this frees up the memory allocated on the heap, the pointer referring to this in the original object will become a **dangling pointer**

> Similar problems will occur if the object reserves resources of
   • Networks
   • Databases

20.6.1.3 Rules of the Copy Constructor

❖ Like the default constructor,

> The name must be the same as the name of the class
> It cannot return anything

❖ The input is of type Reference-to-theClass

❖ Thus a copy constructor of the class Employee will be as follows

```
class Employee
{
        // Copy Constructor
        Employee( Employee& ref_Employee_IN )
        {
                // Functionalities of the Copy Constructor
        }

};//END...Class Employee
```

20.6.1.3.1 Input of a Copy Constructor

❖ The copy constructor must receive its input argument by **reference**, not by value

➢ If it received the object by value,

- The copy constructor would have to make a copy of the input object
- But any time a copy of an object is created, the copy constructor is called
- The copy constructor would thus recursively call itself infinite times, in order to make a copy of the input!

20.6.1.4 Constructor vs. Copy Constructor

	CONSTRUCTOR	**COPY CONSTRUCTOR**
❖	Called when a new object is created	Called when a copy of an object is created i.e. whenever passing an object by value
❖	Used to put the newly created object in a default state	Used to put the newly created copy in the same state as an existing object
❖	Must have the same name as the class it is in	
❖	Cannot return anything	
❖	Can have any inputs	Can have only one input, which must be of type Reference-to-theClass it is in

20.6.2 IMPLEMENTING AND USING THE COPY CONSTRUCTOR

❖ In our case, we will be creating a copy constructor which

➢ Takes a reference to an employee object as its input

➢ Copies all the data from the this object into the current Employee object

➢ In addition, a message displays the object whose copy constructor has been invoked

❖ The declaration is as follows:

```
Employee( Employee& ref_Employee_IN );
```

❖ The definition is as follows:

```
Employee::Employee( Employee& ref_Employee_IN )
{
        name = ref_Employee_IN.name;
        nI_Number = ref_Employee_IN.nI_Number;
        salary_Gross = ref_Employee_IN.salary_Gross;
        tax_Rate = ref_Employee_IN.tax_Rate;
        tax_Amount = ref_Employee_IN.tax_Amount;
        salary_Net = ref_Employee_IN.salary_Net;

        cout << "Copy Constructor called for "<< name<< endl;

}// END...COPY Constructor
```

❖ The full program is as follows:

```
#include <iostream>
#include <string>
using namespace std;

#ifndef EMPLOYEE_H
#define EMPLOYEE_H


const double TAXRATE_THRESHOLD_LOWER = 20000.0;
const double TAXRATE_THRESHOLD_INTERMEDIATE = 25000.0;


const double TAXRATE_RATE_LOWER = 0.3;
const double TAXRATE_RATE_INTERMEDIATE = 0.4;
const double TAXRATE_RATE_HIGHER = 0.5;


// Class DECLARATION
 class Employee
 {
       // DATA MEMBERS
       private:
               string name;
               string nI_Number;
               double salary_Gross;
               double tax_Rate;
               double tax_Amount;
```

```
                    double salary_Net;

                    // ---------- STATIC Data Member(s) ------------
                    static int num_Employees;

    // MEMBER FUNCTIONS
    public:
            // ---------- CONSTRUCTOR(S) ------------
            Employee( string );
            Employee( string, string );
            Employee( string name_IN,
                      string nI_Number,
                      double salary_Gross_IN );

            // ------- COPY Constructor ---------
            Employee( Employee& ref_Employee_IN );

            // ---------- DESTRUCTOR ------------
            ~Employee(  );

            // ---------- MUTATORS ------------
            double get_salary_Gross(  ) const;
            void set_salary_Gross( double salary_Gross_IN );

            // ---------- STATIC MEMBER FUNCTION(S) ------------
            // ...also a Mutator
            static int get_num_Employees();

            // ---------- "NORMAL Member Functions " ------------
            void greet();
            void greet_Name();
            void greet_Name_NINumber();
            void greet_Name_SalaryGross();
            void greet_AllAttributes();


};//END...Class Declaration Employee

#endif
```

```
#include "Employee.h"

// ---------- STATIC Data Member(s) ------------
int Employee::num_Employees = 0;


// ---------- CONSTRUCTOR(S) ------------
Employee::Employee( string name_IN )
{
        name = name_IN;
```

```
        // Keeping track of the number of objects
        ++num_Employees;

}//END...Constructor Employee( string name_IN )


Employee::Employee( string name_IN, string nI_Number_IN )
{
        name = name_IN;
        nI_Number = nI_Number_IN;

        // Keeping track of the number of objects
        ++num_Employees;

}//END...Constructor Employee( string name_IN )

 Employee::Employee( string name_IN,
                     string nI_Number_IN,
                     double salary_Gross_IN )
{
        name = name_IN;
        nI_Number = nI_Number_IN;
        salary_Gross = salary_Gross_IN;
        //----------------------------------------------------
        // Calculating the Tax Rate
        //
        if( salary_Gross < TAXRATE_THRESHOLD_LOWER )
        {
                tax_Rate = TAXRATE_RATE_LOWER;
        }
        else
        {
                if( salary_Gross < TAXRATE_THRESHOLD_INTERMEDIATE )
                {
                        tax_Rate = TAXRATE_RATE_INTERMEDIATE;
                }
                else
                {
                        tax_Rate = TAXRATE_RATE_HIGHER;
                }
         }

    //----------------------------------------------------
    //    Calculating the Tax Amount & Net Salary
    //
    tax_Amount = tax_Rate * salary_Gross;
    salary_Net = salary_Gross - tax_Amount;


    // Incrementing the Number of Objects/Employees

    // Keeping track of the number of objects
```

```
        ++num_Employees;

}/* END...Constructor Employee( string name_IN,
                                string nI_Num_IN,
                                double salary_Gross_IN )
  */

// ------- COPY Constructor ---------
Employee::Employee( Employee& ref_Employee_IN )
{
        name = ref_Employee_IN.name;
        nI_Number = ref_Employee_IN.nI_Number;
        salary_Gross = ref_Employee_IN.salary_Gross;
        tax_Rate = ref_Employee_IN.tax_Rate;
        tax_Amount = ref_Employee_IN.tax_Amount;
        salary_Net = ref_Employee_IN.salary_Net;

        cout << "Copy Constructor called for "<< name<< endl;

}// END...COPY Constructor

// ---------- DESTRUCTOR ------------
Employee::~Employee(  )
{
        cout << "Goodbye, I, "<< name
             << " am  going out of scope now."
             << endl;

      // Keeping track of the number of objects
        --num_Employees;

}// END...Destructor \~{}Employee

// ---------- MUTATORS ------------
// const keyword is required here as well
double Employee::get_salary_Gross(  ) const
{
        // tax_Rate = 0.6;
              // ILLEGAL - const member function cannot
              //                  modify any data members

        return salary_Gross;

}//END...Mutator get_salary_Gross


void Employee::set_salary_Gross( double salary_Gross_IN )
{
        salary_Gross = salary_Gross_IN;

        //----------------------------------------------------
        // Calculating the NEW Tax Rate
        //
```

```
        if( salary_Gross < TAXRATE_THRESHOLD_LOWER )
        {
              tax_Rate = TAXRATE_RATE_LOWER;
        }
        else
        {
              if( salary_Gross < TAXRATE_THRESHOLD_INTERMEDIATE )
              {
                    tax_Rate = TAXRATE_RATE_INTERMEDIATE;
              }
              else
              {
                    tax_Rate = TAXRATE_RATE_HIGHER;
              }
        }

  //----------------------------------------------------
  //    Calculating the NEW Tax Amount & Net Salary
  //
  tax_Amount = tax_Rate * salary_Gross;
  salary_Net = salary_Gross - tax_Amount;

}//END...Mutator set_salary_Gross


  // ---------- STATIC MEMBER FUNCTION(S) ------------
  // ...also a Mutator
  int Employee::get_num_Employees()
  {
        return num_Employees;

  }//END...Mutator get_num_Employees


  // ---------- "NORMAL Member Functions " ------------
  void Employee::greet()
  {
        cout << "Hello, I'm an Employee"<< endl;

  }//END...Member Function greet

  void Employee::greet_Name()
  {
        cout << "Hello, I'm an Employee, "
              << "my name is "<< name
              << endl;

  }//END...Member Function greet_Name

  void Employee::greet_Name_NINumber()
  {
        cout << "Hello, I'm an Employee, "
              << "my name is "<< name
```

```
        << " and my NI Number is "<< nI_Number
        << endl;

}//END...Member Function greet_Name_NINumber


void Employee::greet_Name_SalaryGross()
{
      cout << "Hello, I'm an Employee, my name is "<< name
           << ", my Gross Salary is "<< salary_Gross
           << endl;

}//END...Member Function greet_Name_SalaryGross()


void Employee::greet_AllAttributes()
{
      cout << endl
           << "====================="
           << endl
           << "Hello, I am an employee, my name is "<< name<< ", "
           << endl
           << "my NI Number is "<< nI_Number<< ","
           << endl
           << "my Gross Salary is "<< salary_Gross
           << endl
           << "my tax Rate is "<< tax_Rate<< "%"
           << endl
           << "so the amount I have to pay in tax is "<< tax_Amount
           << endl
           << "thus my Net Salary is "<< salary_Net
           << endl
           << "====================="<< endl;
}// END...MF greet_AllAttributes()
```

Program Listing/Output 20.15: Employee.cpp

```
#include <iostream>
#include <string>
using namespace std;

#include "Employee.h"

int main( )
{
      //»»»»»»»»»»»»»»»»»»»»»»»»»»»»»»»»»»»»»»»»»»»»»»»»»»»»»»»
      //
      Employee joeBloggs( "Joseph Bloggs", "PH 45 67 34 D", 17000.0 );
      Employee steveRichards = joeBloggs;

      //»»»»»»»»»»»»»»»»»»»»»»»»»»»»»»»»»»»»»»»»»»»»»»»»»»»»»»»
        cout << endl<< endl<< "==========="<< endl
```

```
                    << "END...Demo C++ Copy Constructor"<< endl;
while(true){}
return 0;

}//END...Function MAIN
```

```
Copy Constructor called for Joseph Bloggs

===========

END...Demo C++ Copy Constructor
```

Program Listing/Output 20.16: CopyConstructor_Demo_Client.cpp

20.6.3 DEFAULT COPY CONSTRUCTOR

❖ A copy constructor is always there

❖ If the programmer (i.e. you!:-)) do(es) not implement it explicitly, it becomes a member function which

➢ Takes no inputs

➢ Performs a

• Bitwise copy for all primitive variables
• Copy for all objects, and calls their default constructors

➢ Like all constructors, it cannot return anything

❖ The reinterpret cast is used to convert a pointer from one class type to another, including an unrelated one, at compile time

20.7 THE "THIS" POINTER

20.7.1 WHAT IS THE THIS POINTER?

❖ The `this` pointer is a special pointer through which an object can access its address

➢ It is not part of the object itself

➢ It is passed implicitly by the compiler to all non-static member functions

❖ This is essential in object oriented programming

- ➤ As we have been discussing, member functions have access to the data members of the class
- ➤ It is by using the "this" pointer that member functions know which object's data members to manipulate?

❖ Note that the this pointer is always there

- ➤ The programmer (i.e. you! :-)) ha(s)(ve) no control over this

20.7.2 USING THE THIS POINTER

❖ In our case, the following member function has been added to the Employee class:

```
void Employee::display_salary_Net()
{
    cout << "Accessing the Data Member salary_Net of :"
        << get_Name()<< endl;

    cout << "-\tImplicitly\t  salary_Net = "
        << salary_Net<< endl;

    cout << "-\tExplicitly\t  this -> salary_Net = "
        << this -> salary_Net<< endl;

    cout << "-\tExplicitly\t  (*this).salary_Net = "
        << (*this).salary_Net<< endl;

}// END...MF display_salary_Net()
```

Program Listing/Output 20.17: Member Function which displays data implicitly and explicitly

❖ What it is showing is how individual data members can accessed using the this pointer

- ➤ The command
    ```
    this -> dataMemberName
    ```
    allows explicit access to the data member
- ➤ In addition, for demonstration purposes, it has been shown that the following can also be used:
    ```
    (*this).dataMemberName
    ```

❖ The full program has been shown below:

```cpp
#include <iostream>
#include <string>
using namespace std;

#ifndef EMPLOYEE_H
#define EMPLOYEE_H


const double TAXRATE_THRESHOLD_LOWER = 20000.0;
const double TAXRATE_THRESHOLD_INTERMEDIATE = 25000.0;


const double TAXRATE_RATE_LOWER = 0.3;
const double TAXRATE_RATE_INTERMEDIATE = 0.4;
const double TAXRATE_RATE_HIGHER = 0.5;


// Class DECLARATION
class Employee
{
      // DATA MEMBERS
      private:
            string name;
            string nI_Number;
            double salary_Gross;
            double tax_Rate;
            double tax_Amount;
            double salary_Net;

            // ---------- STATIC Data Member(s) ------------
            static int num_Employees;

  // MEMBER FUNCTIONS
  public:
        // ---------- CONSTRUCTOR(S) ------------
        Employee( string );
        Employee( string, string );
        Employee(    string name_IN,
                     string nI_Number,
                     double salary_Gross_IN );

        // ------- COPY Constructor ---------
        Employee( Employee& ref_Employee_IN );

        // ---------- DESTRUCTOR ------------
        ~Employee(  );

        // ---------- MUTATORS ------------
        double get_salary_Gross(  ) const;
        void set_salary_Gross( double salary_Gross_IN );

        // ---------- STATIC MEMBER FUNCTION(S) ------------
```

```
        // ...also a Mutator
        static int get_num_Employees();

        // ---------- "NORMAL Member Functions " ------------
        void greet();
        void greet_Name();
        void greet_Name_NINumber();
        void greet_Name_SalaryGross();
        void greet_AllAttributes();
        void display_salary_Net();


};//END...Class Declaration Employee

#endif
```

Program Listing/Output 20.18: Employee.h

```
#include "Employee.h"

// ---------- STATIC Data Member(s) ------------
int Employee::num_Employees = 0;


// ---------- CONSTRUCTOR(S) ------------
Employee::Employee( string name_IN )
 {
      name = name_IN;

     // Keeping track of the number of objects
     ++num_Employees;

 }//END...Constructor Employee( string name_IN )


 Employee::Employee( string name_IN, string nI_Number_IN )
 {
      name = name_IN;
      nI_Number = nI_Number_IN;

     // Keeping track of the number of objects
     ++num_Employees;

 }//END...Constructor Employee( string name_IN )

 Employee::Employee( string name_IN,
                     string nI_Number_IN,
                     double salary_Gross_IN )
 {
      name = name_IN;
      nI_Number = nI_Number_IN;
```

```
        salary_Gross = salary_Gross_IN;
        //---------------------------------------------------
        //     Calculating the Tax Rate
        //
        if( salary_Gross < TAXRATE_THRESHOLD_LOWER )
        {
                tax_Rate = TAXRATE_RATE_LOWER;
        }
        else
        {
                if( salary_Gross < TAXRATE_THRESHOLD_INTERMEDIATE )
                {
                        tax_Rate = TAXRATE_RATE_INTERMEDIATE;
                }
                else
                {
                        tax_Rate = TAXRATE_RATE_HIGHER;
                }
    }

    //---------------------------------------------------
    //    Calculating the Tax Amount & Net Salary
    //
    tax_Amount = tax_Rate * salary_Gross;
    salary_Net = salary_Gross - tax_Amount;


    // Incrementing the Number of Objects/Employees

    // Keeping track of the number of objects
    ++num_Employees;

}/* END...Constructor Employee( string name_IN,
                                string nI_Num_IN,
                                double salary_Gross_IN )
 */

// ------- COPY Constructor ---------
Employee::Employee( Employee& ref_Employee_IN )
{
        name = ref_Employee_IN.name;
        nI_Number = ref_Employee_IN.nI_Number;
        salary_Gross = ref_Employee_IN.salary_Gross;
        tax_Rate = ref_Employee_IN.tax_Rate;
        tax_Amount = ref_Employee_IN.tax_Amount;
        salary_Net = ref_Employee_IN.salary_Net;

        cout << "Copy Constructor called for "<< name<< endl;

}// END...COPY Constructor

// ---------- DESTRUCTOR ------------
Employee::~Employee(   )
```

```
{
        cout << "Goodbye, I, "<< name
             << " am  going out of scope now."
             << endl;

        // Keeping track of the number of objects
        --num_Employees;

}// END...Destructor ~Employee

// ---------- MUTATORS ------------
// const keyword is required here as well
double Employee::get_salary_Gross( ) const
{
        // tax_Rate = 0.6;
            // ILLEGAL - const member function cannot
            //                     modify any data members

        return salary_Gross;

}//END...Mutator get_salary_Gross


void Employee::set_salary_Gross( double salary_Gross_IN )
{
        salary_Gross = salary_Gross_IN;

        //----------------------------------------------------
        //    Calculating the NEW Tax Rate
        //
        if( salary_Gross < TAXRATE_THRESHOLD_LOWER )
        {
              tax_Rate = TAXRATE_RATE_LOWER;
        }
        else
        {
              if( salary_Gross < TAXRATE_THRESHOLD_INTERMEDIATE )
              {
                    tax_Rate = TAXRATE_RATE_INTERMEDIATE;
              }
              else
              {
                    tax_Rate = TAXRATE_RATE_HIGHER;
              }
    }

    //----------------------------------------------------
    //    Calculating the NEW Tax Amount & Net Salary
    //
    tax_Amount = tax_Rate * salary_Gross;
    salary_Net = salary_Gross - tax_Amount;

}//END...Mutator set_salary_Gross
```

```
// ---------- STATIC MEMBER FUNCTION(S) ------------
// ...also a Mutator
int Employee::get_num_Employees()
{
      return num_Employees;

}//END...Mutator get_num_Employees


// ---------- "NORMAL Member Functions " ------------
void Employee::greet()
{
      cout << "Hello, I'm an Employee"<< endl;

}//END...Member Function greet

void Employee::greet_Name()
{
      cout << "Hello, I'm an Employee, "
           << "my name is "<< name
           << endl;

}//END...Member Function greet_Name

void Employee::greet_Name_NINumber()
{
      cout << "Hello, I'm an Employee, "
           << "my name is "<< name
           << " and my NI Number is "<< nI_Number
           << endl;

}//END...Member Function greet_Name_NINumber


void Employee::greet_Name_SalaryGross()
{
     cout << "Hello, I'm an Employee, my name is "<< name
          << ", my Gross Salary is "<< salary_Gross
          << endl;

}//END...Member Function greet_Name_SalaryGross()


void Employee::greet_AllAttributes()
{
      cout << endl
           << "====================="
           << endl
           << "Hello, I am an employee, my name is "<< name<< ", "
           << endl
           << "my NI Number is "<< nI_Number<< ","
```

```
                << endl
                << "my Gross Salary is "<< salary_Gross
                << endl
                << "my tax Rate is "<< tax_Rate<< "%"
                << endl
                << "so the amount I have to pay in tax is "<< tax_Amount
                << endl
                << "thus my Net Salary is "<< salary_Net
                << endl
                << "====================="<< endl;
}// END...MF greet_AllAttributes()


void Employee::display_salary_Net()
{
        cout << "Accessing the Data Member salary_Net of :"
             << name<< endl;

        cout << "-\tImplicitly\t              salary_Net = "
             << salary_Net<< endl;

        cout << "-\tExplicitly\t              this -> salary_Net = "
             << this -> salary_Net<< endl;

        cout << "-\tExplicitly\t              (*this).salary_Net = "
             << (*this).salary_Net<< endl;

}// END...MF display_salary_Net()
```

Program Listing/Output 20.19: Employee.cpp

```
/*  ******************************************************************
    THIS POINTER DEMO
    ******************************************************************
    -      This program demonstrates using the this pointer
    -      The value of a data member is displayed in 3 ways
           -         Implicitly
           -         using this, and the -> operator
           -         using this, and the dot . operator
*/
#include <iostream>
#include <string>
#include <list>

using namespace std;

#include "Employee.h"


int main( )
{
        //»»»»»»»»»»»»»»»»»»»»»»»»»»»»»»»»»»»»»»»»»»»»»»»»»»»»»»»»»»»»»»»»»»»
```

```
        Employee amandaTaylor( "Amanda Taylor", "RG 35 23 56 T",
        27000.0 );

        //»»»»»»»»»»»»»»»»»»»»»»»»»»»»»»»»»»»»»»»»»»»»»»»»»»»»»»»»»»»»»»»
        //    Using the THIS pointer to access Data Members
        //
        amandaTaylor.display_salary_Net();


        //»»»»»»»»»»»»»»»»»»»»»»»»»»»»»»»»»»»»»»»»»»»»»»»»»»»»»»»»»»»»»»»
        cout  << endl<< endl<< "==========="
              << endl
              << "END...Demo C++ This Pointer"
              << endl;
        while(true){}
        return 0;

}//END...Function MAIN
```

```
Accessing the Data Member salary_Net of :Amanda Taylor
-       Implicitly                      salary_Net = 13500
-       Explicitly                      this -> salary_Net = 13500
-       Explicitly                      (*this).salary_Net = 13500

===========
END...Demo C++ This Pointer
```

Program Listing/Output 20.20: This_Pointer_Demo_Client.cpp

20.8 ADDRESS-OF OPERATOR

❖ The Address-Of operator is used when equating a pointer to the address of an object e.g.

```
Employee *ptr_FredJohnson = &fredJohnson;
```

❖ It is to similar to the assignment operator in that it also controls behaviour when used with the = symbol:

> The assignment operator is invoked when equating two existing objects

> The Address-of operator is invoked when equating a pointer to the address of an object

❖ The default behaviour is to return the `this` pointer

20.9 INNER CLASSES

❖ Inner classes are classes which are implemented inside other entities

➢ The most popular example is the "Nested Class," - a class within a class

➢ There is also the "Local Class", a class within a function

❖ They are of fundamental importance in object oriented programming

➢ New levels of organisation and management can be implemented

❖ To a first approximation,

➢ Inner classes allow the same kind of organisation to data members and member functions

➢ Which directories/folders allow for files in IT

❖ A new level of scope is also established

❖ Hierarchies can, and often are implemented

➢ i.e. It is quite common to have inner classes containing inner classes, which in turn contain inner classes...

20.9.1 NESTED CLASSES

20.9.1.1 What are Nested Classes

❖ A nested class is a class within a class

❖ It provides an often invaluable means of creating further object oriented management within a class, including:

➢ Further organisation

• A hierarchy of class organisation can be created

• This is similar to the practice of having directories within directories in IT

➢ Further scope of variable

20.9.1.2 Rules of Nested Classes

❖ There are a few things to bear in mind when creating nested classes:

❖ There are no special access features:

➢ An inner nested class does not have access to the private members of the outer class[1]

---

[1]Note that this is not the case in other OOP Languages such as Java. There, an Inner Class has access to private members of the Outer Class.

> Likewise, the outer class does not have access to private members of the inner class

❖ In order to organise our program, we need the inner class to work with data from the outer class

> A common practice, is to create
  • A pointer to the outer class...
  • ...as a data member of the inner class

20.9.1.3 Creating and Using Nested Classes

❖ In our Employee example, it has been seen so far that the set of tax calculations, calculating the

> Tax Rate
> Tax Amount
> Net Salary

have been performed twice so far, when the Gross Salary has been specified

> Once when initialising it, in the constructor
> Once when modifying it in the mutator

❖ It will be far better to have an Inner Class where these are implemented; advantages include:

> These calculations will almost certainly be needed in future, this inner class can be accessed any time
> The structure of the program will be much more organised
> When these calculations have to be changed, they can be changed just once in the Inner class, thus preventing any errors due to disorganisation

❖ Thus the Nested Class `TaxCalculations` will be created

> The declaration will be as follows:

```
class Employee
{
    public:
        // ---------- INNER CLASS ------------
        class TaxCalculations
        {
```

```
                    private:
                           Employee *ptr_Employee;

                    public:
                           TaxCalculations( Employee * );

                           void calc_tax_Rate();
                           void calc_tax_Amount();
                           void calc_salary_Net();

    };//END...Inner Class Declaration: TaxCalculations

//Rest of Members: CLASS Employee


};//END...Class Declaration Employee
```

❖ An essential part of this Inner Class is that it

➢ Has a Pointer-to-Employee as a data member

➢ Has a Constructor which initialises it

  • The declaration is as above

```
        TaxCalculations( Employee * );
```

  • The definition is as follows

```
Employee::TaxCalculations::TaxCalculations( Employee
                                         *ptr_Employee_IN
                                         )
{
      ptr_Employee = ptr_Employee_IN;

}//END...Constructor, Inner Class TaxCalculations
```

❖ When using this Inner Class, the outer class will pass the `this` pointer

➢ This fulfils the requirements of having a pointer which points to the current object in question

➢ It is through this pointer that the Inner Class will access the data members of the Outer Class

❖ Thus in order to use the Inner class

➢ First an object of its type must be created:

```
TaxCalculations tax_Results( this );
```

➢ Now the member functions can be invoked, just like any outer class

```
        tax_Results.calc_tax_Rate();
        tax_Results.calc_tax_Amount();
        tax_Results.calc_salary_Net();
```

❖ The full program is as follows:

```
#include <iostream>
#include <string>
using namespace std;

#ifndef EMPLOYEE_H
#define EMPLOYEE_H

const double TAXRATE_THRESHOLD_LOWER = 20000.0;
const double TAXRATE_THRESHOLD_INTERMEDIATE = 25000.0;

const double TAXRATE_RATE_LOWER = 0.3;
const double TAXRATE_RATE_INTERMEDIATE = 0.4;
const double TAXRATE_RATE_HIGHER = 0.5;

// Class DECLARATION
class Employee
{
    public:
            // ---------- INNER CLASS ------------
            class TaxCalculations
            {
                private:
                    Employee *ptr_Employee;

                public:
                    TaxCalculations( Employee * );

                    void calc_tax_Rate();
                    void calc_tax_Amount();
                    void calc_salary_Net();

            };//END...Inner Class Declaration: TaxCalculations

    // DATA MEMBERS
    private:
            string name;
            string nI_Number;
            double salary_Gross;
            double tax_Rate;
            double tax_Amount;
            double salary_Net;

            // ---------- STATIC Data Member(s) ------------
            static int num_Employees;

            // MEMBER FUNCTIONS
```

```
     public:
             // ---------- CONSTRUCTOR(S) ------------
             Employee( string );
             Employee( string, string );
             Employee(   string name_IN,
                         string nI_Number,
                         double salary_Gross_IN );

             // ------- COPY Constructor ---------
             Employee( Employee& ref_Employee_IN );

             // ---------- DESTRUCTOR ------------
             ~Employee(  );

             // ---------- MUTATORS ------------
             double get_salary_Gross(  ) const;
             void set_salary_Gross( double salary_Gross_IN );

             // ---------- STATIC MEMBER FUNCTION(S) ------------
             // ...also a Mutator
             static int get_num_Employees();

             // ---------- "NORMAL Member Functions " ------------
             void greet();
             void greet_Name();
             void greet_Name_NINumber();
             void greet_Name_SalaryGross();
             void greet_AllAttributes();
             void display_salary_Net();


};//END...Class Declaration Employee

#endif
```

Program Listing/Output 20.21: Employee.h

```
#include "Employee.h"

// ---------- STATIC Data Member(s) ------------
int Employee::num_Employees = 0;


// ---------- CONSTRUCTOR(S) ------------
Employee::Employee( string name_IN )
{
     name = name_IN;

     // Keeping track of the number of objects
     ++num_Employees;

}//END...Constructor Employee( string name_IN )
```

```cpp
Employee::Employee( string name_IN, string nI_Number_IN )
{
      name = name_IN;
      nI_Number = nI_Number_IN;

      // Keeping track of the number of objects
      ++num_Employees;

}//END...Constructor Employee( string name_IN )

Employee::Employee( string name_IN,
                    string nI_Number_IN,
                    double salary_Gross_IN )
{
      name = name_IN;
      nI_Number = nI_Number_IN;
      salary_Gross = salary_Gross_IN;

      //---------------------------------------------------
      //    Using the Inner Class TaxCalculations
      //    -       Object required, initialised with this pointer
      //    -           This holds the data of the current object
      //
      TaxCalculations tax_Results( this );

      tax_Results.calc_tax_Rate();
      tax_Results.calc_tax_Amount();
      tax_Results.calc_salary_Net();


      // Keeping track of the number of objects
      ++num_Employees;

}/* END...Constructor Employee( string name_IN,
                                string nI_Num_IN,
                                double salary_Gross_IN )
  */

// ------- COPY Constructor ---------
Employee::Employee( Employee & ref_Employee_IN )
{
      name = ref_Employee_IN.name;
      nI_Number = ref_Employee_IN.nI_Number;
      salary_Gross = ref_Employee_IN.salary_Gross;
      tax_Rate = ref_Employee_IN.tax_Rate;
      tax_Amount = ref_Employee_IN.tax_Amount;
      salary_Net = ref_Employee_IN.salary_Net;

      cout  << "Copy Constructor called for "<< name<< endl;

}// END...COPY Constructor

// ---------- DESTRUCTOR ------------
```

```
Employee::~Employee( )
{
      cout << "Goodbye, I, "<< name
           << " am  going out of scope now."
           << endl;

      // Keeping track of the number of objects
      --num_Employees;

}// END...Destructor ~Employee

// ---------- MUTATORS ------------
// const keyword is required here as well
double Employee::get_salary_Gross( ) const
{
      // tax_Rate = 0.6;
            // ILLEGAL - const member function cannot
            //                   modify any data members

      return salary_Gross;

}//END...Mutator get_salary_Gross


void Employee::set_salary_Gross( double salary_Gross_IN )
{
      salary_Gross = salary_Gross_IN;

      //----------------------------------------------------
      //      Calculating the NEW Tax Rate...
      //         ...Using the Inner Class TaxCalculations
      //         -    Object required, initialised with this pointer
      //         -        This holds the data of the current object
      //
      TaxCalculations tax_Results( this );

      tax_Results.calc_tax_Rate();
      tax_Results.calc_tax_Amount();
      tax_Results.calc_salary_Net();

}//END...Mutator set_salary_Gross


// ---------- STATIC MEMBER FUNCTION(S) ------------
// ...also a Mutator
int Employee::get_num_Employees()
{
      return num_Employees;

}//END...Mutator get_num_Employees


// ---------- "NORMAL Member Functions " ------------
```

```
void Employee::greet()
{
     cout   << "Hello, I'm an Employee"<< endl;

}//END...Member Function greet

void Employee::greet_Name()
{
     cout   << "Hello, I'm an Employee, "
            << "my name is "<< name
            << endl;

}//END...Member Function greet_Name

void Employee::greet_Name_NINumber()
{
     cout   << "Hello, I'm an Employee, "
            << "my name is "<< name
            << " and my NI Number is "<< nI_Number
            << endl;

}//END...Member Function greet_Name_NINumber


void Employee::greet_Name_SalaryGross()
{
     cout   << "Hello, I'm an Employee, my name is "<< name
            << ", my Gross Salary is "<< salary_Gross
            << endl;

}//END...Member Function greet_Name_SalaryGross()


void Employee::greet_AllAttributes()
{
     cout   << endl
            << "====================="
            << endl
            << "Hello, I am an employee, my name is "<< name<< ", "
            << endl
            << "my NI Number is "<< nI_Number<< ","
            << endl
            << "my Gross Salary is "<< salary_Gross
            << endl
            << "my tax Rate is "<< tax_Rate<< "%"
            << endl
            << "so the amount I have to pay in tax is "<< tax_Amount
            << endl
            << "thus my Net Salary is "<< salary_Net
            << endl
            << "====================="<< endl;
}// END...MF greet_AllAttributes()
```

```
void Employee::display_salary_Net()
{
     cout   << "Accessing the Data Member salary_Net of :"
            << name<< endl;

     cout   << "-\tImplicitly\t          salary_Net = "
            << salary_Net<< endl;

     cout   << "-\tExplicitly\t            this -> salary_Net = "
            << this -> salary_Net<< endl;

     cout   << "-\tExplicitly\t            (*this).salary_Net = "
            << (*this).salary_Net<< endl;

}// END...MF display_salary_Net()


//#####################################################################

/*    INNER CLASS: TaxCalculations
                   void calc_tax_Rate();
                   void calc_tax_Amount();
                   void calc_salary_Net();
*/
Employee::TaxCalculations::TaxCalculations( Employee *ptr_Employee_IN )
{
      ptr_Employee = ptr_Employee_IN;

}//END...Constructor, Inner Class TaxCalculations


void Employee::TaxCalculations::calc_tax_Rate()
{
     ptr_Employee -> salary_Gross;

     if( ptr_Employee-> salary_Gross < TAXRATE_THRESHOLD_LOWER )
     {
          ptr_Employee -> tax_Rate = TAXRATE_RATE_LOWER;
     }
     else
     {
          if(ptr_Employee->salary_Gross <TAXRATE_THRESHOLD_INTERMEDIATE)
          {
                 ptr_Employee -> tax_Rate = TAXRATE_RATE_INTERMEDIATE;
          }
          else
          {
                 ptr_Employee -> tax_Rate = TAXRATE_RATE_HIGHER;
          }
     }
}//END...MF calc_tax_Rate
```

```
void Employee::TaxCalculations::calc_tax_Amount()
{
     ptr_Employee -> tax_Amount = ptr_Employee -> salary_Gross *
                                  ptr_Employee -> tax_Rate;


}//END...MF calc_tax_Amount


void Employee::TaxCalculations::calc_salary_Net()
{
      ptr_Employee -> salary_Net =  ptr_Employee -> salary_Gross -
                                    ptr_Employee -> tax_Amount;


}//END...MF calc_tax_Amount

//END...INNER CLASS TaxCalculations
    ################################################
```

Program Listing/Output 20.22: Employee.cpp

```
/*    **************************************************************
      INNER (NESTED) CLASS
      **************************************************************
*/
#include <iostream>
#include <string>
using namespace std;

#include "Employee.h"

int main()
{
    //»»»»»»»»»»»»»»»»»»»»»»»»»»»»»»»»»»»»»»»»»»»»»»»»»»»»»»»»»
    // Creating OBJECTS


    Employee joeBloggs( "Joseph Bloggs", "PH 45 67 34 D", 18000.0 );
    joeBloggs.greet_AllAttributes();

    cout  << "Joe Bloggs is getting promoted..."<< endl;

    joeBloggs.set_salary_Gross( 22000.0 );
    joeBloggs.greet_AllAttributes();


    //»»»»»»»»»»»»»»»»»»»»»»»»»»»»»»»»»»»»»»»»»»»»»»»»»»»»»»»»»
    cout << endl<< "================"<< endl
         << "END...C++ Demo: Inner Classes "
         << endl;
    while(true){}
    return 0;
```

```
}//END...Function MAIN


======================
Hello, I am an employee, my name is Joseph Bloggs,
my NI Number is PH 45 67 34 D,
my Gross Salary is 18000
my tax Rate is 0.3%
so the amount I have to pay in tax is 5400
thus my Net Salary is 12600
======================
Joe Bloggs is getting promoted...


======================
Hello, I am an employee, my name is Joseph Bloggs,
my NI Number is PH 45 67 34 D,
my Gross Salary is 22000
my tax Rate is 0.4%
so the amount I have to pay in tax is 8800
thus my Net Salary is 13200
======================


================
END...C++ Demo: Inner Classes
```

Program Listing/Output 20.23: InnerClass_Demo_Client.cpp

❖ Notice the improvement in organisation of this program, with the Inner class

❖ Whenever two or more parts of your class are doing the same thing, like the tax calculations in this case, that is the cue an Inner Class may be called for

20.9.2 LOCAL CLASSES

❖ A local class is a class defined and instantiated within a function

➢ Just like a local variable is a variable within a function

➢ A local class is a class within a function

❖ Nested functions, the practice of having a function within a function,

➢ Was something supported by some C compilers, such as GCC

➢ Are not supported in C++

❖ Bjarne Stroustrup states

➢ "Most often, the use of a local class is a sign that a function is too large

➤ Beware that a local class cannot be a valid template argument "

❖ **Rule of Thumb:** Avoid Local Classes whenever possible; use the management enabled from class structures

20.10 EXPLICIT CONSTRUCTORS

❖ A constructor can be forced to be `explicit`, meaning it will work only as stated, no flexibility is allowed

❖ This is usually advisable, as without them, Implicit Conversions can take place

➤ These are usually subtle, with no indication from the Compiler

➤ All too often, the only indication is runtime errors...

❖ This section

➤ First looks at the potential problems of Implicit conversion

➤ It then shows how it can be fixed by using an explicit Constructor

20.10.1 IMPLICIT CONVERSIONS

❖ We have our Employee Class as before,

➤ This time with a default Constructor hard coded

❖ We have a new class Office

➤ This has a

• Data Member `personInCharge`, of type Employee
• A Constructor which sets this Data Member

❖ In our Client program,

➤ We have a Function `display_OfficeName`

• This takes in an **Office** object as a formal parameter

➤ The main function calls this, but passes an **Employee** object as the actual parameter

❖ The result, often surprising, is that

➤ The compiler ALLOWS it

• Reason: The Office Class has a Constructor which takes in an Employee object

- It is attempting an IMPLICIT Conversion

➢ We have a runtime error

- The function `display_OfficeName` is unable to perform its functionality

```cpp
#include <iostream>
#include <string>
using namespace std;

#ifndef EMPLOYEE_H
#define EMPLOYEE_H


const double TAXRATE_THRESHOLD_LOWER = 20000.0;
const double TAXRATE_THRESHOLD_INTERMEDIATE = 25000.0;

const double TAXRATE_RATE_LOWER = 0.3;
const double TAXRATE_RATE_INTERMEDIATE = 0.4;
const double TAXRATE_RATE_HIGHER = 0.5;


// Class DECLARATION
class Employee
{
        // DATA MEMBERS
        private:
                string name;
                string nI_Number;
                double salary_Gross;
                double tax_Rate;
                double tax_Amount;
                double salary_Net;

        // MEMBERS FUNCTIONS
        public:
                // ---------- CONSTRUCTOR(S) ------------
                Employee();
                Employee( string );
                Employee( string, string );
                Employee( string, string, double );

                // ---------- "NORMAL" Member Funcrtions ------------
                void greet();
                void greet_Name();
                void greet_Name_NINumber();
                void greet_Name_SalaryGross();
                void greet_AllAttributes();

};//END...Class Declaration Employee
```

```
#endif
```

```
#include "Employee.h"

// ---------- CONSTRUCTOR(S) ------------
Employee::Employee(  )
{
}//END...Constructor Employee( ) - Default

Employee::Employee( string name_IN )
{
      name = name_IN;

}//END...Constructor Employee( string name_IN )


Employee::Employee( string name_IN, string nI_Number_IN )
{
      name = name_IN;
      nI_Number = nI_Number_IN;

}//END...Constructor Employee( string name_IN )

Employee::Employee( string name_IN,
                    string nI_Number_IN,
                    double salary_Gross_IN )
{
      name = name_IN;
      nI_Number = nI_Number_IN;
      salary_Gross = salary_Gross_IN;

      //----------------------------------------------------
      //      Calculating the Tax Rate
      //
      if( salary_Gross < TAXRATE_THRESHOLD_LOWER )
      {
            tax_Rate = TAXRATE_RATE_LOWER;
      }
      else
      {
            if( salary_Gross < TAXRATE_THRESHOLD_INTERMEDIATE )
            {
                  tax_Rate = TAXRATE_RATE_INTERMEDIATE;
            }
            else
            {
                  tax_Rate = TAXRATE_RATE_HIGHER;
            }
      }
```

```
        //-------------------------------------------------------
        //     Calculating the Tax Amount & Net Salary
        //
        tax_Amount = tax_Rate * salary_Gross;
        salary_Net = salary_Gross - tax_Amount;


}/* END...Constructor Employee( string name_IN,
                                string nI_Num_IN,
                                double salary_Gross_IN )
 */
// ---------- "NORMAL Member Functions " ------------
void Employee::greet()
{
        cout  << "Hello, I'm an Employee"<< endl;

}//END...Member Function greet

void Employee::greet_Name()
{
        cout  << "Hello, I'm an Employee, "
              << "my name is "<< name
              << endl;

}//END...Member Function greet_Name

void Employee::greet_Name_NINumber()
{
        cout  << "Hello, I'm an Employee, "
              << "my name is "<< name
              << " and my NI Number is "<< nI_Number
              << endl;

}//END...Member Function greet_Name_NINumber


void Employee::greet_Name_SalaryGross()
{
        cout  << "Hello, I'm an Employee, my name is "<< name
              << ", my Gross Salary is "<< salary_Gross
              << endl;

}//END...Member Function greet_Name_SalaryGross()


void Employee::greet_AllAttributes()
{
        cout  << endl
              << "====================="
              << endl
              << "Hello, I am an employee, my name is "<< name<< ", "
              << endl
```

```
                        << "my NI Number is "<< nI_Number<< ","
                        << endl
                        << "my Gross Salary is "<< salary_Gross
                        << endl
                        << "my tax Rate is "<< tax_Rate<< "%"
                        << endl
                        << "so the amount I have to pay in tax is "<< tax_Amount
                        << endl
                        << "thus my Net Salary is "<< salary_Net
                        << endl
                        << "===================="<< endl;
}// END...MF greet_AllAttributes()
```

Program Listing/Output 20.25: Employee.cpp

```cpp
#include <iostream>
#include <string>
using namespace std;

#ifndef OFFICE_H
#define OFFICE_H

#include "Employee.h"


class Office
{
        private:
                string name;
                string location;
                Employee personInCharge;

        public:
                // ---------- CONSTRUCTOR(S) -----------
                explicit Office( Employee );
                Office( string );
                Office( string, string );

                string get_name();
                string get_location();
                void set_name( string );
                void set_location( string );

};//END...Class Office

#endif
```

Program Listing/Output 20.26: Office.h

```cpp
#include "Office.h"

// Explicit Constructor
```

```
// - Cannot have the keyword explicit here again
Office::Office( Employee personInCharge_IN )
{
      personInCharge = personInCharge_IN;

}//END...Constructor Office( Employee  )

Office::Office( string name_IN )
{
      name = name_IN;

}//END...Constructor Office( string )

 Office::Office( string name_IN, string location_IN )
{
      name = name_IN;
      location = location_IN;

}//END...Constructor Office( Employee  )

string Office::get_name()
{
      return name;

}//END...Mutator get_name

string Office::get_location()
{
      return location;

}//END...Mutator get_location

void Office::set_name( string name_IN )
{
      name = name_IN;

}//END...Mutator set_name

void Office::set_location( string location_IN )
{
      location = location_IN;

}//END...Mutator set_location
```

Program Listing/Output 20.27: Office.cpp

```
/*    ************************************************************
      EXPLICIT CONSTRUCTORS - PROBLEMS, Lack of Them
      ************************************************************
      -    This program demonstrates
           -    Implicit Conversions which might take place without
                -    The Programmer's (i.e. your!:-)) intention
```

```
                        -        Any Compiler feedback
*/
#include <iostream>
#include <string>
using namespace std;

#include "Employee.h"
#include "Office.h"

void display_OfficeName( Office );

int main( )
{
        //»»»»»»»»»»»»»»»»»»»»»»»»»»»»»»»»»»»»»»»»»»»»»»»»»»»»»»»»»»»»»»»»»
        Employee joeBloggs("Joseph Bloggs" );

        //      Calling using an Employee object, not Office
        //      -          Compiler allows it since Office has a
    Constructor
        //                 which takes an Employee input
        //
        display_OfficeName( joeBloggs );

        //»»»»»»»»»»»»»»»»»»»»»»»»»»»»»»»»»»»»»»»»»»»»»»»»»»»»»»»»»»»»»»»»»
    cout   << endl<< endl<< "==========="
           << endl
           << "END...C++ Demo: IMPLICIT Conversions"
           << " "
           << endl;
    while(true){}
    return 0;

}//END...Function MAIN

void display_OfficeName( Office office_IN )
{
        cout   << "The Name of the Office: "
               << office_IN.get_name()<< endl;

}//END...Function display_OfficeLocation
```

```
The Name of the Office:

===========
END...C++ Demo: IMPLICIT Conversions
```

Program Listing/Output 20.28: ImplicitConversions_Client.cpp

20.10.2 PREVENTING IMPLICIT CONVERSIONS USING EXPLICIT CONSTRUCTORS

❖ The remedy to this potential problem is very simple: place the keyword explicit in front of the Constructor in its Declaration

  ➢ Note that when you define this separately, the compiler will not allow the keyword to be duplicated

❖ In our program, if we change the declaration of the Constructor which takes the Employee object as its formal parameter to

```
explicit Office( Employee );
```

  ➢ Then attempting to run the same program will produce the following compiler error

```
"No user-defined-conversion operator available that can
perform this conversion, or the operator cannot be
called"
```

❖ Whether or not all Constructors should be made explicit is down to the individual programmer

  ➢ In some cases, when classes are very simple, Implicit conversion is desirable

  ➢ In other cases, when the results would be erroneous, making Constructors explicit is a good idea

20.11 WHAT A CLASS CREATES BEHIND THE SCENES

❖ Bringing together previous sections, when you define a class, the following member functions have been "silently" created for you

  ➢ Default Constructor
  ➢ Default Copy Constructor
  ➢ Default Destructor
  ➢ Assignment Operator
  ➢ Address-of Operator

❖ This does indeed mean that if you just type the following:

```
class MyClass{};
```

all the above enitites are created

20.12 FRIENDSHIP

❖ As we have been discussing, the `private` access specifier is a means of implementing confidentiality

❖ However, in the real world, confidentiality usually does not mean 100% secrecy

➤ Certain authorised people have privileged access

• Your accountant has privileged access to your financial data
• Your doctor has privileged access to your medical data

❖ Friendship in C++ is a means of implementing privileged access to private and protected members of a class

➤ The class must specify these entities in its declaration
➤ The keyword `friend` is used

20.12.1 FRIEND FUNCTIONS

❖ A friend function is a function which has access to all private and protected elements of a class, as well as the public

➤ In order to implement this privileged access, the class must declare this function as a friend

❖ The example used here will simulate a Blood Pressure Monitor

➤ It will comprise

• A class "Patient" which will have blood pressure as a private data member
• The function `check_IsBloodPressure_OK` which will analyse the blood pressure data of the Patient
• It will have access to this private data since the Patient class will declare it a Friend

❖ The program is as follows

```
#include <iostream>
#include <string>
using namespace std;

#ifndef PATIENT_H
#define PATIENT_H

class Patient
```

```
{
      private:
            string name;
            int bloodPressure;


      public:
            Patient( string name_IN, double bloodPressure_IN  );
            void greet();

      friend bool check_IsBloodPressure_OK( Patient myPatient_IN );

};


#endif
```

Program Listing/Output 20.29: Patient.h

```
#include "Patient.h"

Patient::Patient( string name_IN, double bloodPressure_IN )
{
      name = name_IN;
      bloodPressure = bloodPressure_IN;
}

void Patient::greet()
{
      cout << "Hello, I am a Patient"<< endl
            << "name:\t\t\t"<< name
            << ", Blood Pressure:\t"<< bloodPressure<< endl;
}
```

Program Listing/Output 20.30: Patient.cpp

```
#include <iostream>
#include <string>
using namespace std;

#include "Patient.h"

bool check_IsBloodPressure_OK( Patient myPatient_IN );


int main()
{

      //»»»»»»»»»»»»»»»»»»»»»»»»»»»»»»»»»»»»»»»»»»»»»»»»»»»»»»»»»»»»»»»»»
      Patient abigailMitchellmore( "Abigail Mitchellmore", 115 );
      Patient jonathanOLeary( "Jonathan O'Leary", 90 );
```

```
        Patient fredJohnson( "Fred Johnson", 125 );

        //»»»»»»»»»»»»»»»»»»»»»»»»»»»»»»»»»»»»»»»»»»»»»»»»»»»»»»»»»»»»»»
        check_IsBloodPressure_OK( abigailMitchellmore );
        check_IsBloodPressure_OK( jonathanOLeary );
        check_IsBloodPressure_OK( fredJohnson );

        //»»»»»»»»»»»»»»»»»»»»»»»»»»»»»»»»»»»»»»»»»»»»»»»»»»»»»»»»»»»»»»
        cout << endl<< "================"<< endl
             << "END...C++ Demo: Friend Functions "
             << endl;
        while(true){}
        return 0;

}//END...Function MAIN

/*      Function check_IsBloodPressure_OK
          -       Friend of Class Patient
*/
bool check_IsBloodPressure_OK( Patient myPatient_IN )
{

     if( ( myPatient_IN.bloodPressure > 95 ) &&
           ( myPatient_IN.bloodPressure < 120 ))
     {
                cout << endl<< "The Blood Pressure of "
                     << myPatient_IN.name<< " is OK"
                     << endl;
                return true;
     }

     cout << endl<< "The Blood Pressure of "
          << myPatient_IN.name<< " is not OK"
          << endl;
     return false;

}//END...Function check_IsBloodPressure_OK
```

```
The Blood Pressure of Abigail Mitchellmore is OK

The Blood Pressure of Jonathan O'Leary is not OK

The Blood Pressure of Fred Johnson is not OK

================
END...C++ Demo: Friend Functions
```

Program Listing/Output 20.31: FriendFunctionsDemo_Client.cpp

20.12.2 FRIEND CLASSES

❖ A friend class is a class which has access to all private and protected elements of another class, as well as the public

➢ In order to implement this privileged access, the latter class must declare the former class as a friend

❖ Our example will once again use the Employee class (simple version)

➢ It will now declare a new class "Accountant" as a friend class

❖ The program is as follows:

```cpp
#include <iostream>
#include <string>
using namespace std;

#ifndef EMPLOYEE_H
#define EMPLOYEE_H

// Class DECLARATION
class Employee
{
    // Data Members
    private:
            string name;
            string nI_Number;
            double salary_Gross;
            double tax_Rate;
            double tax_Amount;
            double salary_Net;

            // ---------- STATIC Data Member(s) -----------
            static int num_Employees;


    public:
            // ---------- CONSTRUCTORS ------------
            Employee( string name_IN );

            Employee( string name_IN,
                        string nI_Number );

            Employee( string name_IN,
                        string nI_Number,
                        double salary_Gross_IN );

            // ---------- DESTRUCTOR ------------
            ~Employee( );

            // ---------- MUTATORS ------------
```

```
              double get_salary_Gross(  ) const;
              void set_salary_Gross( double salary_Gross_IN );

              // ---------- STATIC MEMBER FUNCTION(S) ------------
              // ...also a Mutator
              static int get_num_Employees();


              //------- "Normal" Member Functions -------
              void greet();
              void greet_Name();
              void greet_Name_SalaryGross();

       friend class Accountant;

};

#endif
```

```
#include <iostream>
#include <string>
using namespace std;

#include "Employee.h"

#ifndef ACCOUNTANT_H
#define ACCOUNTANT_H

class Accountant
{
      private:
            string name;

      public:
            Accountant( string name_IN );
            void display_Employee_Salary( Employee employee_IN );
};
#endif
```

```
#include "Employee.h"

// ---------- STATIC Data Member(s) ------------
int Employee::num_Employees = 0;


// ---------- CONSTRUCTORS ------------
Employee::Employee( string name_IN )
{
        name = name_IN;
```

```
          ++num_Employees;


}//END...Constructor Employee( string )


Employee::Employee( string name_IN, string nI_Number_IN )
{
      name = name_IN;
      nI_Number = nI_Number_IN;

      ++num_Employees;

}//END...Constructor Employee( string, string )

Employee::Employee( string name_IN,
                              string nI_Number_IN,
                              double salary_Gross_IN )
{
     name = name_IN;
     nI_Number = nI_Number_IN;
     salary_Gross = salary_Gross_IN;

     ++num_Employees;

}//END...Constructor Employee( string, string, double )

// ---------- DESTRUCTOR ------------
Employee::~Employee(  )
{
       cout << "Goodbye, I, "<< name
            << " am going out of scope now."
            << endl;

}//END...Destructor ~Employee( )


// ---------- MUTATORS ------------
double Employee::get_salary_Gross(  ) const
{
      return salary_Gross;

}//END...Mutator get_salary_Gross( )


void Employee::set_salary_Gross( double salary_Gross_IN )
{
      salary_Gross = salary_Gross_IN;

}//END...Mutator set_salary_Gross( )
```

```
// ---------- STATIC MEMBER FUNCTION(S) ------------
// ...also a Mutator
int Employee::get_num_Employees()
{
      return num_Employees;

}//END...Static Mutator get_num_Employees( )


// Member Function DEFINITIONS:
void Employee::greet()
{
     cout  << endl
           << "Hello, I'm an Employee"<< endl;

}//END...Member Function greet( )


void Employee::greet_Name()
{
     cout << endl
           << "Hello, I'm an Employee, my name is "<< name
           << endl;

}//END...Member Function greet_Name( )

void Employee::greet_Name_SalaryGross()
{
    cout << endl
           << "Hello, I'm an Employee, my name is "<< name
           << ", my Gross Salary is "<< get_salary_Gross()
           << endl;

}//END...Member Function greet_Name_SalaryGross( )
```

Program Listing/Output 20.34: Employee.cpp

```
#include "Accountant.h"

Accountant::Accountant( string name_IN )
{
      name = name_IN;
}//END...Constructor Accountant( string name_IN )

void Accountant::display_Employee_Salary( Employee employee_IN )
{
      cout   << endl<< "==================="<< endl
             << "As an Accountant, I, "<< name
             << " am in a privileged position to report the Gross
    Salary of "
             << employee_IN.name << " as "<< employee_IN.salary_Gross
             << endl;
```

```
}//END...Member Function display_Employee_Salary
```

Program Listing/Output 20.35: Accountant.cpp

```cpp
#include <iostream>
#include <string>
using namespace std;

#include "Employee.h"
#include "Accountant.h"

int main()
{
    //»»»»»»»»»»»»»»»»»»»»»»»»»»»»»»»»»»»»»»»»»»»»»»»»»»»»»»»»»»»»»»»»»»»
    Employee joeBloggs( "Joseph Bloggs", "PH 45 67 34 D", 17000.0 );
    Employee johnSmith( "John Smith", "GH 67 34 67 D", 18000.0 );
    Employee amandaTaylor( "Amanda Taylor", "SG 36 78 24 F", 22000.0 );

    //»»»»»»»»»»»»»»»»»»»»»»»»»»»»»»»»»»»»»»»»»»»»»»»»»»»»»»»»»»»»»»»»»»»
    Accountant peterJohnson( "Peter Johnson" );
    peterJohnson.display_Employee_Salary( joeBloggs );
    peterJohnson.display_Employee_Salary( johnSmith );
    peterJohnson.display_Employee_Salary( amandaTaylor );

    //»»»»»»»»»»»»»»»»»»»»»»»»»»»»»»»»»»»»»»»»»»»»»»»»»»»»»»»»»»»»»»»»»»»
    cout << endl<< "==============="<< endl
         << "END...C++ Demo: Friend Classes "
         << endl;
    while(true){}
    return 0;
}//END...Function MAIN
```

```
==================
As an Accountant, I, Peter Johnson am in a priviledged position to
    report the Gross Salary of Joseph Bloggs as 17000
Goodbye, I, Joseph Bloggs am going out of scope now.
==================
As an Accountant, I, Peter Johnson am in a priviledged position to
    report the Gross Salary of John Smith as 18000
Goodbye, I, John Smith am going out of scope now.
==================
As an Accountant, I, Peter Johnson am in a priviledged position to
    report the Gross Salary of Amanda Taylor as 22000
Goodbye, I, Amanda Taylor am going out of scope now.
===============
END...C++ Demo: Friend Classes
```

Program Listing/Output 20.36: FriendClassDemo.cpp

20.12.3 PROPERTIES OF FRIENDSHIP

❖ Friendship is NOT REFLEXIVE

➢ If B is a Friend of A, that does not imply that A is a Friend of B
➢ Reflects the non-programming world
  • Your accountant knows your bank details,
  • But you don't know your accountant's bank details

❖ Friendship is NOT TRANSITIVE

➢ If
  • B is a Friend of A
  • C is a Friend of B

  That does not imply that C is a Friend of A
➢ Reflects the non-programming world
  • Your accountant knows your bank details,
  • Your accountant's doctor knows your acountant's medical details
  • But your accountant's doctor does not know anything confidential about you

20.12.4 PROSPECTIVE EVILS OF FRIENDSHIP

❖ What friendship is doing is exposing the private and protected members

➢ If this is done excessively,
  • Private data members become like global variables
    ○ Debugging, finding which friend function/member function of a friend class caused the problem, becomes like a treasure hunt i.e.
  • It undoes the encapsulation so carefully and painstakingly engineered into object oriented programming

❖ **Rule of Thumb:** Use friendship with caution, if at all

➢ Other object oriented programming languages such as Java, C# and Swift do not have friendship at all
➢ In order to access private data members, use Mutator Member functions whenever possible

20.12.5 REFINED DEFINITIONS OF ACCESS SPECIFIERS

❖ Having seen how the concept of friendship works, we can now refine the definitions of `private` and `public`

➤ `private`

Accessible only by

- Member functions of the same class
- Its friends
  - ○ Friend Functions
  - ○ Friend Classes

➤ `public`

Accessible by

- Anything

❖ These are still not full definitions

❖ Further refinements of these, along with the further access qualifier "protected" will be encountered when looking at Inheritance, section 22 onwards

21. OPERATOR OVERLOADING

21.1 WHAT IS OPERATOR OVERLOADING?

❖ Operator Overloading is a facility which enables the programmer (i.e. YOU! :-)) to implement new functionalities for existing operators in C++

➤ The consequences can be far reaching and so it is used to implement sophisticated entities (section 21.5 and Cybernetics in C++ - Expert)

- It is also supported in C#

➤ However, its misuse can be catastrophic,

- Most languages, including Java, C, VBA, Python, Perl, Web Programming Languages, do not support it

❖ In C++, as long as there is sufficient caution, operator overloading can be used to produce high performance sophistication (section 21.5 & Cybernetics in C++ - Expert), otherwise not possible

21.1.1 C++ OPERATORS - MOST BUT NOT ALL CAN BE OVERLOADED

❖ Revision: Operators are the entities denoted by symbols as opposed to words which perform operations

➤ They can be thought of as special kinds of functions

- They also comprise
  - Input(s)
  - Process
  - Output
- ➢ Functions are denoted by words, operators are denoted by symbols

❖ There are primarily 3 categories of operator in C++, two of which are looked at in Operator Overloading

- ➢ Unary Operators
  - These take in ONE operand
  - Examples:
    - Pre Increment and Post Increment ++
    - Pre Decrement and Post Decrement− −
- ➢ Binary Operators
  - These take in TWO operands
  - Examples:
    - Arithmetic Operators:
      - + - / * % || &&
    - Logical Operators:
      - | &
- ➢ The Ternary Operator
  - Takes in THREE Operands
  - Only ONE example
    - The conditional i.e. If, Then Else operator ? :
  - This cannot be overloaded

❖ Operator Overloading means that new meanings can be assigned only to existing operators

- ➢ You cannot create new operators and assign meanings to them

❖ The majority of operators can be overloaded

- ➢ Appendix B contains a list of the C++ operators which can and can't be overloaded

21.1.1.1 Prohibited Overloads

❖ Overloading any of the following operators is prohibited:

- ➢ Membership

- .
- .*
- ::

➤ Conditional

- ?:

➤ Memory Management

- new
- delete
- sizeof

➤ OOP Dynamic Polymorphism (Cybernetics in C++ - Expert)

- typeid

➤ Casting Operators (Cybernetics in C++ - Expert)

- static_cast
- dynamic_cast
- const_cast
- reinterpret_cast

21.1.2 IMPLEMENTING OPERATOR OVERLOADING

❖ In order to overload an operator, it needs to be written as a function:

```
ReturnType operator +( Inputs with Types )
{
    // Functionalities of overloaded +
}
```

➤ The keyword operator is used

➤ This is followed by the symbol used for the operator

➤ After this comes the parentheses brackets

- Between these, the inputs with their types are placed, just like with member functions

❖ It is strongly advised, usually instructed, to implement this as a member function of a class

➤ This way, the new meaning for the operator will be used only for objects of that Class type

➤ If it is done globally, it will mean every time the operator is used anywhere in the project, the new meaning will be used

21.2 OVERLOADING UNARY OPERATORS

❖ The examples here show how to overload the two increment operators

➢ Note that these have been included for demonstration purposes only

➢ You are advised to avoid overloading these operators in most programs

❖ For the Employee class, the increment operator will be overloaded to increment the data member `salary_gross`

➢ If there is an object `richardSharp` with a `salary_gross` of 23000

➢ Either

```
richardSharp++

++ richardSharp
```

will return an object of type Employee with the `salary_gross` data member incremented

21.2.1 OVERLOADING THE INCREMENT OPERATORS

❖ In order to distinguish between pre-increment and post-increment overloading, the keyword `int` is used, as follows

➢ Declarations

```
Employee operator ++();      // PRE Increment
Employee operator ++( int ); // POST Increment
```

➢ Definitions:

```
// ---------- OPERATOR OVERLOADERS ------------

// PRE Increment

Employee Employee::operator ++()

{
     return ( Employee( name, nI_Number, ++salary_Gross ) );

}//END...Operator Overloader Pre ++

// POST Increment

Employee Employee::operator ++( int )

{
     return ( Employee( name, nI_Number, salary_Gross++ ) );

}//END...Operator Overloader Post ++
```

❖ C++ designers have been criticised heavily for this counter-intuitive syntax

➢ The word `int` in between the parentheses in the post-increment form makes it look like

- A member function taking an integer input, especially in the declaration
- An entity which works with only ints
  ○ It actually works with all primitive types

➢ The fact that there is no input name after the word int is the cue to the programmer that its presence means something different

❖ The full program is as follows:

```cpp
#include <iostream>
#include <string>
using namespace std;

#ifndef EMPLOYEE_H
#define EMPLOYEE_H


const double TAXRATE_THRESHOLD_LOWER = 20000.0;
const double TAXRATE_THRESHOLD_INTERMEDIATE = 25000.0;


const double TAXRATE_RATE_LOWER = 0.3;
const double TAXRATE_RATE_INTERMEDIATE = 0.4;
const double TAXRATE_RATE_HIGHER = 0.5;


// Class DECLARATION
class Employee
{
      // DATA MEMBERS
      private:
            string name;
            string nI_Number;
            double salary_Gross;
            double tax_Rate;
            double tax_Amount;
            double salary_Net;

      // MEMBERS FUNCTIONS
      public:
      // ---------- CONSTRUCTOR(S) ------------
        Employee( string );
        Employee( string, string );
        Employee( string name_IN,
                        string nI_Number,
                        double salary_Gross_IN );
```

```
        // ---------- DESTRUCTOR ------------
        ~Employee(  );

        // ---------- OPERATOR OVERLOADERS ------------
        Employee operator ++();  // PRE Increment
        Employee operator ++( int );// POST Increment


        // ---------- "NORMAL Member Functions " ------------
        void greet();
        void greet_Name();
        void greet_Name_NINumber();
        void greet_Name_SalaryGross();
        void greet_AllAttributes();


};//END...Class Declaration Employee

#endif
```

Program Listing/Output 21.1: Employee.h

```
#include "Employee.h"

// ---------- CONSTRUCTOR(S) ------------
Employee::Employee( string name_IN )
{
     name = name_IN;

}//END...Constructor Employee( string name_IN )


Employee::Employee( string name_IN, string nI_Number_IN )
{
     name = name_IN;
     nI_Number = nI_Number_IN;

}//END...Constructor Employee( string name_IN )

Employee::Employee( string name_IN,
                                string nI_Number_IN,
                                double salary_Gross_IN )
{
     name = name_IN;
     nI_Number = nI_Number_IN;
     salary_Gross = salary_Gross_IN;
     //----------------------------------------------------
     // Calculating the Tax Rate
     //
```

```
        if( salary_Gross < TAXRATE_THRESHOLD_LOWER )
        {
                tax_Rate = TAXRATE_RATE_LOWER;
        }
        else
        {
                if( salary_Gross < TAXRATE_THRESHOLD_INTERMEDIATE )
                {
                        tax_Rate = TAXRATE_RATE_INTERMEDIATE;
                }
                else
                {
                        tax_Rate = TAXRATE_RATE_HIGHER;
                }
        }

    //----------------------------------------------------
    // Calculating the Tax Amount & Net Salary
    //
    tax_Amount = tax_Rate * salary_Gross;
    salary_Net = salary_Gross - tax_Amount;


    // Incrementing the Number of Objects/Employees
}/* END...Constructor Employee( string name_IN,
                                        string nI_Num_IN,
                                        double salary_Gross_IN )
*/

// ---------- DESTRUCTOR ------------
 Employee::~Employee(  )
{/*
        cout << "Goodbye, I, "<< name
                << " am  going out of scope now."
                << endl;
*/
}// END...Destructor ~Employee



// ---------- "NORMAL Member Functions " ------------
void Employee::greet()
{
        cout << "Hello, I'm an Employee"<< endl;

}//END...Member Function greet

void Employee::greet_Name()
{
        cout << "Hello, I'm an Employee, "
                << "my name is "<< name
                << endl;
```

```
}//END...Member Function greet_Name

void Employee::greet_Name_NINumber()
{
        cout << "Hello, I'm an Employee, "
            << "my name is "<< name
            << " and my NI Number is "<< nI_Number
            << endl;

}//END...Member Function greet_Name_NINumber


void Employee::greet_Name_SalaryGross()
{
        cout << "Hello, I'm an Employee, my name is "<< name
            << ", my Gross Salary is "<< salary_Gross
            << endl;

}//END...Member Function greet_Name_SalaryGross()


void Employee::greet_AllAttributes()
{
        cout << endl
            << "====================="
            << endl
            << "Hello, I am an employee, my name is "<< name<< ", "
            << endl
            << "my NI Number is "<< nI_Number<< ","
            << endl
            << "my Gross Salary is "<< salary_Gross
            << endl
            << "my tax Rate is "<< tax_Rate<< "%"
            << endl
            << "so the amount I have to pay in tax is "<< tax_Amount
            << endl
            << "thus my Net Salary is "<< salary_Net
            << endl
            << "====================="<< endl;
}// END...MF greet_AllAttributes()

// ---------- OPERATOR OVERLOADERS ------------
// PRE Increment
Employee Employee::operator ++()
{
    return ( Employee( name, nI_Number, ++salary_Gross ) );

}//END...Operator Overloader Pre ++

// POST Increment
Employee Employee::operator ++( int )
{
```

```
        return ( Employee( name, nI_Number, salary_Gross++ ) );

}//END...Operator Overloader Post ++
```

```
/*     ***********************************************************
       OPERATOR OVERLOADING DEMO: UNARY OPERATORS
       ***********************************************************
       -           This program demonstrates overloading unary operators
*/
#include <iostream>
#include <string>
using namespace std;

#include "Employee.h"


int main()
{
       //»»»»»»»»»»»»»»»»»»»»»»»»»»»»»»»»»»»»»»»»»»»»»»»»»»»»»»»»»»»»
       // Creating OBJECTS

       Employee joeBloggs("Joseph Bloggs", "PH 24 25 67 G", 18000.0 );
       joeBloggs.greet_Name();
       joeBloggs.greet_AllAttributes();

       cout << endl
            << "============"<< endl
            << "Object creation: Employee joeBloggs_01 = joeBloggs++"
            << endl;
     Employee joeBloggs_01 = joeBloggs++;
     joeBloggs_01.greet_Name_SalaryGross();

       cout << endl
            << "============"<< endl
            << "Object creation: Employee joeBloggs_02 = joeBloggs"
            << endl;
     Employee joeBloggs_02 = joeBloggs;
     joeBloggs_02.greet_Name_SalaryGross();

    cout << endl
            << "============"<< endl
            << "Object creation: Employee joeBloggs_03 = ++joeBloggs"
            << endl;
    Employee joeBloggs_03 = ++joeBloggs;
    joeBloggs_03.greet_Name_SalaryGross();



    //»»»»»»»»»»»»»»»»»»»»»»»»»»»»»»»»»»»»»»»»»»»»»»»»»»»»»»»»»»»»
    cout    << endl<< "==============="<< endl
```

```
            << "END...C++ Demo: Overloading Increment Operators "
            << endl;
    while(true){}
    return 0;

}//END...Function MAIN
```

```
Hello, I'm an Employee, my name is Joseph Bloggs

======================
Hello, I am an employee, my name is Joseph Bloggs,
my NI Number is PH 24 25 67 G,
my Gross Salary is 18000
my tax Rate is 0.3%
so the amount I have to pay in tax is 5400
thus my Net Salary is 12600
=====================


============
Object creation: Employee joeBloggs_01 = joeBloggs++
Hello, I'm an Employee, my name is Joseph Bloggs, my Gross
Salary is 18000

============
Object creation: Employee joeBloggs_02 = joeBloggs
Hello, I'm an Employee, my name is Joseph Bloggs, my Gross
Salary is 18001

============
Object creation: Employee joeBloggs_03 = ++joeBloggs
Hello, I'm an Employee, my name is Joseph Bloggs, my Gross
Salary is 18002

================
END...C++ Demo: Overloading Increment Operators
```

Program Listing/Output 21.3: OverloadingIncrementEmployee_Client.cpp

21.3 OVERLOADING BINARY OPERATORS

❖ This section will look at overloading the + operator for the Employee
   class so that it will add together the `salary_gross` data members of each
   object

   ➢ i.e. If we have a statement like

```
Employee total = joeBloggs + johnSmith;
```

it will be like saying:

```
Employee total(joeBloggs.get_salary_gross() +

                johnSmith.get_salary_gross()  );
```

❖ The standard means of overloading binary operators is to

➢ Create the operator overloading member function, to have as
- Input: An object of the same class type
- Output: An object of the same class type
- Thus in our case, the declaration will be as follows:

```
Employee operator +( Employee  );
```

○ ...and the prototype of the definition will be:

```
Employee Employee::operator +( Employee employee_In )
```

➢ Next, the data member in question needs to be extracted from the input object, and added to the data member of the current class

➢ In our case, since the data member is `salary_gross`, the required commands will be:

```
double salary_Gross_Total = 0.0;

salary_Gross_Total = salary_Gross +

                        employee_In.get_salary_Gross();
```

➢ Now, a new temporary object, often called a dummy object, needs to be created, with this new value for its data member
- In our case, we require an object to have `salary_Gross_Total` as its value for `salary_Gross`

```
Employee employee_Temp( salary_Gross_Total );
```

- There is a constructor which initialises with the input value `salary_Gross`, as shown in

➢ Finally, this object needs to be returned

❖ When using this new operator, the following lines show how an Employee object can be created, by using the addition operator of two existing Employee objects

```
Employee susanWhitmore( 12000.0 );
Employee scottHamilton(1.0);


Employee employee_Total = susanWhitmore + scottHamilton;
```

❖ The full program is as follows:

```cpp
#include <iostream>
#include <string>
using namespace std;

#ifndef EMPLOYEE_H
#define EMPLOYEE_H


const double TAXRATE_THRESHOLD_LOWER = 20000.0;
const double TAXRATE_THRESHOLD_INTERMEDIATE = 25000.0;


const double TAXRATE_RATE_LOWER = 0.3;
const double TAXRATE_RATE_INTERMEDIATE = 0.4;
const double TAXRATE_RATE_HIGHER = 0.5;


// Class DECLARATION
class Employee
{
     // DATA MEMBERS
     private:
            string name;
            string nI_Number;
            double salary_Gross;
            double tax_Rate;
            double tax_Amount;
            double salary_Net;

    // MEMBERS FUNCTIONS
    public:
            // ---------- CONSTRUCTOR(S) ------------
            Employee( string );
            Employee( string, string );
            Employee( string name_IN,
            string nI_Number,
            double salary_Gross_IN );
            Employee( double );


    // ---------- DESTRUCTOR ------------
    ~Employee(  );

    // ---------- MUTATOR(S) ------------
    double get_salary_Gross();

    // ---------- OPERATOR OVERLOADERS ------------
    Employee operator +( Employee  );
```

```
    // ---------- "NORMAL Member Functions " ------------
    void greet();
    void greet_Name();
    void greet_Name_NINumber();
    void greet_salaryGross();
    void greet_Name_SalaryGross();
    void greet_AllAttributes();



};//END...Class Declaration Employee

#endif
```

Program Listing/Output 21.4: Employee.h

```
#include "Employee.h"

// ---------- CONSTRUCTOR(S) ------------
Employee::Employee( string name_IN )
{
      name = name_IN;

}//END...Constructor Employee( string name_IN )


Employee::Employee( string name_IN, string nI_Number_IN )
{
      name = name_IN;
      nI_Number = nI_Number_IN;

}//END...Constructor Employee( string name_IN )

Employee::Employee( string name_IN,
      string nI_Number_IN,
      double salary_Gross_IN )
{
      name = name_IN;
      nI_Number = nI_Number_IN;
      salary_Gross = salary_Gross_IN;
      //--------------------------------------------------
      //      Calculating the Tax Rate
      //
      if( salary_Gross < TAXRATE_THRESHOLD_LOWER )
      {
            tax_Rate = TAXRATE_RATE_LOWER;
      }
      else
      {
            if( salary_Gross < TAXRATE_THRESHOLD_INTERMEDIATE )
            {
                  tax_Rate = TAXRATE_RATE_INTERMEDIATE;
```

```
                }
                else
                {
                      tax_Rate = TAXRATE_RATE_HIGHER;
                }
        }

        //----------------------------------------------------
        // Calculating the Tax Amount & Net Salary
        //
        tax_Amount = tax_Rate * salary_Gross;
        salary_Net = salary_Gross - tax_Amount;



}/* END...Constructor Employee( string name_IN,
                                        string nI_Num_IN,
                                        double salary_Gross_IN )
*/

Employee::Employee( double salary_Gross_IN )
{
        salary_Gross = salary_Gross_IN;
        //----------------------------------------------------
        // Calculating the Tax Rate
        //
        if( salary_Gross < TAXRATE_THRESHOLD_LOWER )
        {
                tax_Rate = TAXRATE_RATE_LOWER;
        }
        else
        {
                if( salary_Gross < TAXRATE_THRESHOLD_INTERMEDIATE )
                {
                        tax_Rate = TAXRATE_RATE_INTERMEDIATE;
                }
                else
                {
                        tax_Rate = TAXRATE_RATE_HIGHER;
                }
    }

  //----------------------------------------------------
  // Calculating the Tax Amount & Net Salary
  //
  tax_Amount = tax_Rate * salary_Gross;
  salary_Net = salary_Gross - tax_Amount;

}//END...Constructor Employee( double salary_Gross_IN )


// ---------- DESTRUCTOR ------------
```

```
Employee::~Employee( )
{/*
        cout   << "Goodbye, I, "<< name
               << " am  going out of scope now."
               << endl;
*/
}// END...Destructor ~Employee

// ---------- MUTATOR(S) ------------
double Employee::get_salary_Gross()
{
        return salary_Gross;

}// END...Mutator get_salary_Gross


// ---------- "NORMAL Member Functions " ------------
void Employee::greet()
{
       cout    << "Hello, I'm an Employee"<< endl;

}//END...Member Function greet

void Employee::greet_Name()
{
      cout     << "Hello, I'm an Employee, "
               << "my name is "<< name
               << endl;

}//END...Member Function greet_Name

void Employee::greet_Name_NINumber()
{
      cout     << "Hello, I'm an Employee, "
               << "my name is "<< name
               << " and my NI Number is "<< nI_Number
               << endl;

}//END...Member Function greet_Name_NINumber


void Employee::greet_salaryGross()
{
      cout     << "Hello, I'm an Employee, "
               << "my Gross Salary is "<< salary_Gross
               << endl;

}//END...Member Function greet_salaryGross()
```

```
void Employee::greet_Name_SalaryGross()
{
    cout    << "Hello, I'm an Employee, my name is "<< name
            << ", my Gross Salary is "<< salary_Gross
            << endl;

}//END...Member Function greet_Name_SalaryGross()




void Employee::greet_AllAttributes()
{
    cout    << endl
            << "======================"
            << endl
            << "Hello, I am an employee, my name is "<< name<< ", "
            << endl
            << "my NI Number is "<< nI_Number<< ","
            << endl
            << "my Gross Salary is "<< salary_Gross
            << endl
            << "my tax Rate is "<< tax_Rate<< "%"
            << endl
            << "so the amount I have to pay in tax is "<< tax_Amount
            << endl
            << "thus my Net Salary is "<< salary_Net
            << endl
            << "======================"<< endl;
}// END...MF greet_AllAttributes()




// ---------- OPERATOR OVERLOADERS -----------
// This is HYPOTHETICAL...in the real world,
// this is a bad idea - here for demo purposes
//
Employee Employee::operator +( Employee employee_In )
{
        double salary_Gross_Total = 0.0;
        salary_Gross_Total = salary_Gross +
                            employee_In.get_salary_Gross();
        Employee employee_Temp( salary_Gross_Total );
        return employee_Temp;

}//END...Operator Overloader:Binary Addition +
```

Program Listing/Output 21.5: Employee.cpp

```
/*    ****************************************************************
      OPERATOR OVERLOADING DEMO: BINARY OPERATORS
      ****************************************************************
      -       This program demonstrates overloading binary operators
*/
#include <iostream>
#include <string>
using namespace std;

#include "Employee.h"


int main()
{
      //»»»»»»»»»»»»»»»»»»»»»»»»»»»»»»»»»»»»»»»»»»»»»»»»»»»»»»»»»»»»»»»»»»
      // Creating OBJECTS
      Employee susanWhitmore( 15000.0 );
      susanWhitmore.greet_Name_SalaryGross();

      Employee scottHamilton( 13000.0 );
      scottHamilton.greet_Name_SalaryGross();

      cout    << endl
              << "============"<< endl
              << "Object creation: Employee employee_Total = "
              << "susanWhitmore + scottHamilton:"
              << endl;
      Employee employee_Total = susanWhitmore + scottHamilton;
      employee_Total.greet_salaryGross();


      //»»»»»»»»»»»»»»»»»»»»»»»»»»»»»»»»»»»»»»»»»»»»»»»»»»»»»»»»»»»»»»»»»»
      cout    << endl<< "==============="<< endl
              << "END...C++ Demo: Overloading Binary Operators "
              << endl;
      while(true){}
      return 0;

}//END...Function MAIN
```

```
Hello, I'm an Employee, my name is , my Gross Salary is 15000
Hello, I'm an Employee, my name is , my Gross Salary is 13000

============
Object creation: Employee employee_Total = susanWhitmore +
    scottHamilton:
Hello, I'm an Employee, my Gross Salary is 28000

===============
```

```
END...C++ Demo: Overloading Binary Operators
```

21.4 OVERLOADING THE ASSIGNMENT OPERATOR

❖ As we have seen,

➢ the assignment operator is invoked whenever two existing objects are equated to each other e.g.

```
robertFletcher = joeBloggs;
```

➢ It is created "silently" whenever a class is created - there is no further work for the programmer (i.e. YOU!:-)) to do, in order to use its default functionality

❖ Overloading this operator is possible, as the following program shows:

➢ Note that this is also for demonstration purposes only:

```cpp
#include <iostream>
#include <string>
using namespace std;

#ifndef EMPLOYEE_H
#define EMPLOYEE_H


const double TAXRATE_THRESHOLD_LOWER = 20000.0;
const double TAXRATE_THRESHOLD_INTERMEDIATE = 25000.0;


const double TAXRATE_RATE_LOWER = 0.3;
const double TAXRATE_RATE_INTERMEDIATE = 0.4;
const double TAXRATE_RATE_HIGHER = 0.5;


// Class DECLARATION
class Employee
{
        // DATA MEMBERS
        private:
                string name;
                string nI_Number;
                double salary_Gross;
                double tax_Rate;
                double tax_Amount;
                double salary_Net;
```

```cpp
        // MEMBERS FUNCTIONS
        public:
                // ---------- CONSTRUCTOR(S) ------------
                Employee( string );
                Employee( string, string );
                Employee( string name_IN,
                          string nI_Number,
                          double salary_Gross_IN );


                // ---------- OPERATOR OVERLOADER(S) ------------
                Employee operator =( Employee );


                // ---------- "NORMAL Member Functions " ------------
                void greet();
                void greet_Name();
                void greet_Name_NINumber();
                void greet_Name_SalaryGross();
                void greet_AllAttributes();


};//END...Class Declaration Employee

#endif
```

Program Listing/Output 21.7: Employee.h

```cpp
#include "Employee.h"

// ---------- CONSTRUCTOR(S) ------------
Employee::Employee( string name_IN )
{
     name = name_IN;

}//END...Constructor Employee( string name_IN )


Employee::Employee( string name_IN, string nI_Number_IN )
{
     name = name_IN;
     nI_Number = nI_Number_IN;

}//END...Constructor Employee( string name_IN )

Employee::Employee( string name_IN,
                    string nI_Number_IN,
                    double salary_Gross_IN )
{
     name = name_IN;
     nI_Number = nI_Number_IN;
     salary_Gross = salary_Gross_IN;
```

```
    //----------------------------------------------------
    // Calculating the Tax Rate
    //
    if( salary_Gross < TAXRATE_THRESHOLD_LOWER )
    {
          tax_Rate = TAXRATE_RATE_LOWER;
    }
    else
    {
              if( salary_Gross < TAXRATE_THRESHOLD_INTERMEDIATE )
              {
                    tax_Rate = TAXRATE_RATE_INTERMEDIATE;
              }
              else
              {
                    tax_Rate = TAXRATE_RATE_HIGHER;
              }
    }

    //----------------------------------------------------
    // Calculating the Tax Amount & Net Salary
    //
    tax_Amount = tax_Rate * salary_Gross;
    salary_Net = salary_Gross - tax_Amount;


    // Incrementing the Number of Objects/Employees

}/* END...Constructor Employee( string name_IN,
                               string nI_Num_IN,
                               double salary_Gross_IN )
 */

// ---------- "NORMAL Member Functions " ------------
void Employee::greet()
{
      cout << "Hello, I'm an Employee"<< endl;

}//END...Member Function greet

void Employee::greet_Name()
{
      cout << "Hello, I'm an Employee, "
           << "my name is "<< name
           << endl;

}//END...Member Function greet_Name

void Employee::greet_Name_NINumber()
{
      cout << "Hello, I'm an Employee, "
           << "my name is "<< name
           << " and my NI Number is "<< nI_Number
```

```
                << endl;

}//END...Member Function greet_Name_NINumber


void Employee::greet_Name_SalaryGross()
{
        cout << "Hello, I'm an Employee, my name is "<< name
             << ", my Gross Salary is "<< salary_Gross
             << endl;

}//END...Member Function greet_Name_SalaryGross()


void Employee::greet_AllAttributes()
{
      cout << endl
           << "====================="
           << endl
           << "Hello, I am an employee, my name is "<< name<< ", "
           << endl
           << "my NI Number is "<< nI_Number<< ","
           << endl
           << "my Gross Salary is "<< salary_Gross
           << endl
           << "my tax Rate is "<< tax_Rate<< "%"
           << endl
           << "so the amount I have to pay in tax is "<< tax_Amount
           << endl
           << "thus my Net Salary is "<< salary_Net
           << endl
           << "====================="<< endl;
}// END...MF greet_AllAttributes()

// ---------- OPERATOR OVERLOADER(S) ------------
Employee Employee::operator =( Employee employee_IN )
{
      cout << "Assignment Operator called..."<< endl;
      return( employee_IN );

}//END...Operator Overloader: Assignment =
```

Program Listing/Output 21.8: Employee.cpp

```
/*     *************************************************************
       DEMO: OVERLOADING THE ASSIGNMENT OPERATOR
       *************************************************************
*/
#include <iostream>
#include <string>
using namespace std;
```

```
#include "Employee.h"

int main()
{
    //»»»»»»»»»»»»»»»»»»»»»»»»»»»»»»»»»»»»»»»»»»»»»»»»»»»»»»»»»»»»»»»»»»»»»»»»»»
    // Creating OBJECTS

        Employee joeBloggs("Joseph Bloggs", "PH 24 25 67 G", 18000.0 );
        joeBloggs.greet_Name();

        Employee richardBradford( "Richard Bradford" );

        richardBradford = joeBloggs;


    //»»»»»»»»»»»»»»»»»»»»»»»»»»»»»»»»»»»»»»»»»»»»»»»»»»»»»»»»»»»»»»»»»»»»»»»»»»
    cout    << endl<< "================"<< endl
            << "END...C++ Demo: Overloading the Assignment Operator "
            << endl;
    while(true){}
    return 0;

}//END...Function MAIN
```

```
Hello, I'm an Employee, my name is Joseph Bloggs
Assignment Operator called...

================
END...C++ Demo: Overloading the Assignment Operator
```

Program Listing/Output 21.9: OverloadAssignmentOperator_Client.cpp

21.4.1 ARGUMENTS FOR & AGAINST OVERLOADING =

❖ One of the major reasons for Overloading the Assignment Operator is to deal with resource control

➢ Suppose one data member is a pointer to some heap memory

➢ If this object is equated with another, the counterpart pointer in the latter object will point to the same part of the heap

➢ We now have the problem of potentially too many pointers to the same part of the heap (section 14.5), possibly allowing for dangling pointers

➢ In cases like these, overloading the assignment operator can enable dealing with special cases

❖ However, it is essential to do it with caution

  ➢ Since the assignment operator is one of the few entities guaranteed to be used frequently, any runtime bug can have potentially disastrous consequences

   • It could mean objects not being created properly

❖ **Rule of Thumb**: Avoid overloading the Assignment Operator whenever possible

  ➢ While overloading the Assignment Operator can be useful in a few cases, doing so excessively can lead to malfunction

  ➢ Whenever objects need to be equated, it will be called - overlooking special cases can cause problems

21.5 ADVANCED USES OF OPERATOR OVERLOADING

❖ One of the major reasons Operator Overloading is used in C++ is to implement the Proxy Design Pattern

  ➢ This comprises implementing a class, the objects of which will mimic other software entities

❖ Two famous examples are

  ➢ **Functors:** These comprise overloading the parentheses operators () in order to make an object behave like a function

   • Several further functionalities can be implemented
   • These are widely used in libraries such as the STL and Boost
   • You will meet these in Cybernetics in C++ - Expert

  ➢ **Smart Pointers:** These comprise overloading the * and -> operators in order to create objects which behave like Pointers

   • These also have several further functionalities built in
   • These are widely used in libraries such as the STL and Boost
   • You will meet these in Cybernetics in C++ - Expert

21.6 PROSPECTIVE EVILS OF OPERATOR OVERLOADING

❖ Operator Overloading is another practice in C++ which gives you a lot of power...

  ➢ You can effectively redefine a language by redefining operators

❖ ...with which comes great responsibility

➤ It is essential to make sure that overloading operators do not make things worse, or unworkable

❖ There are many potential problems, some of which can have catastrophic consequences

➤ Operator Overloading facilitates vandalism of programs

• For operators which can be overloaded, there are no restrictions for what can be done
• It is perfectly possible to
  ○ Overload the + operator to perform subtraction!
    ■ In fact, any mathematical operator can be redefined!

❖ Once redefined, the new meaning is the only one the operator can take on, the "old" meaning has been lost

➤ If this is done globally, then the entire program can only ever use the new meaning

• You are effectively instructed never to implement an operator over-loader as a global function

➤ If this is implemented as a member function of a class, only objects of that class will be affected

❖ Common advice in C++ is never to overload at least the following operators

➤ &&   Logical
➤ || Logical OR
➤ ,   Comma/Separator

❖ Whether Operator Overloading is an asset or a liability is a subject of considerable debate

➤ Most languages, including Java, Visual Basic, C, Pascal, Web programming languages, do not support it at all
➤ However, along with C++, it is supported in the Microsoft .NET languages C# and F#
➤ The view in this book is it is an asset provided it is used responsibly:

❖ **Rules of Thumb:**

➤ Use Operator Overloading only when you know it will do a lot of good

> Make sure the overloaded functionality resembles the original
> If in doubt, or if it is not essential, avoid it
> Only ever implement it as a member function of a class, so it applies only to objects of that class
  - Do not implement it as a global function

22. INHERITANCE

22.1 WHAT IS INHERITANCE?

❖ Inheritance is the practice of building one entity by extending (an)other(s)

❖ In Object Oriented Programming, it is about creating one class by extending (an)other class(es)

❖ It is often described as implementing the **IS-A** relationship

> It is all about creating categories, with progressively increased refinements

> Note that while this model is common, Inheritance is not limited to this

❖ Even when one level of inheritance is implemented, but particularly with more than one, an "Inheritance Hierarchy" is said to build up

22.1.1 TERMINOLOGY

❖ The class at the lowest level is called the "Base class"

> Alternative terms include:
  - Super class
  - Parent class
  - Ancestor class

❖ A class which inherits from this one is called the "Derived" class

> Alternative terms include:
  - Sub-class
  - Child class
  - Descendant class

22.2 MODELLING INHERITANCE IN UML

❖ In UML, Inheritance is shown using an arrow with a hollow head

❖ It is referred to as "Generalisation"

➤ As one moves down the hierarchy, there is increased specialisation
➤ As one moves up the hierarchy, there is increased generalisation

❖ The examples below show it in action

22.2.1 INHERITANCE OUTSIDE THE PROGRAMMING WORLD

❖ Inheritance is another powerful means of thinking which is used abundantly outside as well as inside the programming world

❖ **Nature**. These ideas can be seen abundantly, Figure 22.1, since

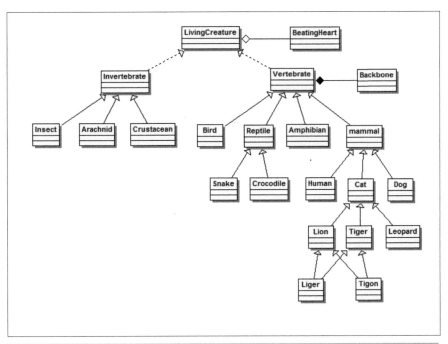

Figure 22.1: UML Diagram, showing Inheritance in Nature

➤ A mammal IS-A vertebrate
➤ A dog IS-A mammal
➤ Note that from Cat, **multiple** inheritance is used, showing that one type can belong to more than one parent type
   • A Tigon IS-A Tiger, as well as a Lion
   • A Liger IS-A Tiger, as well as a Lion

> The relationship between beating heart and animal is not one of inheritance

- It does not make sense to say an "Animal" is a beating heart!
- It does make sense to say an "Animal" **has a** beating heart
  - Consequently, an "Aggregation " relationship is used

❖ **Vehicles**. Likewise, Inheritance ideas can be seen abundantly here:

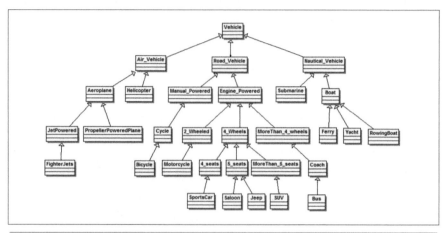

Figure 22.2: UML Diagram, showing Inheritance in Vehicles

EXERCISE: FURTHER INHERITANCE EXAMPLES

❖ These lists can be extended almost indefinitely

❖ *Come up with 3 further examples of Inheritance outside the computer world*

22.2.2 INHERITANCE INSIDE THE COMPUTER WORLD

❖ In Object Oriented Programming, Inheritance is the practice of building one class by extending another

❖ It is used abundantly since a common need is to build sophisticated types

> Starting with simple types and developing them is generally considered significantly advantageous over starting with complex ones

❖ One example is GUI programming:

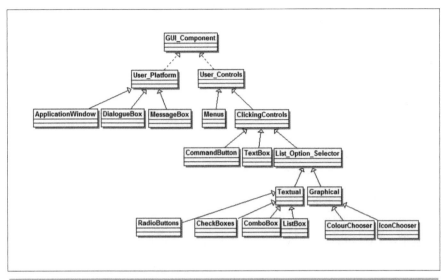

Figure 22.3: UML Diagram, showing Inheritance in GUI Programming

➢ While there are several technologies/libraries for GUI programming, the fundamental structure in all cases is similar to the above

VENDOR	TECHNOLOGY	OOP LANGUAGE
• Microsoft	.NET	C++, VB and C#
	MFCs	C++
• Apple	Carbon	Objective C, Swift
	Cocoa	
• Nokia/QT Technologies	QT	Mainly C++, also Python and Java
• Oracle/Sun Microsystems	Java Swing Java AWT	Java
• FSF Free Software Foundation	wxWidgets	Several languages including C++, Python

22.3 C++ SYNTAX: INHERITANCE

22.3.1 SINGLE INHERITANCE

❖ The Colon : Operator is used to implement Inheritance in C++, as follows

```
class  className  :  Access Category  Base Class Name
{
       Contents of Derived Class
}
```

➢ Examples

```
class  Manager    :  public     Employee {}
class  Car        :  public     Vehicle  {}
class  Vertebrate :  protected  Animal   {}
class  Surgeon    :  private    Doctor   {}
class  Programmer :  public     Engineer {}
```

❖ Nothing further needs to be written to specify whether the Inheritance is Implementation or Interface

➢ The Compiler will be able to tell, from the Class name, whether the Class is

• Concrete
  ○ All the Member Functions have their "bodies" implemented
• Abstract
  ○ At least one of the Member Functions has no body

22.3.2 MULTIPLE INHERITANCE

❖ In order to implement Multiple Inheritance,

➢ Put the name of each Base Class

➢ Each separated by a Comma

➢ With each having its Access Category specified Separately

```
class  className  :  Access      Base
                     Category    Class 1  ,
                     Access      Base
                     Category    Class 2  ...
{
        Contents of Derived Class
}
```

➢ Examples

```
class Liger : public Lion, public Tiger{};
class Mule : public Horse, protected Donkey{};
class ArchtopGuitar: private ElectricGuitar, public
    AcousticGuitar{};
class Director : private Manager, private Employee{};
```

22.4 THE ACCESS SPECIFIER "PROTECTED"

❖ "`Protected`," the 3rd access specifier in Encapsulation means that members are accessible within

➤ That class
➤ Derived Classes

❖ Let us look at all 3 principal (but in C++, not their full) definitions and compare them:

➤ `private`
  • Accessible only within
    ○ that class
➤ `protected`
  • Accessible only within
    ○ that class
    ○ derived classes
➤ `public`
  • Accessible
    ○ anywhere

22.4.1 PROSPECTIVE EVILS OF PROTECTED

❖ While the protected qualifier has its uses, it has many problems associated with it

➤ The very person who initially persuaded Bjarne Stroustrup to include `protected` in C++, Mark Linton, regrets it in retrospect[2]

❖ **Potentially Excessive Exposure**

➤ Revision: Global Variables and Public Data Members are considered evil because of their excessive exposure: anything anywhere can write data to them
  • When they have incorrect runtime values, debugging can become like a treasure hunt
➤ `Protected` is exposing data members to all member functions of all classes below in the inheritance hierarchy
  • This could potentially be thousands/millions of member functions

---

[2]The Design and Evolution of C++, Bjarne Stroustrup, 1994 Addison-Wesley.

- • When software is sold to companies, in order for them to inherit from the given classes in their development, there is no way to tell how much software will depend on the base classes

❖ Assigning invalid values

- ➤ It follows from above, that being able to access data members without the need for member functions of that class, means that validity checks can be a problem

❖ Software Fragility/Brittleness

- ➤ The principal role of the base class is to determine the structure of the entire inheritance hierarchy
- ➤ It is advisable for derived class implementations to depend only on base class services i.e. non-private members
- ➤ Any changes made to protected data in the base class will instantly require all calls/references to them in derived classes to be changed accordingly
- ➤ Software is said to be fragile/brittle - a small change at one level can "break" the overall structure

❖ **Rule of Thumb:** Use `protected` with caution if at all;

- ➤ If in doubt, Avoid it

22.5 CATEGORIES OF INHERITANCE

- ❖ There are several kinds of Inheritance:
  - ➤ Single, Multiple
  - ➤ Implementation, Interface
  - ➤ Public, Protected, Private

- ❖ C++, being very powerful, supports all of these
  - ➤ There are potentially 12 categories of Inheritance
  - ➤ However, only a few are commonly used, some are advised as "best avoided"

	IMPLEMENTATION		INTERFACE	
	Single	Multiple	Single	Multiple
Public	Very Common & useful	Advised against - can be dangerous	Very Common & useful	Becoming more common, fairly safe
Protected	Rare	Rare - advised against	Rare	Rare
Private	Rare but can be useful	Advised against	Rare	Rare

22.5.1 SINGLE, MULTIPLE INHERITANCE

❖ Single Inheritance is inheriting from one class

❖ Multiple Inheritance is inheriting from more than one class

22.5.2 IMPLEMENTATION, INTERFACE

❖ Implementation Inheritance is inheriting from a class which is fully built and so can have objects based on it

❖ Interface Inheritance means inheriting from a "framework" class, which has been built for that sole purpose, and cannot be instantiated

22.5.2.1 UML Realisation

❖ In UML, Interface Inheritance is referred to as "Realisation"

❖ It is shown as

> ➤ An arrow with a hollow head, just like Generalisation
> ➤ The line of the arrow is dotted

❖ It is widely used

> ➤ Nature: looking back at Figure 22.1
> - The base class `Living_Creature` cannot be instantiated i.e. we cannot have an object that is just a `Living_Creature`, it must be more specific
> - Both Vertebrate and Invertebrate inherit from it
> ➤ GUI programming: looking back at Figure 22.3
> - The base class `GUI_Component` cannot be instantiated i.e. we cannot have an object that is just a GUI component, it must be more specific

❖ The idea is that interface inheritance enables "realisation" of a base class

> ➤ i.e. creates (a) class(es) which can be instantiated

22.5.3 PUBLIC, PROTECTED, PRIVATE INHERITANCE

❖ If the inheritance is **Public**

> ➤ Public members of the base class become public members of the derived class

➢ Protected members of the base class become protected members of the derived class

➢ Private members of the base class remain inaccessible

➢ i.e. Public Inheritance implements an IS-A relationship

❖ If the inheritance is **Protected**

➢ Public and Protected members of the base class become **Protected** members of the derived class

➢ Private members of the base class remain inaccessible

➢ i.e. Protected Inheritance does not implement an IS-A relationship

❖ If the inheritance is **Private**

➢ Public and Protected members of the base class become **Private** members of the derived class

➢ Private members of the base class remain inaccessible

➢ i.e. Private Inheritance does not implement an IS-A relationship either

22.6 CONSTRUCTOR & DESTRUCTOR CALLS IN INHERITED CLASSES

❖ Whenever an object based on a derived class is created, BOTH

➢ The Constructor for the Derived Class

➢ AND the Constructor for the Base Class
are invoked

❖ Likewise whenever an object based on a derived class goes out of scope, BOTH

➢ The Destructor for the Derived Class

➢ AND the Destructor for the Base Class
are invoked

❖ The idea is the following:

➢ When a car has to be created:
• All initial set up work for a car, the derived class, must take place
  ○ e.g. filling up with petrol, checking the oil level
• All initial set up work for a vehicle in general, the base class, must also take place

○ e.g. registering the vehicle, getting it insured

➤ Likewise, when a car has to be destroyed,

- All final clean up work for a car must take place
    - ○ e.g. ensuring it gets disposed of safely, any reusable parts are reused
- All final clean up required for vehicles in general must also take place
    - ○ e.g. Informing local authorities that registration is no longer used, cancelling insurance policies

22.7 SINGLE IMPLEMENTATION INHERITANCE

❖ The program used here to demonstrate Single Implementation Inheritance will be Inheriting from the base Class Employee to create the derived class Manager:

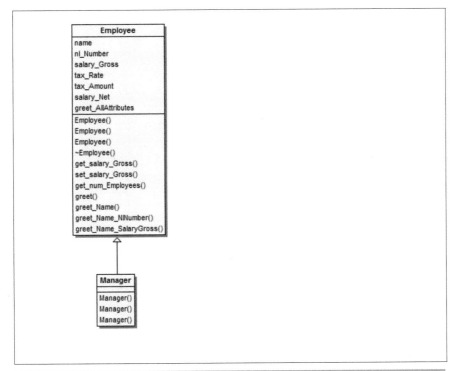

Figure 22.4: Design of the Employee - Manager Inheritance Hierarchy

```
#include <iostream>
#include <string>
using namespace std;

#ifndef EMPLOYEE_H
#define EMPLOYEE_H


const double TAXRATE_THRESHOLD_LOWER = 20000.0;
const double TAXRATE_THRESHOLD_INTERMEDIATE = 25000.0;


const double TAXRATE_RATE_LOWER = 0.3;
const double TAXRATE_RATE_INTERMEDIATE = 0.4;
const double TAXRATE_RATE_HIGHER = 0.5;


// Class DECLARATION
class Employee
{
      // DATA MEMBERS
      protected:
            string name;
            string nI_Number;
            double salary_Gross;
            double tax_Rate;
            double tax_Amount;
            double salary_Net;

      // MEMBERS FUNCTIONS
      public:
      // ---------- CONSTRUCTOR(S) ------------
            Employee( );
            Employee( string );
            Employee( string, string );
            Employee( string name_IN,
                          string nI_Number,
                          double salary_Gross_IN );


            // ---------- DESTRUCTOR ------------
            ~Employee(   );


            // ---------- "NORMAL Member Functions " ------------
            void greet();
            void greet_Name();
            void greet_Name_NINumber();
            void greet_Name_SalaryGross();
            void greet_AllAttributes();
```

```
};//END...Class Declaration Employee

#endif
```

```
#include "Employee.h"

// ---------- CONSTRUCTOR(S) ------------
Employee::Employee(  )
{
     cout << "Base Class Employee constructor has been called"
          << endl;

}//END...Constructor Employee - Default


Employee::Employee( string name_IN )
{
     cout << "Base Class Employee constructor has been called"
          << endl;

     name = name_IN;

}//END...Constructor Employee( string name_IN )


Employee::Employee( string name_IN, string nI_Number_IN )
{
     cout << "Base Class Employee constructor has been called"
          << endl;

     name = name_IN;
     nI_Number = nI_Number_IN;

}//END...Constructor Employee( string name_IN, string nI_Number_IN )

Employee::Employee( string name_IN,
                              string nI_Number_IN,
                              double salary_Gross_IN )
{
     cout << "Base Class Employee constructor has been called"
          << endl;

     name = name_IN;
     nI_Number = nI_Number_IN;
     salary_Gross = salary_Gross_IN;
     //---------------------------------------------------
     //     Calculating the Tax Rate
     //
     if( salary_Gross < TAXRATE_THRESHOLD_LOWER )
```

```
        {
                tax_Rate = TAXRATE_RATE_LOWER;
        }
        else
        {
                if( salary_Gross < TAXRATE_THRESHOLD_INTERMEDIATE )
                {
                        tax_Rate = TAXRATE_RATE_INTERMEDIATE;
                }
                else
                {
                        tax_Rate = TAXRATE_RATE_HIGHER;
                }
        }

    //----------------------------------------------------
    // Calculating the Tax Amount & Net Salary
    //
    tax_Amount = tax_Rate * salary_Gross;
    salary_Net = salary_Gross - tax_Amount;


    // Incrementing the Number of Objects/Employees
}/* END...Constructor Employee( string name_IN,
                                        string nI_Num_IN,
                                        double salary_Gross_IN )
*/

// ---------- DESTRUCTOR ------------
Employee::~Employee(   )
{
    cout  << "Base Class Employee destructor has been called:"
          << endl;

    cout  << "\tGoodbye, I, "<< name
          << " am  going out of scope now."
          << endl;

}// END...Destructor ~Employee



// ---------- "NORMAL Member Functions " ------------
void Employee::greet()
{
        cout << "Hello, I'm an Employee"<< endl;

}//END...Member Function greet

void Employee::greet_Name()
{
        cout << "Hello, I'm an Employee, "
```

```
                    << "my name is "<< name
                    << endl;

}//END...Member Function greet_Name

void Employee::greet_Name_NINumber()
{
            cout   << "Hello, I'm an Employee, "
                    << "my name is "<< name
                    << " and my NI Number is "<< nI_Number
                    << endl;

}//END...Member Function greet_Name_NINumber


void Employee::greet_Name_SalaryGross()
{
            cout   << "Hello, I'm an Employee, my name is "<< name
                    << ", my Gross Salary is "<< salary_Gross
                    << endl;

}//END...Member Function greet_Name_SalaryGross()


void Employee::greet_AllAttributes()
{
            cout   << endl
                    << "====================="
                    << endl
                    << "Hello, I am an employee, my name is "<< name<< ","
                    << endl
                    << "my NI Number is "<< nI_Number<< ","
                    << endl
                    << "my Gross Salary is "<< salary_Gross
                    << endl
                    << "my tax Rate is "<< tax_Rate<< "%"
                    << endl
                    << "so the amount I have to pay in tax is "<<
                       tax_Amount
                    << endl
                    << "thus my Net Salary is "<< salary_Net
                    << endl
                    << "====================="<< endl;
}// END...MF greet_AllAttributes()
```

Program Listing/Output 22.2: Employee.cpp

```
#include "Employee.h"

#ifndef MANAGER_H
#define MANAGER_H
```

```
class Manager : public Employee
{
    private:

    public:
            // ---------- CONSTRUCTOR(S) ------------
            Manager( );
            Manager( string name_IN );
            Manager( string name_IN,
                            string nI_Number_IN,
                            double salary_Gross_IN );

        // ---------- DESTRUCTOR ------------
        ~Manager();

};//END...Class Declaration Manager

#endif
```

```
#include "Manager.h"

// ---------- CONSTRUCTORS ------------
Manager::Manager(   )
{
    cout    << "Inherited Class Manager constructor has been called"
            << endl;

}//END...Constructor Manager - Default



Manager::Manager( string name_IN )
{
    cout    << "Inherited Class Manager constructor has been called"
            << endl;
    name = name_IN;

}//END...Constructor Manager( string name_IN )

Manager::Manager( string name_IN,
                                string nI_Number_IN,
                                double salary_Gross_IN )

        : Employee( name_IN,
                            nI_Number_IN,
                            salary_Gross_IN   )
{
    cout    << "Inherited Class Manager constructor has been called"
    <<   endl;
```

```
}/* END...Constructor Manager( string name_IN,
                                        string nI_Number_IN,
                                        double salary_Gross_IN )
*/

Manager::~Manager()
{
        cout  << endl
              << "Inherited Class Manager destructor has been called"
              << endl;

}// END...Destructor ~Manager
```

Program Listing/Output 22.4: Manager.cpp

```
#include <iostream>
#include <string>
using namespace std;

#include "Employee.h"
#include "Manager.h"

void createOffice();

int main()
{
     createOffice();

     //»»»»»»»»»»»»»»»»»»»»»»»»»»»»»»»»»»»»»»»»»»»»»»»»»»»»»»»»»»»»»»»
     cout    << endl<< "================"<< endl
             << "END...C++ Demo: Single Implementation Inheritance"
             << endl;
     while(true){}
     return 0;

}//END...Function MAIN


void createOffice()
{
     //»»»»»»»»»»»»»»»»»»»»»»»»»»»»»»»»»»»»»»»»»»»»»»»»»»»»»»»»»»»»»»»
     cout    << "====== Creating Employee joeBloggs ========"<< endl;
     Employee joeBloggs( "Joseph Bloggs", "PH 45 67 34 D", 17000.0 );
     joeBloggs.greet_Name_SalaryGross();
     //joeBloggs.name = "Joseph Bloggs Snr";
          // ILLEGAL: name is protected; like private,
          //   it cannot be accessed from here

     cout  << endl
           << endl<< "===== Creating Manager pierreHancock ======="
           << endl;
     Manager pierreHancock( "Pierre Hancock" );
```

```
        pierreHancock.greet_Name();

        //»»»»»»»»»»»»»»»»»»»»»»»»»»»»»»»»»»»»»»»»»»»»»»»»»»»»»»»»»»»»»»»»»»»
        cout    << endl
                << endl<< "===== Creating Manager sheilaHemingway ======="
                << endl;
        Manager sheilaHemingway( "Sheila Hemingway",
                                                    "FG 56 34 67 H",
                                                    23000.0 );
        sheilaHemingway.greet_Name_SalaryGross();


}//END...Function createOffice
```

```
====== Creating Employee joeBloggs ========
Base Class Employee constructor has been called
Hello, I'm an Employee, my name is Joseph Bloggs, my Gross
Salary is 17000


===== Creating Manager pierreHancock =======
Base Class Employee constructor has been called
Inherited Class Manager constructor has been called
Hello, I'm an Employee, my name is Pierre Hancock




===== Creating Manager sheilaHemingway =======
Base Class Employee constructor has been called
Inherited Class Manager constructor has been called
Hello,
I'm an Employee, my name is Sheila Hemingway, my
Gross Salary is 23000

Inherited Class Manager destructor has been called
Base Class Employee destructor has been called:
        Goodbye, I, Sheila Hemingway am  going out of scope now.

Inherited Class Manager destructor has been called
Base Class Employee destructor has been called:
        Goodbye, I, Pierre Hancock am  going out of scope now.
Base Class Employee destructor has been called:
        Goodbye, I, Joseph Bloggs am  going out of scope now.

===============
END...C++ Demo: Single Implementation Inheritance
```

Program Listing/Output 22.5: Implementation_Single_Client.cpp

22.8 SINGLE INTERFACE INHERITANCE

❖ Interface Inheritance is the practice of Building a Framework from which specific types can be created

❖ In Object oriented programming, this is achieved by creating a base class where some or all of the member functions have no body, only the prototype

22.8.1 ABSTRACT CLASS

❖ An Abstract Class is a class where one or more member function definition(s) have no body

❖ Note that there is a slight discrepancy in C++ terminology compared to other OOP languages:

		Terminology		
		UML	**Java, C#**	**C++**
Member Functions with NO BODY	All	Interface Class	Interface Class	Abstract Class
	At Least One	Abstract Class	Abstract Class	
	None	Concrete Class	Concrete Class	Concrete Class

22.8.2 USING INTERFACE INHERITANCE

❖ In our case, we will create an Abstract Base Class, Bear

❖ From this, Inherited classes will include

➢ Black_Bear
➢ Sloth_Bear

❖ The full program is as follows:

```
#include <iostream>
#include <string>
using namespace std;

/*    ABSTRACT CLASS:
          -     All of the member functions have no body
*/
class Bear
{
    protected:
```

```
                string name;

        public:
                void greet()
                {
                }

                void greet_withName()
                {
                }
};


class BlackBear : public Bear
{
/*
        private:
                string name;
*/
        public:
                BlackBear( string name_In )
                {
                        name = name_In;
                }

                void greet()
                {
                        cout    << "Hello, I'm a Black Bear"<< endl;
                }

                void greet_withName()
                {
                        cout    << "Hello, I'm a Black Bear and my name is "
                                << name<< endl;
                }
};

class SlothBear : public Bear
{
// private:
//          string name;

    public:
            SlothBear( string name_In )
            {
                    name = name_In;
            }

            void greet()
            {
                    cout << "Hello, I'm a Brown Bear"<< endl;
            }
```

```
        void greet_withName()
        {
                cout << "Hello, I'm a Sloth Bear and my name is "
                     << name<< endl;
        }


};

int main()
{
    //»»»»»»»»»»»»»»»»»»»»»»»»»»»»»»»»»»»»»»»»»»»»»»»»»»»»»»»»»»»»»»»»»»»»»»»»»»»»»
    BlackBear gentleBen( "Gentle Ben" );
    gentleBen.greet_withName();


    SlothBear baloo( "Baloo" );
    baloo.greet_withName();


    //»»»»»»»»»»»»»»»»»»»»»»»»»»»»»»»»»»»»»»»»»»»»»»»»»»»»»»»»»»»»»»»»»»»»»»»»»»»»»
    cout    << endl<< "================"<< endl
            << "END...C++ Demo: Interface Single Inheritance "
            << endl;
    while(true){}
    return 0;
}//END...Function MAIN
```

```
Hello, I'm a Black Bear and my name is Gentle Ben
Hello, I'm a Sloth Bear and my name is Baloo
================
 END...C++ Demo: Interface Single Inheritance
```

Program Listing/Output 22.6: InterfaceInheritance_Single_Client.cpp

22.8.3 BEAUTIES OF INTERFACE INHERITANCE

❖ Interface Inheritance is ideal in teamwork

➢ The bulk of software projects have teams working on different parts

- One of the many beauties of OOP is that often, these parts can be directly represented by classes, so the coding of different classes can be allocated to different teams

➢ It is important to ensure that an overall company structure/style is maintained

- One of the beauties of software engineering, the opportunities for self expression, could become a victim of its own success

❖ The team leader/software architect can implement the base abstract classes

❖ These can then be handed over to the various teams, for them to inherit from these

❖ Consequently, a consistent style is ensured

❖ This is an example of the benefits of OOP representing the real world directly

➢ This is the basis of the Abstract Factory Design Pattern (Cybernetics in C++ - Expert)

22.9 MULTIPLE IMPLEMENTATION INHERITANCE

22.9.1 REASONS FOR MULTIPLE IMPLEMENTATION INHERITANCE

❖ The main motivation for Multiple Inheritance is that entities often belong to more than one category e.g.

➢ An office IS BOTH A building and a workplace, so will possess all the attributes of both

➢ A Liger IS BOTH A Lion and a Tiger

❖ Multiple Inheritance facilitates this in a program by allowing a class to inherit from more than one class

22.9.2 CREATING MULTIPLE IMPLEMENTATION INHERITANCE

❖ In looking at multiple inheritance, the following hierarchy will be implemented:

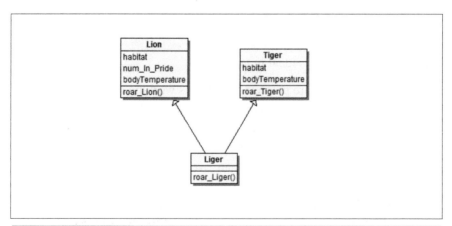

Figure 22.5: UML Representation of Multiple Implementation Inheritance

❖ The program is as follows:

```cpp
#include <iostream>
#include <string>
using namespace std;

#ifndef LION_H
#define LION_H

class Lion
{
      private:
              string habitat;
              int num_In_Pride;
              double bodyTemperature;

      public:
              void greet();
              void roar_Lion();

};//END...Class Declaration Lion

#endif
```

Program Listing/Output 22.7: Lion.h

```cpp
#include "Lion.h"

void Lion::roar_Lion()
{
      cout << "Lion Roar RRRR"<< endl;

}//END...MF roar_Lion

void Lion::greet()
{
      cout << "Hello from the Lion World"<< endl;

}//END...MF roar_Lion
```

Program Listing/Output 22.8: Lion.cpp

```cpp
#include <iostream>
#include <string>
using namespace std;

#ifndef TIGER_H
#define TIGER_H

 class Tiger
 {
  private:
```

```
        string habitat;
        double bodyTemperature;

  public:
        void roar_Tiger();
        void greet();

};//END...Class Declaration Tiger

#endif
```

Program Listing/Output 22.9: Tiger.h

```
#include "Tiger.h"

void Tiger::roar_Tiger()
{
        cout << "Tiger Roar RRRR"<< endl;

}//END...MF roar_Tiger

void Tiger::greet()
{
        cout << "Hello from the Tiger World"<< endl;

}//END...MF roar_Tiger
```

Program Listing/Output 22.10: Tiger.cpp

```
#include <iostream>
#include <string>
using namespace std;

#include "Lion.h"
#include "Tiger.h"

#ifndef LIGER_H
#define LIGER_H

// MULTIPLE Inheritance from Lion & Tiger
class Liger : public Lion, public Tiger
{
        private:

        public:
                void roar_Liger();

};//END...Class Declaration Liger

#endif
```

Program Listing/Output 22.11: Liger.h

```
#include "Liger.h"

void Liger::roar_Liger()
{
      cout << "Liger Roar RRRR"<< endl;

}//END...MF roar_Liger
```

```
#include <iostream>
#include <string>
using namespace std;

#include "Lion.h"
#include "Tiger.h"
#include "Liger.h"


int main()
{
      //»»»»»»»»»»»»»»»»»»»»»»»»»»»»»»»»»»»»»»»»»»»»»»»»»»»»»»»»»»»»
      cout    << endl
              << "============"<< endl
              << "Object creation: Liger, hercules"
              << endl;
      Liger hercules;

      hercules.roar_Lion();
      hercules.roar_Tiger();
      hercules.roar_Liger();

      //»»»»»»»»»»»»»»»»»»»»»»»»»»»»»»»»»»»»»»»»»»»»»»»»»»»»»»»»»»»»
      cout    << endl<< "==============="<< endl
              << "END...C++ Demo: Multiple Implementation Inheritance"
              << endl;
      while(true){}
      return 0;

}//END...Function MAIN
```

```
============
Object creation: Liger, hercules
Lion Roar RRRR
Tiger Roar RRRR
Liger Roar RRRR
===============
END...C++ Demo: Multiple Implementation Inheritance
```

22.9.3 PROSPECTIVE EVILS OF MULTIPLE IMPLEMENTATION INHERITANCE

❖ While multiple Implementation Inheritance can provide flexibility, it can also cause several potential problems

22.9.3.1 Ambiguity

❖ The current program shows one potential problem: both the Lion and Tiger class each have a member function with the same signature: greet()

❖ Common Question:

(Q)　　　So when we have the command

hercules.greet();

Will it call the greet from `Lion`, or from `Tiger`?

(A)　　　There is ambiguity, so this will result in a compile-time error:

22.9.3.1.1 Resolving Using the Scope Resolution Operator

❖ This problem can be solved using the scope resolution operator, as follows:

hercules.Lion::greet();

hercules.Tiger::greet();

❖ The "client part" of the program is as follows:

➢ The rest is the same as Program Listing/Output 22.9 - Program Listing/Output 22.14

```
#include <iostream>
#include <string>
using namespace std;

#include "Lion.h"
#include "Tiger.h"
#include "Liger.h"


int main()
{
    //»»»»»»»»»»»»»»»»»»»»»»»»»»»»»»»»»»»»»»»»»»»»»»»»»»»»»»»»»»»»»»»»
    cout    << endl
            << "============"<< endl
            << "Object creation: Liger, hercules"
            << endl;
    Liger hercules;

    // hercules.greet();
    // ERROR: error C2385: ambiguous access of 'greet'
```

```
        // Resolving the Ambiguity
        hercules.Lion::greet();
        hercules.Tiger::greet();

        //»»»»»»»»»»»»»»»»»»»»»»»»»»»»»»»»»»»»»»»»»»»»»»»»»»»»»»»»»»»»»»»
        cout    << endl<< "================"<< endl
                << "END...C++ Demo: Multiple Implementation Inheritance"
                << endl;
        while(true){}
        return 0;

}//END...Function MAIN
```

```
============
Object creation: Liger, hercules
Hello from the Lion World
Hello from the Tiger World

================
END...C++ Demo: Multiple Implementation Inheritance
```

Program Listing/Output 22.14: Implementation_Multiple_Ambiguity_Client.cpp

22.9.3.2 Diamond Shaped Multiple Inheritance

❖ Diamond Shaped Multiple Inheritance occurs

> When two or more classes inherit from the same class

> Then another class inherits from them

❖ In our example of the Cat family, the following could happen:

> i.e. Cat is the name of a class

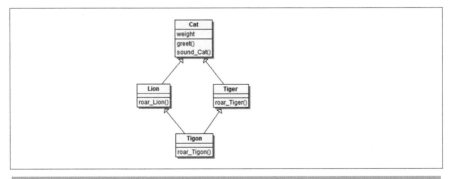

Figure 22.6: Diamond Shaped Multiple Implementation Inheritance

❖ The problem is that

➤ Lion and Tiger will each have their own copy of the data member
`weight`

➤ Tigon will have 2 copies

22.9.3.2.1 Resolving Using Virtual Inheritance/Virtual Classes

❖ The problem can be solved by making the inheritance of both Lion and Tiger, a Virtual Inheritance:

```
class Lion : virtual public Cat{};

class Tiger : virtual public Cat{};
```

❖ This will ensure only one copy of any data members of Cat will be inherited by Tigon

❖ The full program is as follows:

```
#include <iostream>
#include <string>
using namespace std;

#ifndef CAT_H
#define CAT_H

class Cat
{
    protected:
        double weight;

    public:
        Cat( );

};//END...Class Declaration Lion

#endif
```
Program Listing/Output 22.15: Cat.h

```
#include "Cat.h"

Cat::Cat( )
        : weight(180.0)
{

}//END...Constructor Cat( double weight_IN )
```
Program Listing/Output 22.16: Cat.cpp

```
#include "Cat.h"

#ifndef LION_H
#define LION_H

class Lion : virtual public Cat
{
      private:

      public:


};//END...Class Declaration Lion

#endif
```

Program Listing/Output 22.17: Lion.h

```
#include "Cat.h"

#ifndef TIGER_H
#define TIGER_H

class Tiger : virtual public Cat
{
      private:

      public:


};//END...Class Declaration Tiger

#endif
```

Program Listing/Output 22.18: Tiger.h

```
#include "Lion.h"
#include "Tiger.h"

#ifndef TIGON_H
#define TIGON_H

class Tigon : public Lion, public Tiger
{
      public:
            void display_Weight();

};//END...Class Declaration Tiger

#endif
```

Program Listing/Output 22.19: Tigon.h

```
#include "Tigon.h"

void Tigon::display_Weight()
{
        cout << "The weight of this Tigon is "
             << weight
             << endl;

}//END...MF roar_Liger
```

Program Listing/Output 22.20: Tigon.cpp

```
#include <iostream>
#include <string>
using namespace std;

#include "Lion.h"
#include "Tiger.h"
#include "Tigon.h"


int main()
{
    //»»»»»»»»»»»»»»»»»»»»»»»»»»»»»»»»»»»»»»»»»»»»»»»»»»»»»»»»»»»»»»»»
    Tigon shiva;
    shiva.display_Weight();


    //»»»»»»»»»»»»»»»»»»»»»»»»»»»»»»»»»»»»»»»»»»»»»»»»»»»»»»»»»»»»»»»»
    cout    << endl<< "==============="<< endl
            << "END...C++ Demo: Fixing Diamond Multiple "
            << "Implementation Inheritance using Virtual Classes"
            << endl;
    while(true){}
    return 0;

}//END...Function MAIN
```

```
The weight of this Tigon is 180

===============
END...C++ Demo: Fixing Diamond Multiple Implementation Inheritance
using Virtual
 Classes
```

Program Listing/Output 22.21: VirtualClasses_Demo_Client.cpp

❖ Note: The word `virtual` has duplicate meanings in C++

➢ Do not confuse this with the `virtual` member function, used in Polymorphism (section 23)

  • This also involves Inheritance

22.9.3.3 Potential Lack of Organisation

❖ A major problem with Multiple Inheritance is that

➢ It quickly becomes cumbersome

➢ Disorganisation can easily go out of hand, as shown in Figure 22.5

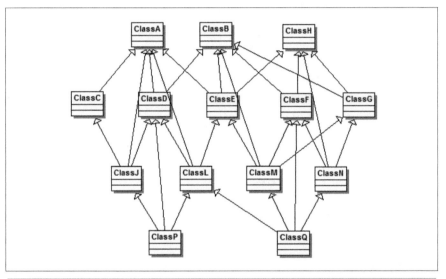

Figure 22.7: Prospective Chaos in Multiple Implementation Inheritance

❖ This is a particularly large problem when project teamwork spans more than one company

➢ If one company has a large project which heavily uses multiple implementation inheritance

➢ They then merge this project with another which also does this

➢ The result of merging two projects with different policies may well be chaotic

22.9.3.4 Multiple Inheritance in other OOP Languages

❖ Several programmers argue that the immense power and flexibility of C++ is a liability rather than an asset

  ➢ The fact that so many different kinds of inheritance are allowed gives too much scope for error

❖ For these reasons

  ➢ Java supports multiple inheritance only with interface inheritance

  ➢ C# does not support multiple inheritance at all

    • In fact, the whole of .NET does not support multiple inheritance at all

22.10 MULTIPLE INTERFACE INHERITANCE

❖ Multiple Interface Inheritance eliminates much of the problems encountered in Multiple Implementation Inheritance

❖ In our example, we will have

  ➢ The Abstract Base Classes

    • Bear

    • CartoonCharacter

  ➢ The Derived Class

    • CartoonBear

```
#include <iostream>
#include <string>
using namespace std;

#ifndef BEAR_H
#define BEAR_H

class Bear
{
      private:

      public:
            void greet_Bear();

};//END...Class Declaration Bear

#endif
```

Program Listing/Output 22.22: Bear.h

```
#include "Bear.h"

void Bear::greet_Bear()
{

}//END...MF greet_Bear
```

```
#include <iostream>
#include <string>
using namespace std;

#ifndef CARTOONCHARACTER_H
#define CARTOONCHARACTER_H

class CartoonCharacter
{
      private:

      public:
            void greet_Cartoon();

};//END...Class Declaration CartoonCharacter

#endif
```

```
#include "CartoonCharacter.h"

void CartoonCharacter::greet_Cartoon()
{

}//END...MF greet_Cartoon
```

```
#include "Bear.h"
#include "CartoonCharacter.h"

#ifndef CARTOONBEAR_H
#define CARTOONBEAR_H

class CartoonBear
      : public Bear, public CartoonCharacter
{
      private:
            string name;
```

```
      public:
            CartoonBear( string );
            void greet_CartoonBear();
            void greet_Bear();
            void greet_Cartoon();

};//END...Class Declaration CartoonBear

#endif
```

Program Listing/Output 22.26: CartoonBear.h

```
#include "CartoonBear.h"

CartoonBear::CartoonBear( string name_IN )
{
      name = name_IN;

}//END...Cponstructor CartoonBear( string name_IN )

void CartoonBear::greet_Bear()
{
    cout << "Hello, I'm a Bear "
        << endl;

}//END...MF greet_Bear

void CartoonBear::greet_Cartoon()
{
    cout << "Hello, I'm a Cartoon Character "
        << endl;

}//END...MF greet_Bear

void CartoonBear::greet_CartoonBear()
{
    greet_Bear();
    greet_Cartoon();
    cout << "Hello, I'm a Cartoon Bear, "
        << "my name is "
        << name
        << endl;

}//END...MF greet_CartoonBear
```

Program Listing/Output 22.27: CartoonBear.cpp

```
#include <iostream>
#include <string>
using namespace std;

#include "Bear.h"
```

```
#include "CartoonCharacter.h"
#include "CartoonBear.h"


int main()
{
    //»»»»»»»»»»»»»»»»»»»»»»»»»»»»»»»»»»»»»»»»»»»»»»»»»»»»»»»»»»»»»»»»»»»»»»»»»
    cout    << endl
            << "============"<< endl
            << "Object creation: Yogi Bear"
            << endl;
    CartoonBear yogi( "Yogi Bear" );
    yogi.greet_CartoonBear();


    //»»»»»»»»»»»»»»»»»»»»»»»»»»»»»»»»»»»»»»»»»»»»»»»»»»»»»»»»»»»»»»»»»»»»»»»»»
    cout    << endl<< "================"<< endl
            << "END...C++ Demo: Multiple Interface Inheritance"
            << endl;
    while(true){}
    return 0;

}//END...Function MAIN
```

```
============
Object creation: Yogi Bear
Hello, I'm a Bear
Hello, I'm a Cartoon Character
Hello, I'm a Cartoon Bear, my name is Yogi Bear

================
END...C++ Demo: Multiple Interface Inheritance
```

Program Listing/Output 22.28: InterfaceInheritance_Multiple_Client.cpp

22.11 PRIVATE & PROTECTED INHERITANCE

❖ As discussed, Private and Protected Inheritance are means of controlling the access specifiers in derived classes

❖ If the inheritance is **Public**

➢ Public members of the base class become public members of the derived class

➢ Protected members of the base class become protected members of the derived class

➢ Private members of the base class remain inaccessible

➢ i.e. Public Inheritance implements an IS-A relationship

❖ If the inheritance is **Protected**

  ➢ Public and Protected members of the base class become **Protected** members of the derived class

  ➢ Private members of the base class remain inaccessible

  ➢ i.e. Protected Inheritance does not implement an IS-A relationship

❖ If the inheritance is **Private**

  ➢ Public and Protected members of the base class become **Private** members of the derived class

  ➢ Private members of the base class remain inaccessible

  ➢ i.e. Private Inheritance does not implement an IS-A relationship either

❖ The main reason for these is facilitating the possibility of hiding implementation details

  ➢ This is commonly encountered in special purpose data structures such as the Stack and the Queue (Cybernetics in C++ - Expert)

  ➢ These can be implemented in several ways, e.g. using

    • Arrays
    • Linked Lists

  ➢ More often than not, you do not want the programmers who use your stack/queue to

    • Know how you have implemented them
    • Have access to the implementation
      ○ They may make unwanted/unintended changes

❖ The following program uses animals, to demonstrate Private, Protected and Public Inheritance:

  ➢ The following Classes and Objects are used:

CLASS		OBJECT
❖ Base Class	`Bear`	`winnieThePooh`
➢ Public Inheritance	`SlothBear`	`baloo`
➢ Protected Inheritance	`PetBear`	`gentleBen`
• Public Inheritance	`TeddyBear`	`paddingtonBear`
➢ Private Inheritance	`Panda`	`masterPoPing`

- A Panda IS-NOT a Bear (at least not all kinds:-))
  - However, it is similar
    - Most of the attributes of Bear can be reused in Panda
  - We do not want users of the Panda class to know/care that we are using the Bear class

```
/*      *********************************************************
        PRIVATE, PROTECTED & PUBLIC INHERITANCE
        *********************************************************
*/
#include <iostream>
using namespace std;

//######################################################################
class Bear
{
        private:
                int private_Data_Bear;
        protected:
                int protected_Data_Bear;
        public:
                int public_Data_Bear;

                void greet()
                {
                        private_Data_Bear = 6; // All OK, in same class
                        protected_Data_Bear = 7; //
                        public_Data_Bear = 8; //

                }//end...MF greet

};//END...CLASS Bear

//######################################################################
/*      PUBLIC INHERITANCE
        Public and Protected members of Bear are unchanged here
*/
class SlothBear : public Bear
{

        public:
                void greet_SlothBear()
                {
                    // private_Data_Bear = 8;
                        //ILLEGAL - cannot access private data in
                        //inherited classes

                    protected_Data_Bear = 9;
                        //OK, can access protected data of Base Classes

                    public_Data_Bear = 10;
```

```
                              //OK, can access public data of Base Classes

            }//END...MF greet_SlothBear

};//END...CLASS SlothBear

//#####################################################################
/*    PROTECTED INHERITANCE
      Public and Protected members of Bear are Protected members here
*/
class PetBear : protected Bear
{
      public:
      void greet_PetBear()
      {
            // private_Data_Bear = 8;
                  //ILLEGAL - cannot access private data in
                  //Inherited classes

            protected_Data_Bear = 9;
                  //OK, can access protected data of Base Classes
                  //- It is Protected here

            public_Data_Bear = 10;
                  // OK, can access public data of Base Classes
                  // - It is Protected here

      }//END...MF greet_SlothBear

};//END...CLASS PetBear

//#####################################################################
/*    PUBLIC i.e. PROTECTED then PUBLIC INHERITANCE
      Public and Protected members of Bear are Protected members here
*/
class TeddyBear : public PetBear
{
      public:
            void greet_TeddyBear()
            {
                  // private_Data_Bear = 8;
                        // ILLEGAL - cannot access private data in
                        // inherited classes

                  protected_Data_Bear = 9;
                        // OK, can access protected data of Base Class
                        // - Protected data of PetBear

                  public_Data_Bear = 10;
                        // OK, can access public data of Base Class
                        // - Protected data of PetBear

            }//END...MF greet_SlothBear
```

```
};//END...CLASS TeddyBear

 //##################################################################
/*    PRIVATE INHERITANCE
       Public and Protected members of Bear are Private members here
*/
class Panda : private Bear
{
            void greet_Panda()
            {
                // private_Data_Bear = 8;
                    // ILLEGAL - cannot access private data in
                    // Inherited classes

                protected_Data_Bear = 9;
                    //OK, can access protected data of Base Classes

                public_Data_Bear = 10;
                    //OK, can access private data of Base Classes

            }//END...MF greet_Panda

};//END...CLASS Panda


int main()
{
    //»»»»»»»»»»»»»»»»»»»»»»»»»»»»»»»»»»»»»»»»»»»»»»»»»»»»»»»»»
    Bear winnieThePooh;
    // winnieThePooh.private_Data_Bear = 9;
        // ILLEGAL: cannot access private data in Bear

    // winnieThePooh.protected_Data_Bear = 10;
        // ILLEGAL: cannot access protected data in Bear

    winnieThePooh.public_Data_Bear = 5;
        // OK but inadvisable

    //»»»»»»»»»»»»»»»»»»»»»»»»»»»»»»»»»»»»»»»»»»»»»»»»»»»»»»»»»
    //     PUBLIC INHERITANCE:
    //     Protected and Public data members of Bear are unchanged
    //
    SlothBear baloo;

    // baloo.private_Data_Bear = 8;
        // ILLEGAL: cannot access private data in SlothBear/Bear

    // baloo.protected_Data_Bear = 9;
        // ILLEGAL: cannot access protected data in SlothBear/Bear

    baloo.public_Data_Bear = 10;
        // OK but inadvisable
```

```
//»»»»»»»»»»»»»»»»»»»»»»»»»»»»»»»»»»»»»»»»»»»»»»»»»»»»»»»»»»»»»»»»»»»
//      PROTECTED INHERITANCE:
//      Protected and Public data members of Bear are Protected
//
PetBear gentleBen;

// gentleBen.private_Data_Bear = 5;
//      // ILLEGAL: cannot access private data in SlothBear/Bear

// gentleBen.protected_Data_Bear = 9;
//      // ILLEGAL: protected data in PetBear

// PetBear.public_Data_Bear = 10;
//      // ILLEGAL: protected data in PetBear


//»»»»»»»»»»»»»»»»»»»»»»»»»»»»»»»»»»»»»»»»»»»»»»»»»»»»»»»»»»»»»»»»»»»
//   PROTECTED then PUBLIC INHERITANCE
//   Protected and Public data members of Bear are Protected
//
TeddyBear paddingtonBear;
//   paddingtonBear.private_Data_Bear = 7;
//      // ILLEGAL: cannot access private data in SlothBear/Bear

// paddingtonBear.protected_Data_Bear = 9;
//      // ILLEGAL: cannot access protected data in SlothBear/
Bear

// paddingtonBear.public_Data_Bear = 10;
//      // ILLEGAL: public_Data_Bear is protected in TeddyBear


//»»»»»»»»»»»»»»»»»»»»»»»»»»»»»»»»»»»»»»»»»»»»»»»»»»»»»»»»»»»»»»»»»»»
//   PRIVATE INHERITANCE:
//   All data members in Bear are private
//
Panda masterPoPing;

// masterPoPing.private_Data_Bear = 9;
//      // ILLEGAL: cannot access private data in Panda/Bear

// masterPoPing.protected_Data_Bear = 10;
//      // ILLEGAL: cannot access protected data in Panda/Bear

// masterPoPing.public_Data_Bear = 11;
//      // ILLEGAL: cannot access public data in Bear

//»»»»»»»»»»»»»»»»»»»»»»»»»»»»»»»»»»»»»»»»»»»»»»»»»»»»»»»»»»»»»»»»»»»
cout    << endl<< "===================="
        << endl<< "END...C++ Demo: Private, Protected Inheritance"
        << endl;
```

```
      while(true){}
      return 0;

}//END...Function main
```

Program Listing/Output 22.29: Private_Protected_Inheritance_Client.cpp

Base Class Access Specifier	TYPE OF INHERITANCE		
	Public	**Protected**	**Private**
Private	Hidden	Hidden	Hidden
Protected	Protected	Protected	Private
Public	Public	Protected	Private

22.12 FULL DEFINITIONS OF ACCESS SPECIFIERS

❖ private

Accessible only by

➢ Member functions of the same class
➢ Its friends
  • Friend Functions
  • Friend Classes

❖ protected

Accessible only by

➢ Member functions of the same class
➢ Its friends
  • Friend Functions
  • Friend Classes
➢ Member functions of Derived classes
➢ Their friends
  • Friend Functions
  • Friend Classes

❖ public

Accessible by

➢ Anything
➢ Except member functions of privately inherited classes

22.13 SUGGESTED RULES OF THUMB

❖ While OOP Inheritance is a young and evolving field, programmers suggest the following practices:

➢ Use Single Inheritance whenever possible

➢ Interface Inheritance is often a good idea

➢ Use the access qualifier `protected` with caution if at all;

• If in doubt, do not use it

➢ Multiple Inheritance

• Should be used with Interface Inheritance

• Multiple Implementation Inheritance should be avoided whenever possible

23. POLYMORPHISM

23.1 WHAT IS POLYMORPHISM?

❖ Polymorphism is the practice of associating **multiple behaviours** with a **single type**

❖ It is

➢ Based on the Greek word "polumorphos (πολυμορπηοσ) " meaning many shapes

➢ Made up of 2 words

• *"Poly"* meaning many

• *"Morph"* meaning change

❖ This is another powerful paradigm of thinking which is used abundantly outside as well as inside the programming world

23.1.1 POLYMORPHISM OUTSIDE THE PROGRAMMING WORLD

❖ Rooms

➢ There are many different sorts of room e.g. living room, dining room, bedroom, bathroom, kitchen

➢ Vacuum cleaning is one activity which is performed in rooms in general

❖ Tuning guitars

➢ There are several different sorts of guitar: classical, acoustic, 12 string acoustic, archtop, semi-acoustic electric, solid body electric...

➤ An electronic tuner should be able to work on all of them

❖ Biology: two or more phenotypes can exist in the same species

➤ Leopards: panthers, snow leopards, clouded leopards

➤ Jaguars: Dark-morph/melanistic jaguars

❖ Chemistry: the ability of a solid material to exist in more than one form or crystal structure; examples include:

➤ Allotropy of carbon: exists as diamond or graphite...or the recently discovered Buckminster Fullerene

➤ Allotropy of oxygen: exists as dioxygen O_2, ozone O_3, tetraoxygen O_4, octaoxygen O_8

❖ People speaking different languages

➤ The same story can be told in many different languages

❖ IT

➤ You are organising your files in a hierarchy of directories

➤ There are several actions you need to perform on all of them, including

• Copying files, creating backups
• Carrying out virus scans

23.2 IMPLEMENTING POLYMORPHISM IN C++

23.2.1 REQUIREMENTS

❖ Cue that polymorphism is needed: Having a data structure, containing different but related objects, on which similar actions need to be performed

❖ There are three essential components in the C++ implementation of OOP polymorphism

➤ An Inheritance Hierarchy with member functions which have the SAME signature

➤ Virtual member functions

➤ Pointers to different classes in the inheritance hierarchy

• References can also be used but we will discuss in section 23.4 why this is a less common practice

23.2.2 OOP INHERITANCE HIERARCHY

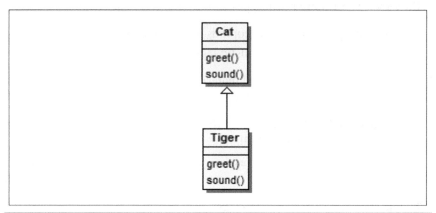

Figure 23.1: Inheritance/Generalisation Hierarchy, to be used in polymorphism

❖ Performing the usual check of whether or not inheritance is appropriate...

> Inheritance is appropriate here since a Tiger IS-A Cat

❖ Each class has the following member functions, each with the **same signature**

> `greet()`

> `sound()`

❖ This is a common and important practice

> The reasons will be explained in section 23.4

❖ Our objective is to create a data structure based on the above inheritance hierarchy, representing a zoo

> For this example, an array will be used

  • This is not the best data structure for objects

> An STL data structure, such as a Vector or Deque, is much better

  • These are covered in Cybernetics in C++ - Expert

❖ We want to traverse this array, then for each element, call the appropriate `sound()` member function, a mixture of

➢ Cat sounds and

➢ Tiger sounds

❖ Notice immediately, we cannot have an array of objects, some of type Cat, some of type Tiger

➢ A fundamental rule of data structures is that every element must be of the same type

➢ It is Polymorphism which gives a powerful solution to this ubiquitous problem

23.2.3 VIRTUAL MEMBER FUNCTIONS

❖ The `virtual` qualifier in a member function determines the behaviour at runtime (demonstrated section 23.2.5)

➢ Which class to bind to in the inheritance hierarchy

❖ In this example, in the class `Cat`

➢ the member function `greet()` is non-virtual

➢ the member function `sound()` is virtual

❖ When a class contains one or more virtual functions, it becomes a **polymorphic** class

➢ i.e. A class on which type flexibility can be conducted

❖ When a member function with the same signature is encountered anywhere lower down in the inheritance hierarchy, it is also virtual

➢ Thus in our example, in the class `Tiger`

• `sound()` is also virtual

• `greet()` is non-virtual

➢ It is acceptable to put the keyword `virtual` again, it just becomes redundant

➢ It is up to the programmer (i.e. you! :-)) to decide whether to put it in

• It is a common practice to put it in again, since this reminds everyone maintaining and referring to the program that it is also virtual

• Without it, another written reminder in a comment, would be a useful reminder in debugging

23.2.4 POINTERS IN INHERITANCE HIERARCHIES

❖ In our main program, we will create the following objects and pointers:

```
Cat oMalley;
Cat *ptr_oMalley = &oMalley;

Tiger shereKhan;
Cat *ptr_shereKhan = &shereKhan;
```

❖ When looking at the fourth line,

```
Cat *ptr_shereKhan = &shereKhan;
```

you may be thinking there will be a type mismatch

➢ After all, the following will cause a type mismatch:

```
int myInt_A = 8;
double *ptrDbl_A;
ptrDbl_A = &myInt_A; // ILLEGAL: Cannot have a Pointer-to-double
                     // containing the address of an int
```

➢ Why then is it acceptable to have a Pointer-to-Cat containing the address of an object of type Tiger?

   • This is a major aspect of object oriented programming:

❖ **It is acceptable for a pointer-to-class to contain the address of an object of type any class below it in the inheritance hierarchy**

➢ This shows the power of inheritance in object oriented programming: the relationship between classes is strong enough to allow this flexibility

➢ In our example, Tiger comes below Cat belong in the same inheritance hierarchy

   • Since a Tiger IS-A Cat, the strength of the relationship is further illustrated

❖ In contrast, with primitive variables, the types are independent of each other, thus there is no question of pointer type flexibility

➢ A double is independent of type int

23.2.5 POLYMORPHISM ON THE RUNTIME STACK

❖ At runtime, any object or pointer/reference to an object must decide which class to BIND-TO at runtime

➢ When there is no inheritance hierarchy, this is predetermined

➢ When there is an inheritance hierarchy, this is determined by the nature of the pointers and whether or not the member functions are virtual

❖ In our example, the question now is what kind of runtime behaviour can we expect?

➢ Will the

- Objects
- Pointers-to-Objects

bind to the

- Base class
- Derived class

at runtime?

```
#include <iostream>
using namespace std;

#ifndef CAT_H
#define CAT_H

class Cat     // Cat is a POLYMORPHIC Base class
              // - it contains a VIRTUAL member function
{
      private:
      public:
              void greet();              // greet() is NON-VIRTUAL
              virtual void sound();      // sound() is VIRTUAL

};//END...Class Cat

#endif
```

Program Listing/Output 23.1: Cat.h Declaration for the Polymorphic Base Class Cat

```
#include "Cat.h"


void Cat::greet()
{
      cout << "Hello from the Cat World"<< endl;
```

```
}//END...Member Function greet

// sound() is VIRTUAL - cannot have the virtual keyword again here
void Cat::sound()
{
      cout << "Miaow"<< endl;

}//END...Member Function sound
```

Program Listing/Output 23.2: Cat.cpp Definition for the Polymorphic Base Class Cat

```
#include "Cat.h"

#ifndef TIGER_H
#define TIGER_H

class Tiger : public Cat // Inheritance because a Tiger IS-A Cat
{
      private:
      public:
            // Member functions with the same name as the Base Class
      Cat
            void greet();            // greet() is NON-VIRTUAL
            virtual void sound();    // sound() is VIRTUAL

};//END...Class Tiger

#endif
```

Program Listing/Output 23.3: Tiger.h Declaration for the Polymorphic Derived Class Tiger

❖ Tiger is a derived class which has member functions with the same signatures as those in its base class Cat

❖ The member function sound() in Tiger is virtual

```
#include "Tiger.h"


void Tiger::greet()
{
      cout << "Hello from the Tiger World"<< endl;

}//END...Member Function greet

// sound() is VIRTUAL - cannot have the virtual keyword again here
void Tiger::sound()
{
      cout << "Tiger Roar RRRR"<< endl;
```

```
}//END...Member Function sound
```

Program Listing/Output 23.4: Tiger.cpp Definition for the Polymorphic Base Class Tiger

```
/*      ****************************************************************
        OOP Polymorphism on the Runtime Stack
        ****************************************************************
        -       This program demonstrates using Objects based on
   Polymorphic
                Classes
                -       Calling Member Functions via OBJECTS
                        -       Polymorphism does not take effect here
                -       Calling Member Functions via POINTERS
                        -       The Pointers are of type Pointer-to-BASEclass
                        -       Polymorphism takes effect here
                        -       The differences between virtual and non-virtual
                                functions becomes apparent
*/

#include "Cat.h"
#include "Tiger.h"

#include <iostream>
using namespace std;


int main()
{
        //===============================================================
        //      Defining Objects and Pointers to Objects
        //
        Cat oMalley;
        Cat *ptr_oMalley = &oMalley;

        Tiger shereKhan;
        Cat *ptr_shereKhan = &shereKhan;


        //===============================================================
        //      ((1)) Calling Member Functions via OBJECTS
        //
        cout    << endl<< "=============="
                << endl<< "Calling Member Functions via Objects"
                << endl<< "-------"
                << endl;
        cout    << "oMalley.greet()\t\t\t";
        oMalley.greet();

        cout    << "oMalley.sound()\t\t\t";
        oMalley.sound();
```

```
       cout    << "shereKhan.greet():\t\t";
       shereKhan.greet();

       cout    << "shereKhan.sound():\t\t";
       shereKhan.sound();

       //============================================================
       //      (2) Calling Member Functions via POINTERS
       //
       cout    << endl<< "==============="
               << endl<< "Calling Member Functions via POINTERS"
               << endl<< "-------"
               << endl;

       cout    << "ptr_oMalley -> greet()\t\t";
       ptr_oMalley -> greet();

       cout    << "ptr_oMalley -> sound()\t\t";
       ptr_oMalley -> sound();

       cout    << "ptr_shereKhan -> greet():\t";
       ptr_shereKhan -> greet();

       cout    << "ptr_shereKhan -> sound():\t";
       ptr_shereKhan -> sound();

       //============================================================
       cout    << endl<< endl<< "============"<< endl
               << "END...OOP Polymorphism, Runtime Stack Demo"
               << endl;
       while(true){}
       return 0;

}//END...Function main
```

```
===============
Calling Member Functions via Objects
-------
oMalley.greet()                 Hello from the Cat World
oMalley.sound()                 Miaow
shereKhan.greet():              Hello from the Tiger World
shereKhan.sound():              Tiger Roar RRRR

===============
Calling Member Functions via POINTERS
-------
ptr_oMalley -> greet()          Hello from the Cat World
ptr_oMalley -> sound()          Miaow
ptr_shereKhan -> greet():       Hello from the Cat World
```

```
ptr_shereKhan -> sound():        Tiger Roar RRRR

===========
END...OOP Polymorphism, Runtime Stack Demo
```

Program Listing/Output 23.5: OOP_Polymorphism_Stack_Client.cpp

23.2.6 RUNTIME BINDING

❖ We observe the following:

  ➢ When we call member functions via objects, the `virtual` qualifier is insignificant

   • i.e. This part of the project was a revision of section 18
   • The runtime behaviour just means binding to the class of the object's type

  ➢ When we call member functions via pointers,

   • When the object is of the BASE class type, here Cat, the runtime binding is to the BASE class
     ○ This is the only option
       ■ both calls via `ptr_oMalley` bind to the base class
   • When the object is of the DERIVED class type
     ○ When the member function is NON-VIRTUAL, the binding takes place to the BASE class
       ■ `ptr_shereKhan -> greet()` binds to the BASE class
     ○ When the member function is VIRTUAL, the binding takes place to the DERIVED class
       ■ `ptr_shereKhan -> sound()` binds to the BASE class

❖ The results can be summarised as follows:

Pointer Type	Object Type	Member Function: Virtual/ Non-Virtual?	Runtime BINDING
Pointer-to- BASE Class	Base Class	Non-Virtual	Base Class
	Base Class	Virtual	Base Class
	Derived Class	Non-Virtual	Base Class
	Derived Class	Virtual	Derived Class

23.2.7 POLYMORPHISM ON THE RUNTIME HEAP

❖ The notation for allocating objects on the runtime heap is as follows:

```
CompileTime   *  pointerName  =  new   Runtime    (  Constructor  ) ;
ClassName                              ClassName  (  Inputs
```

➤ The Left-Hand side of the = is an l-value, a location, known at COMPILE-TIME

➤ The Right-Hand side of the = is an r-value, a location, known at RUNTIME

❖ Reminder: When there is no inheritance/polymorphism involved,

➤ The class bound to at compile-time and the class bound to at runtime are the same, so we see the same class name twice, e.g.

```
Cat *ptr_Duchess = new Cat();
```

❖ As we have been discussing, polymorphism is all about making the compile-time type and runtime type different

```
Cat *ptr_Rajah = new Tiger();
```

❖ Our forthcoming demonstration uses the same Inheritance hierarchy as section 23.2.5:

➤ Cat.h, as Program Listing/Output 23.1

➤ Cat.cpp as Program Listing/Output 23.2

➤ Tiger.h as Program Listing/Output 23.3

➤ Tiger.cpp as Program Listing/Output 23.4

❖ There is now the additional client program, as follows:

```
/*      ****************************************************************
        OOP Polymorphism on the Runtime Heap
        ****************************************************************
        -       This program demonstrates using Objects based on
    Polymorphic
            Classes
        -       Calling Member Functions via POINTERS
            -       The Pointers are of type Pointer-to-BASEclass
            -       Polymorphism takes effect here
            -       The differences between virtual and non-virtual
                    functions becomes apparent
*/
#include "Cat.h"
```

```cpp
#include "Tiger.h"


#include <iostream>
using namespace std;


int main()
{
    //==============================================================
    //     OOP Polymorphism on the Runtime HEAP
    //
    Cat *ptr_Duchess = new Cat(); // Cat twice
    Cat *ptr_Rajah = new Tiger();

     cout   << "==================================="<< endl
            << "POLYMORPHISM on the Runtime HEAP"<< endl
            << "==================================="<< endl;
    //==============================================================
    // (2.1) Call Member Functions via Pointers
    //
    cout   << endl<< "=============="
           << endl<< "Calling Member Functions via POINTERS"
           << endl<< "-------"
           << endl;

    cout   << "ptr_Duchess -> greet()\t\t";
    ptr_Duchess -> greet();

    cout   << "ptr_Duchess -> sound()\t\t";
    ptr_Duchess -> sound();


    cout   << "ptr_Rajah -> greet()\t\t";
    ptr_Rajah -> greet();

    cout   << "ptr_Rajah -> sound()\t\t";
    ptr_Rajah -> sound();


    //
    ================================================================
    cout   << endl<< endl<< "==========="<< endl
           << "END...OOP Polymorphism, Runtime Heap Demo"
           << endl;
    while(true){}
    return 0;

}//END...Function main
```

```
=====================================
POLYMORPHISM on the Runtime HEAP
=====================================


===============
Calling Member Functions via POINTERS
-------
ptr_Duchess -> greet()          Hello from the Cat World
ptr_Duchess -> sound()          Miaow
ptr_Rajah -> greet()            Hello from the Cat World
ptr_Rajah -> sound()            Tiger Roar RRRR



===========
END...OOP Polymorphism, Runtime Heap Demo
```

Program Listing/Output 23.6: OOP_Polymorphism_Heap_Client.cpp

❖ As far as results go, this program confirms the findings:

➢ Both pointers are to the Base class i.e. type Pointer-to-Cat

➢ For `ptr_Duchess`, the runtime type is also the Base Class Cat, so the runtime binding is to the

- The Base class `Cat` for the non-virtual member function `greet()`
- The Base class `Cat` for the virtual member function `sound()`

➢ For `ptr_Rajah`, the runtime type is the Derived Class Tiger, so the runtime binding is to

- The Base class `Cat` for the non-virtual member function `greet()`
- The Derived Class `Tiger` for the virtual member function `sound()`

23.3 PURE VIRTUAL MEMBER FUNCTIONS

23.3.1 INTERFACE INHERITANCE

❖ As a small extension, the inheritance hierarchy often involves

➢ Having an Abstract Class as the Base Class Abstract Class

➢ Creating a Realisation Inheritance hierarchy

❖ If one or more virtual functions in the base class are made pure virtual member function(s), then the base class is forced to be an abstract class

❖ The following Realisation hierarchy will be used in forthcoming polymorphism project:

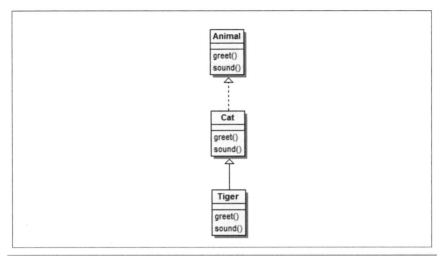

Figure 23.2: Inheritance/Realisation Hierarchy, to be used in polymorphism

❖ Here, the Base class Animal will be used; this is highly appropriate, since

   ➢ A Cat IS-AN Animal
   ➢ A Tiger IS-A Cat which IS-AN Animal

❖ Again, we want to make

   ➢ `greet()` non-virtual
   ➢ `sound()` virtual

23.3.2 NOTATION

❖ In order to make a virtual member function pure, the following notation is used:

```
virtual returnType memberFuntionName() = 0;
```

   ➢ Thus in order to make sound() a pure virtual member function, the following is required:

```
virtual void sound() = 0;
```

❖ C++ is often castigated for this notation

   ➢ The `= 0` makes it look like a variable, possibly a data member of a class, is being set equal to 0

- It is nothing of the sort
➤ This is the notation which is recognised by C++ compilers

23.3.3 CLASS FORCED TO BE ABSTRACT

❖ The important feature is that when a class contains a pure virtual function, it can only ever be an **Abstract** class

➤ i.e. Any attempts to create an object based on it

```
Animal fred;  // ILLEGAL - Animal is an Abstract class,
              // cannot have an object of that type
```

will result in a compile-time error

➤ Like the rest of OOP, this does reflect the real world

- If someone asks the question "What pet do you own?"
  ○ it is inappropriate to answer "An animal!"

23.3.4 USING PURE VIRTUAL FUNCTIONS

❖ Once the abstract base class has been created, concrete classes should be created using inheritance

❖ After that, they should be used in the usual way

❖ The following project shows this:

```
#include <iostream>
using namespace std;

#ifndef ANIMAL_H
#define ANIMAL_H

class Animal // Cat is a POLYMORPHIC Base class - it contains a
    VIRTUAL  member function
{
    private:
    public:
            void greet();              // greet() is NON-VIRTUAL

            // PURE Virtual Function
            virtual void sound( ) = 0; // sound() is VIRTUAL

};//END...Class Animal

#endif
```
Program Listing/Output 23.7: Animal.h - Declaration of the Abstract Base Class Animal

```
#include "Animal.h"

#ifndef CAT_H
#define CAT_H

class Cat : public Animal        // Cat is a POLYMORPHIC Base class - it
     contains a VIRTUAL member function
{
        private:
        public:
        void greet();            // greet() is NON-VIRTUAL
        virtual void sound();    // sound() is VIRTUAL
};//END...Class Cat

#endif
```

Program Listing/Output 23.8: Cat.h Declaration for the Polymorphic Base Class Cat

```
#include "Cat.h"


void Cat::greet()
{
     cout << "Hello from the Cat World"<< endl;

}//END...Member Function greet

// sound() is VIRTUAL - cannot have the virtual keyword again here
void Cat::sound()
{
     cout << "Miaow"<< endl;

}//END...Member Function sound
```

Program Listing/Output 23.9: Cat.cpp Definition for the Polymorphic Base Class Cat

```
#include "Cat.h"

#ifndef TIGER_H
#define TIGER_H

class Tiger : public Cat // Inheritance because a Tiger IS-A Cat
{
     private:
     public:
             // Member functions with the same name as the Base Class
   Cat
             void greet();            // greet() is NON-VIRTUAL
             virtual void sound();    // sound() is VIRTUAL

};//END...Class Tiger
```

```
#endif
```

Program Listing/Output 23.10: Tiger.h Declaration for the Polymorphic Derived Class Tiger

```
#include "Tiger.h"

void Tiger::greet()
{
    cout << "Hello from the Tiger World"<< endl;

}//END...Member Function greet

// sound() is VIRTUAL - cannot have the virtual keyword again here
void Tiger::sound()
{
    cout << "Tiger Roar RRRR"<< endl;

}//END...Member Function sound
```

Program Listing/Output 23.11: Tiger.cpp Definition for the Polymorphic Base Class Tiger

```
/*      ***************************************************************
        OOP Polymorphism
        with Pure Virtual Member Functions and Abstract Classes
        ***************************************************************
        -       This program demonstrates
                -       Creating an Abstract Class with a Pure Virtual
                        Member Function
                -       Building concrete classes in an Inheritance
                        hierarchy
                -       Implementing Polymorphism on the Runtime Stack
                        -       Calling Member Functions via OBJECTS
                        -       Calling Member Functions via POINTERS
                -       Implementing Polymorphism on the Runtime Heap
                        -       Calling Member Functions via POINTERS
*/
#include "Cat.h"
#include "Tiger.h"
#include "Animal.h"

#include <iostream>
using namespace std;


int main()
{
        //============================================================
        //      ((1)) POLYMORPHISM on the Runtime STACK
        //
        cout    << "==================================="<< endl
                << "POLYMORPHISM on the Runtime STACK"<< endl
```

```
                    << "===================================="<< endl;

  //===============================================================
  //          Defining Objects and Pointers to Objects
  //

  // Animal fred;                    // ILLEGAL - Animal is an
Abstract class,
                                     // cannot have an object of
that type
  Cat oMalley;
  Cat *ptr_oMalley = &oMalley;

  Tiger shereKhan;
  Cat *ptr_shereKhan = &shereKhan;


  //===============================================================
  //      (1.1) Calling Member Functions via OBJECTS
  //
  cout    << endl<< "=============="
          << endl<< "Calling Member Functions via Objects"
          << endl<< "-------"
          << endl;
  cout    << "oMalley.greet()\t\t";
  oMalley.greet();

  cout    << "oMalley.sound()\t\t\t";
  oMalley.sound();

  cout    << "shereKhan.greet():\t\t";
  shereKhan.greet();

cout      << "shereKhan.sound():\t\t";
shereKhan.sound();

//===============================================================
//      (1.2) Calling Member Functions via POINTERS
//
  cout    << endl<< "=============="
          << endl<< "Calling Member Functions via POINTERS"
          << endl<< "-------"
          << endl;

  cout    << "ptr_oMalley -> greet()\t\t";
  ptr_oMalley -> greet();

  cout << "ptr_oMalley -> sound()\t\t";
  ptr_oMalley -> sound();

  cout << "ptr_shereKhan -> greet():\t";
  ptr_shereKhan -> greet();
```

```
      cout << "ptr_shereKhan -> sound():\t";
      ptr_shereKhan -> sound();

      //================================================================
      //   (2) POLYMORPHISM on the Runtime HEAP
      //
      cout << endl
           << "==================================="<< endl
           << "POLYMORPHISM on the Runtime HEAP"<< endl
           << "==================================="<< endl;

      //================================================================
      // Defining Objects and Pointers to Objects
      //
      Cat *ptr_Duchess = new Cat(); // Cat twice
      Cat *ptr_Rajah = new Tiger();

      //================================================================
      // (2.1) Call Member Functions via Pointers
      //
      cout    << endl<< "==============="
              << endl<< "Calling Member Functions via POINTERS"
              << endl<< "-------"
              << endl;


      cout    << "ptr_Duchess -> greet()\t\t";
      ptr_Duchess -> greet();          // "Hello from the Cat World"

      cout    << "ptr_Duchess -> sound()\t\t";
      ptr_Duchess -> sound();          // "Miaow"


      cout << "ptr_Rajah -> greet()\t\t";
      ptr_Rajah -> greet();            // "Hello from the Cat World"

      cout << "ptr_Rajah -> sound()\t\t";
      ptr_Rajah -> sound();  // "Tiger Roar RRRR"

      //================================================================
      cout    << endl<< "===================="
              << endl<< "END...OOP Polymorphism Demo"
              << endl<< "...with Pure Virtual Member Functions "
                  << "and Abstract Classes";
      while(true){}
      return 0;

}//END...Function main
```

```
====================================
POLYMORPHISM on the Runtime STACK
====================================



==============
Calling Member Functions via Objects
-------
oMalley.greet()            Hello from the Cat World
oMalley.sound()            Miaow
shereKhan.greet():         Hello from the Tiger World
shereKhan.sound():         Tiger Roar RRRR



==============
Calling Member Functions via POINTERS
-------
ptr_oMalley -> greet()     Hello from the Cat World
ptr_oMalley -> sound()     Miaow
ptr_shereKhan -> greet():  Hello from the Cat World
ptr_shereKhan -> sound():  Tiger Roar RRRR



====================================
POLYMORPHISM on the Runtime HEAP
====================================



==============
Calling Member Functions via POINTERS
-------
ptr_Duchess -> greet()     Hello from the Cat World
ptr_Duchess -> sound()     Miaow
ptr_Rajah -> greet()       Hello from the Cat World
ptr_Rajah -> sound()       Tiger Roar RRRR

===================
END...OOP Polymorphism Demo
...with Pure Virtual Member Functions and Abstract Classes
```

Program Listing/Output 23.12: Pure_Virtual_MemberFunction_Client.cpp

23.4 BEAUTIES OF POLYMORPHISM

23.4.1 THE ESSENTIAL NATURE OF POLYMORPHISM

❖ Polymorphism, the P of A PIE, is often considered the most important building block of object oriented programming

➢ Programmers who are familiar with only abstraction, encapsulation and inheritance can at best call themselves "Object based programmers"

❖ Polymorphism is essential when dealing with data structures of objects

➢ Revision: For all data structures, the TYPE of each element must be the same

❖ We frequently have situations where we want to perform a similar action on several objects which are similar

➢ A common example is using Graphical User Interfaces

• As revision, here is a fairly typical setup of a GUI hierarchy:

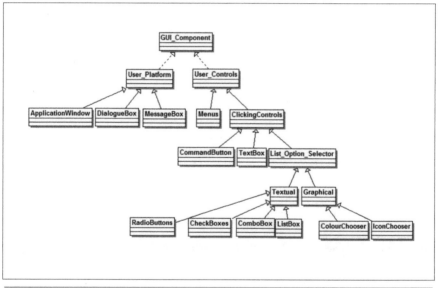

Figure 23.3: Common Inheritance Hierarchy used in GUI programming

➢ You want your user to be able to create their own settings e.g.

• Change the font in the title bar of every GUI component from the default font, to one of their choice, say Tahoma

- We want to have a data structure which we can traverse, and for each element, call the member function

  `change_TitleBarFont( Font desiredFont )`

- Issue: We want to perform this action on several different objects, including Application Windows, Dialogue Boxes, Message Boxes
  - These are all of different types, so we cannot have these objects residing in the same data structure

❖ It is polymorphism which provides a powerful solution to this ubiquitous problem

  ➢ Since it is perfectly acceptable to have the following

  `BaseClass *ptr_ObjectOfTypeDerivedClass`

  We can have a data structure of type `Pointer-to-GUI_Component`
  ➢ It is now possible to traverse this data structure, for each element, the member function `change_TitleBarFont( Font desiredFont )` `is called`

  - This explains the practice of ensuring the member functions have the same signature
    - If they did not, a different member function would have to be called for each component
      - The entire procedure being investigated here would not be possible

  ➢ We ensure it is a VIRTUAL member function, so the DERIVED class will be bound to

  - i.e. the GUI components themselves
    - the dialogue boxes
    - messages boxes
    - application windows
    - menus...
      will be modified

23.4.2 POINTERS PREFERRED OVER REFERENCES

❖ This also shows why it is more common to implement polymorphism using pointers rather than references

  ➢ The main aim of polymorphism is to use data structures
  ➢ Arrays are made up using pointers

➤ As we will see in Cybernetics in C++ - Expert, the same can be said for most, if not all data structures, such as linked lists, trees and graphs

➤ Attempting to have a data structure of references will mean mixing together pointers and references, which is often advised against

❖ Having an array, or any other data structure of pointers, avoids this problem

23.5 DATA STRUCTURES OF POLYMORPHIC OBJECTS

❖ With this in mind, we now look back at our original objective: we want a data structure of different animals

❖ This is possible by making it of type Pointer-to-Animal

➤ The following will be used:

```
Animal *zoo[] = {  ptr_oMalley,
                   ptr_shereKhan,
                   ptr_Duchess,
                   ptr_Rajah

                };
```

➤ Since the array operator [ ] takes precedence over the pointer operator*

  o This is an array of type Pointer-to-`Animal`
  o It is not a Pointer to an array of type `Animal`
    Unscrambling declarations is covered in detail in Cybernetics in C++ - Expert.

❖ The aim is to traverse this array, calling the virtual member function `sound()` for each element

❖ The full program has been shown below:

```
#include <iostream>
using namespace std;

#ifndef ANIMAL_H
#define ANIMAL_H

class Animal // Cat is a POLYMORPHIC Base class - it contains a
VIRTUAL  member function
{
      private:
      public:
            void greet();                  // greet() is NON-VIRTUAL
```

```
                 // PURE Virtual Function
                 virtual void sound( ) = 0;  // sound() is VIRTUAL

};//END...Class Animal

#endif
```

```
#include "Animal.h"

#ifndef CAT_H
#define CAT_H

class Cat : public Animal  // Cat is a POLYMORPHIC Base class - it
    contains a VIRTUAL member function
{
        private:
        public:
                void greet();               // greet() is NON-VIRTUAL
                virtual void sound();        // sound() is VIRTUAL
};//END...Class Cat

#endif
```

```
#include "Cat.h"


void Cat::greet()
{
     cout << "Hello from the Cat World"<< endl;

}//END...Member Function greet

// sound() is VIRTUAL - cannot have the virtual keyword again here
void Cat::sound()
{
     cout   << "Miaow"<< endl;

}//END...Member Function sound
```

```
#include "Cat.h"

#ifndef TIGER_H
#define TIGER_H

class Tiger : public Cat // Inheritance because a Tiger IS-A Cat
```

```
{
      private:
      public:
      // Member functions with the same name as the Base Class Cat
      void greet();            // greet() is NON-VIRTUAL
      virtual void sound();    // sound() is VIRTUAL

};//END...Class Tiger

#endif
```

Program Listing/Output 23.16: Tiger.h Declaration for the Polymorphic Derived Class Tiger

```
#include "Tiger.h"


void Tiger::greet()
{
      cout << "Hello from the Tiger World"<< endl;

}//END...Member Function greet

// sound() is VIRTUAL - cannot have the virtual keyword again here
void Tiger::sound()
{
      cout << "Tiger Roar RRRR"<< endl;

}//END...Member Function sound
```

Program Listing/Output 23.17: Tiger.cpp Definition for the Polymorphic Base Class Tiger

```
/*      ************************************************************
        OOP Polymorphism in Data Structures
        ************************************************************
        -       This program demonstrates
                -       Creating a Data Structure (an array) of type
                        Pointer-to-AbstractBaseClass
                -       Populating this with pointers to objects of
                        different types in the OOP Inheritance Hierarchy
                        -       These objects reside on both the
                                -       Runtime Stack
                                -       Runtime Heap
                -       Traversing this array and calling the same virtual
                        member function for each element
                -       Components in this program:
                        -       Pure Virtual Member Functions
                        -       Abstract Classes

        -       NB. The Array is not the optimal Data Structure
                for Objects
                                -       Better: STL Container, vector or deque
*/
```

```
#include <iostream>
using namespace std;

#include "Cat.h"
#include "Tiger.h"
#include "Animal.h"

int main()
{
    //==============================================================
    //      (1) POLYMORPHISM on the Runtime STACK
    //
    cout    << "===================================="<< endl
            << "POLYMORPHISM on the Runtime STACK"<< endl
            << "===================================="<< endl;


    //==============================================================
    //         Defining Objects and Pointers to Objects
    //

    // Animal fred;         // ILLEGAL - Animal is an Abstract class,
                            // cannot have an object of that type
    Cat oMalley;
    Cat *ptr_oMalley = &oMalley;

    Tiger shereKhan;
    Cat *ptr_shereKhan = &shereKhan;

    //==============================================================
    // (1.1) Calling Member Functions via OBJECTS
    //
    cout   << endl<< "=============="
           << endl<< "Calling Member Functions via Objects"
           << endl<< "-------"
           << endl;
    cout   << "oMalley.greet()\t\t\t";
    oMalley.greet();

    cout   << "oMalley.sound()\t\t\t ";
    oMalley.sound();

    cout   << "shereKhan.greet():\t\t";
    shereKhan.greet();

    cout   << "shereKhan.sound():\t\t";
    shereKhan.sound();

    //==============================================================
    // (1.2) Calling Member Functions via POINTERS
    //
    cout   << endl<< "=============="
           << endl<< "Calling Member Functions via POINTERS"
           << endl<< "-------"
```

```cpp
                << endl;

  cout  << "ptr_oMalley -> greet()\t\t";
  ptr_oMalley -> greet();

  cout  << "ptr_oMalley -> sound()\t\t";
  ptr_oMalley -> sound();

  cout  << "ptr_shereKhan -> greet():\t";
  ptr_shereKhan -> greet();

  cout << "ptr_shereKhan -> sound():\t";
  ptr_shereKhan -> sound();

//==================================================================
// (2) POLYMORPHISM on the Runtime HEAP
//
cout  << endl
      << "===================================="<< endl
      << "POLYMORPHISM on the Runtime HEAP"<< endl
      << "===================================="<< endl;

//==================================================================
//   Defining Objects and Pointers to Objects
//
Cat *ptr_Duchess = new Cat();    // Cat twice
Cat *ptr_Rajah = new Tiger();

//==================================================================
// (2.1) Call Member Functions via Pointers
//
cout  << endl<< "=============="
      << endl<< "Calling Member Functions via POINTERS"
      << endl<< "-------"
      << endl;


  cout  << "ptr_Duchess -> greet()\t\t";
  ptr_Duchess -> greet();  // "Hello from the Cat World"

  cout  << "ptr_Duchess -> sound()\t\t";
  ptr_Duchess -> sound();  // "Miaow"

  cout  << "ptr_Rajah -> greet()\t\t";
  ptr_Rajah -> greet();  // "Hello from the Cat World"


  cout  << "ptr_Rajah -> sound()\t\t";
  ptr_Rajah -> sound();  // "Tiger Roar RRRR"

//==================================================================
//  (3) DATA STRUCTURE of Objects - the reason why
//  Polymorphism is essential
```

```
   //
   cout  << endl
         << "============================================"<< endl
         << "CREATING AN ARRAY of (Pointers-to-) Animals"<< endl
         << "============================================"<< endl;

   // Creating and initialising an array
   //    -    Name: zoo
   //    -    Type: Pointer-to-Animal
   //
   Animal *zoo[] =     {      ptr_oMalley,
                              ptr_shereKhan,
                              ptr_Duchess,
                              ptr_Rajah
                      };

   cout   << endl<< "============="
          << endl<< "Calling each animal's sound() in the array zoo"
          << endl;

   // Not the best way
   //    - Better to use STL Iterators (Cybernetics in C++ - Expert)
          for( int i = 0; i < (sizeof(zoo) / sizeof(Animal *)); ++i )
   {
          zoo[i] -> sound();
   }

   //================================================================
   cout   << endl<< "===================="
          << endl<< "END...OOP Polymorphism Demo"
          << endl<< "...with Data Structures";
   while(true){}
   return 0;

}//END...Function main
```

```
===================================
POLYMORPHISM on the Runtime STACK
===================================


==============
Calling Member Functions via Objects
-------
oMalley.greet()               Hello from the Cat World
oMalley.sound()               Miaow
shereKhan.greet():            Hello from the Tiger World
shereKhan.sound():            Tiger Roar RRRR


==============
Calling Member Functions via POINTERS
```

```
--------
ptr_oMalley -> greet()          Hello from the Cat World
ptr_oMalley -> sound()          Miaow
ptr_shereKhan -> greet():       Hello from the Cat World
ptr_shereKhan -> sound():       Tiger Roar RRRR


====================================
POLYMORPHISM on the Runtime HEAP
====================================


==============
Calling Member Functions via POINTERS
--------
ptr_Duchess -> greet()          Hello from the Cat World
ptr_Duchess -> sound()          Miaow
ptr_Rajah -> greet()            Hello from the Cat World
ptr_Rajah -> sound()            Tiger Roar RRRR


=============================================
CREATING AN ARRAY of (Pointers-to-) Animals
=============================================


=============
Calling each animal's sound() in the array zooMiaow
Tiger Roar RRRRMiaow
Tiger Roar RRRR


====================
END...OOP Polymorphism Demo
...with Data Structures
```

Program Listing/Output 23.18: Polymorphism_DataStructures_Client.cpp

23.6 DYNAMIC CASTING

❖ This section has shown that OOP Polymorphism

➢ Enables type flexibility

➢ Takes place at RUNTIME

❖ Further flexibility is also possible:

➢ At runtime, it is possible to cast one type into another of the same Inheritance Hierarchy

➢ This is part of a wider branch known as RTTI <u>R</u>un <u>T</u>ime <u>T</u>ype <u>I</u>nformation (covered in Cybernetics in C++ - Expert)

23.7 FURTHER POLYMORPHISM

❖ What has been covered here is Object Oriented Polymorphism

➢ This is all about runtime binding

❖ This is one of two major categories of Polymorphism which is supported in C++

❖ The other kind, polymorphism in Generic Programming, will be covered in section 26 onwards

IMPERATIVE PROGRAMMING (2)

❖ This second part of Imperative Programming focuses on 2 concepts which are again, covered to a greater or lesser extent in all programming languages

   ➢ Exception Handling

   • These are like Emergency Protocols - how should the program react when unexpected things happen

   • Failure to deal properly deal with Exceptions is a common reason for program crashes

   ➢ Files and Streams

   • This is all about writing data to files which can be permanently saved on a Hard Drive

      ○ This could be in several possible formats, including a text file, a database, spreadsheet, or even a new type of format created by your program

      ○ In contrast, at runtime, data is placed in runtime memory, mostly RAM, which is lost when the program exits

❖ C++ covers these both with several built-in libraries

   ➢ These are object oriented, so ideal for reuse

24. EXCEPTION HANDLING

24.1 WHAT IS EXCEPTION HANDLING?

❖ Exception Handling can be thought of like "Emergency Protocols"

   ➢ i.e. What to do in the event things do not go according to plan at runtime

❖ A common problem is a program comes to a standstill when a situation that is unaccounted for, is encountered

❖ The idea of exception handling is that this can be dealt with, and the smooth operation of the program continues

> It is essential in all but the most trivial of programs

24.1.1 WHEN TO USE EXCEPTION HANDLING

❖ Exception Handling is designed to deal with synchronous errors

> i.e. Ones which occur when a statement executes
> Common examples include:
  • Attempted division by zero
  • User entering unrecognised/disallowed input
  • Unsuccessful memory allocation/deallocation
    ○ Attempting to write to out of bounds memory, especially with arrays

❖ It is not designed to process synchronous errors

> i.e. Ones which occur due to external events
> Common examples include:
  • User input activities, such as with the mouse and keyboard
  • Messages sent from networks
  • Issues with further external connections, such as to a database

24.1.2 EXCEPTION HANDLING IN C++

❖ C++ includes libraries which deal with exception handling

> This enables the programmer (i.e. YOU!:-)) to deal with several possible exceptions, ensuring your program is robust
  • A widely used case is the `runtime_error` library, the standard base class for dealing with runtime errors
  • It is common to inherit from this class when creating exception handlers, as the forthcoming examples will show

❖ In languages with no support for Exception Handling, the only alternative is to attempt handling all special cases using conditional statements

> Each case would have to be written from scratch every time
> Having a class makes the exception handling much more reusable

24.2 ANATOMY OF AN EXCEPTION HANDLER

❖ An exception handler comprises

> A class which deals with the exception

- The constructor gets called
➢ A "Try-Catch" handler, in the (member) function which is carrying out the operation

```
try
{
        // Normal Operations
}
catch( Exception ClassType and Name )
{
        // Instructions in the event of an Exception
}
```

- There can be, and often are, more than one catch handler(s), to deal with different exceptions
- A special catch statement is one which deals with "all other" exceptions
  ○ This has the syntax of "dots" for the exception name

```
catch( ... )
{
    // Instructions in the event of all other Exception(s)
}
```

❖ (Member) functions where the "emergency" may occur

➢ These include

- A `throw` statement in the prototype:
- A conditional statement where the exception(s) need(s) to be invoked

```
returnType functionName( )
     throw( Exception_Class_Type )
{
     // Action(s) in function

     if( Condition for Invoking Exception )
     {
          throw Exception_Class_Type();
               // Call Constructor of Exception Class
     }
     // Further action(s) in function
}//END...Function functionName
```

24.3 EXCEPTION: DIVISION BY ZERO

❖ One of the most common exceptions is handling division by zero

❖ The components include

➤ The class `Exception_DivisionByZero`, which deals with the Exception Handling
- This is inherited from the class `runtime_error`, the standard base class for dealing with runtime errors
- defined in the header file `<stdexcept>`

➤ The function `calc_Division` carries out the division
- The definition must include any exceptions it may throw
  - In this case, the line `throw( Exception_DivisionByZero )` is included

```
#pragma once
#include <stdexcept>
class Exception_DivisionByZero
      : public runtime_error
{

      private:

      public:
              Exception_DivisionByZero(  )
                    : runtime_error( "Attempted Division by Zero" )
                  {

                  }

};//END...Class Exception_DivisionByZero
```
Program Listing/Output 24.1: Exception_DivisionByZero.h

```
/*    ***********************************************************
      EXCEPTION HANDLING: DIVISION BY ZERO
      ***********************************************************


*/

#include <iostream>
#include <string>
using namespace std;

#include "Exception_DivisionByZero.h"

double calc_Division( double , double );

int main()
{
      //»»»»»»»»»»»»»»»»»»»»»»»»»»»»»»»»»»»»»»»»»»»»»»»»»»»»»»»»»»
      double myDbl_A = 0.0, myDbl_B = 0.0, quotient_A_B = 0.0;
```

```
       char userInput = ' ';

       do
       {

       //===========================================================
             cout   << endl<< "================================"
                          << endl;
             cout   << "Please enter your 1st number: ";
             cin >> myDbl_A;

             cout   << "Please enter your 2nd number: ";
             cin >> myDbl_B;

        //===========================================================
             try
             {
                   quotient_A_B = calc_Division( myDbl_A, myDbl_B );
                   cout   << endl
                      << myDbl_A<< " divided by "<< myDbl_B<< " is "
                          << quotient_A_B
                          << endl;

             }
             catch( Exception_DivisionByZero&
                          ref_Exception_DivisionByZero_IN
                    )
             {
                   cout   << "Exception occurred: "
                          << ref_Exception_DivisionByZero_IN.what()
                          << endl;

             }//End...try, catch DivisionByZero

       //===========================================================
             cout   << "---------"<< endl
                  << "Do you want to try again (Y/N)? ";
             cin >> userInput;

       } while( userInput != 'N' );
       // End...Do Loop, Request for User Input

       //»»»»»»»»»»»»»»»»»»»»»»»»»»»»»»»»»»»»»»»»»»»»»»»»»»»»»»»»»»»»»»»»
       cout << endl<< "================"
            << endl<< "END...C++ Demo: Exception Handling"
                   << "...Division by Zero"
            << endl;
       while(true){}
       return 0;

}//END...Function MAIN

//++++++++++++++++++++++++++++++++++++++++++++++++++++++++++++++++++
 double calc_Division( double myDbl_IN_A, double myDbl_IN_B )
```

```
      throw( Exception_DivisionByZero )
{
      double result = 0.0;
      if( myDbl_IN_B == 0 )
      {
            throw Exception_DivisionByZero();
      }
      else
      {
            result = myDbl_IN_A / myDbl_IN_B;
      }
      return result;

}//END...Function calc\_Division
```

```
==================================
Please enter your 1st number: 45
Please enter your 2nd number: 9

45 divided by 9 is 5
---------
Do you want to try again (Y/N)? Y

==================================
Please enter your 1st number: 23.0
Please enter your 2nd number: 5.0

23 divided by 5 is 4.6
---------
Do you want to try again (Y/N)? Y

==================================
Please enter your 1st number: 25
Please enter your 2nd number: 0

Exception occurred: Attempted Division by Zero
---------
Do you want to try again (Y/N)? N

================
END...C++ Demo: Exception Handling...Division by Zero
```

Program Listing/Output 24.2: Exception_DivisionByZero_Client.cpp

24.4 EXCEPTION: USER INPUT OF INAPPROPRIATE TYPE

❖ Another common runtime problem is the user enters data which is not of
  the required type

➤ A typo could mean that what would otherwise be interpreted as an int or double, is interpreted as a string e.g. 234# instead of 234

➤ The user might enter a currency sign, such as £ or $, thinking it is required, and again, change the type from a numeric type to string

❖ The following example looks at a program which requires the input data to be of type int, but the user enters string data

```cpp
#pragma once

#include <iostream>
using namespace std;
#include <stdexcept>

class Exception_UserInputType
      : public runtime_error
{
      private:

      public:
            Exception_UserInputType(   )
                  : runtime_error( "Type not recognised " )
            {

            }

};//END...Class Exception_DivisionByZero
```

Program Listing/Output 24.3: Exception_UserInputType.h

```cpp
#include <iostream>
#include <string>
using namespace std;

#include "Exception_UserInputType.h"

int request_UserEntersInt( string )
   throw( Exception_UserInputType );

int main()
{
   int num_People = 0, cost_PerTicket = 0, cost_Total = 0;

   try
   {
      num_People = request_UserEntersInt( "the number of people"
);
      cost_PerTicket = request_UserEntersInt( "the cost per ticket" );
      cost_Total = num_People * cost_PerTicket;

      cout  << "The total cost for "<< num_People
```

```
                         << " at $"<< cost_PerTicket<< " per ticket"
                         <<  " will be $"<< cost_Total
                         << endl;

    }
    catch( Exception_UserInputType&
                 ref_Exception_UserInputType
                 )
    {
                 cout << endl<< "Exception occurred: "
                         << ref_Exception_UserInputType.what()
                         << endl;

    }//End...try, catch requesting user input

    //»»»»»»»»»»»»»»»»»»»»»»»»»»»»»»»»»»»»»»»»»»»»»»»»»»»»»»»»»»»»»»»»»»»»
    cout   << endl<< "================"
           << endl<< "END...C++ Demo: Exception Handling"
           << "...User Input wrong type"
           << endl;
    while(true){}
    return 0;

}//END...Function MAIN


// Returns default of 0 or the value entered by the user
int request_UserEntersInt( string theRequest )
     throw( Exception_UserInputType )
{
     int myInt_A = 0;
     cout << "Please enter "<< theRequest
          <<  " then press Return"
          << endl;

     if( ! (cin >> myInt_A ) )
     {
          throw Exception_UserInputType(); // Call Constructor
     }
     return myInt_A;

}//END...Function request_UserEntersInt
```

```
Please enter the number of people then press Return 34
Please enter the cost per ticket then press Return 6
The total cost for 34 at $6 per
ticket will be $204

================
END...C++ Demo: Exception Handling...User Input wrong type
```

```
Please enter the number of people then press Return 23
Please enter the cost per ticket then press Return Hello
Exception occurred: Type not recognised

================
END...C++ Demo: Exception Handling...User Input wrong type
```

Program Listing/Output 24.4: Exception_UserInputType_Client.cpp

24.5 EXCEPTION: OUT OF BOUNDS ARRAY

❖ Exception Handling is the favoured way to deal with attempts to write to non-allocated memory

❖ A common example of this is attempting to write to a data structure outside its boundaries

❖ The example here looks at the array; note that

　➢ This was essential in most C programs

　➢ In C++, you are advised to use the STL containers vector and deque wherever possible; much of this exception handling is covered

　　• i.e. Treat this as reference material

```cpp
#pragma once
#include <iostream>
using namespace std;
#include <stdexcept>

class Exception_Array_OutOfBoundsAccess
      : public runtime_error
{
      private:

      public:
            Exception_Array_OutOfBoundsAccess(  )
                : runtime_error( "Attempting Out of Bounds Access, " )
            {

            }
};//END...Class Exception_DivisionByZero
```

Program Listing/Output 24.5: Exception_Array_OutOfBoundsAccess.h

```
#pragma once
#include "Exception_Array_OutOfBoundsAccess.h"

class ArrayTester
{
      private:
            int myIntArr_A[5];
            int elem_index, elem_value;

      public:
            // ********** CONSTRUCTOR(S) **********
            ArrayTester();

            // ********** NORMAL MEMBER FUNCTION(S) **********
            void enter_Data();

            void insert_Data(  )
                  throw(Exception_Array_OutOfBoundsAccess);

};//END...class ArrayTester
```

Program Listing/Output 24.6: ArrayTester.h

```
#include "ArrayTester.h"
#include "Exception_Array_OutOfBoundsAccess.h"

// ********** CONSTRUCTOR(S) **********
ArrayTester::ArrayTester()
{
      elem_index = 0;
      elem_value = 0;

      for( int i = 0; i < 5; ++i )
      {
            myIntArr_A[i] = 0;
      }

}//END...Constructor: Default.

// ********** NORMAL MEMBER FUNCTION(S) **********
void ArrayTester::enter_Data()
{
      try
      {
            insert_Data();
            myIntArr_A[elem_index] = elem_value;

      }
      catch( Exception_Array_OutOfBoundsAccess&
                  ref_Exception_Array_OutOfBoundsAccess)
{
      cout    << endl<< "Exception occurred: "
```

```
                << ref_Exception_Array_OutOfBoundsAccess.what()
                << endl;

      }//End...try, catch requesting user input


}//END...Member Function enter_Data

void ArrayTester::insert_Data(  )
      throw(Exception_Array_OutOfBoundsAccess)
      {
      cout  << "Please enter an index"<< endl;
      cin >> elem_index;

      cout << "Please enter an element"<< endl;
      cin >> elem_value;

      if( elem_index >= 4 )
      {
            throw Exception_Array_OutOfBoundsAccess();
      }

}//END...Member Function insert_Data
```

Program Listing/Output 24.7: ArrayTester.cpp

```
#include <iostream>
#include <string>
using namespace std;

#include "Exception_Array_OutOfBoundsAccess.h"
#include "ArrayTester.h"

int main()
{
      //»»»»»»»»»»»»»»»»»»»»»»»»»»»»»»»»»»»»»»»»»»»»»»»»»»»»»»»»»»»»
      ArrayTester myArrayTester;
      myArrayTester.enter_Data();

      //»»»»»»»»»»»»»»»»»»»»»»»»»»»»»»»»»»»»»»»»»»»»»»»»»»»»»»»»»»»»
      cout << endl<< "================"
          << endl<< "END...C++ Demo: Exception Handling"
          << "...Out of Bounds Array"
          << endl;
      while(true){}
      return 0;

}//END...Function MAIN


Please enter an index 4
```

```
Please enter an element 5
Exception occurred: Attempting Out of Bounds Access,

===============
END...C++ Demo: Exception Handling...Out of Bounds Array
```

Program Listing/Output 24.8: Exception_OutOfBoundsArray_Client.cpp

24.6 SPECIAL EXCEPTIONS

❖ The C++ Standard specifies a number of further exceptions; common examples include:

> ➢ **bad_typeid:** When typeid is applied to a dereferenced null pointer a bad_typeid exception is thrown
> ➢ **bad_alloc:** This is thrown whenever an attempt to allocate memory on the runtime heap, using `new`, fails

24.7 WARNING: NEVER GET A DESTRUCTOR TO THROW AN EXCEPTION

❖ It is very important to note that an exception should never be thrown inside a destructor

❖ The catastrophic possible events include:

> ➢ The object not being properly destroyed
> ➢ Another exception calling the destructor
>> • i.e. an exception within an exception: this will cause the program to crash

24.8 CATCHING MULTIPLE EXCEPTIONS

❖ Multiple exceptions can be handled by using more than one catch statement

❖ A special catch statement, which says catch all other exceptions, uses 3 dots as the notation ...

25. FILES & STREAMS

❖ Feeding data to external files, where it can be saved, is a pervasive requirement in programming

➢ A stream is required whenever data has to be transferred from one part of a program to another

❖ In normal operation, data is saved to runtime memory

➢ This is erased when the program ends

❖ This section looks at Persistence – the practice of saving data to the hard drive after the program ends

➢ It starts with a look at the I/O Stream Classes, then using this with the Main function

➢ It then looks at working with

• Text files, where the data types where the original data types are converted

• Binary files, where original data types are retained

25.1 THE C++ I/O STREAM CLASSES

❖ Streams are used whenever data needs to be passed from one part of a program to another

➢ Input streams are used when data needs to be passed into a program from an external file

➢ Output streams are used when data needs to be passed from a program to an external file

❖ Two commonly used commands, are

➢ `cout` Used to pass data from the program to the console output screen

➢ `cin` Used to pass data from the console output screen to the program
These are both part of the library

iostream Input, Output stream

❖ ANSI C++, comes with a comprehensive stream library:

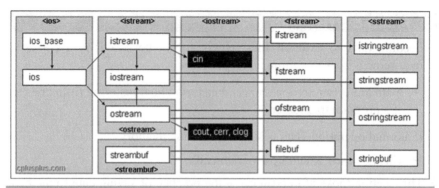

Figure 25.1: Standard Input / Output Streams Library - from
www.cplusplus.com

* ❖ Coding Note: When writing the path of a file, which is a common requirement in most streaming programs, remember to use the double backslash
\\ to divide the directories, as opposed to single

  ➢ e.g. In MS Windows, instead of writing

    ● `C:\Users\John Smith\Documents\Cpp Projects\Files` as you
    would in any IT program such as MS Windows Explorer and the
    MS DOS Prompt
    ● Write `C:\\Users\John Smith\\Documents\\Cpp Projects\\
    Files`

  ➢ The lone backslash means the C++ statement continues on the next line
  (section)

25.2 GIVING INPUTS TO THE MAIN FUNCTION

* ❖ One use of streams is enabling the user of a program to give inputs, of type
string, to the main function]

  ➢ This has to be done via the command line interface
  ➢ In recent times, when graphical IDEs have become common, this
  practice has become less popular, but is still widely used

* ❖ The first line of the definition of the main function is often as follows,
when nothing needs to be input:

```
int main( )
```

❖ However, when the main function does have to take in inputs, the prototype is as follows:

```
int main( int argc, char *argv[] )
```

➢ When calling the main function, from the command line prompt i.e.

- DOS command window in MS Windows
- Terminal in Apple Mac OS X
- Shell scripting terminal in UNIX and Linux the following call is required:

```
programName InputsToMain
```

➢ Thus if you have a program

- Called "greeter"
- Where the main function is to take the inputs "Jonathan" and "Louisa", the command will be as follows:

```
greeter Jonathan Louisa
```

❖ The inputs are placed in the array `argv`

➢ This is of type pointer-to-char i.e. string

- Note that this does not use the string type from the C++ Standard Library, in order to ensure backward compatibility with C

➢ Thus each word can be accessed by accessing elements of this array in the usual way i.e.

```
argv[0]
argv[1]
...
```

❖ The other input of type int, is the number of elements in the array

➢ It is populated automatically by the program, when it knows the number of inputs

➢ i.e. The user does not specify any value for this

❖ The following program shows this in use:

```
/*      ************************************************************
        INPUTS TO THE MAIN FUNCTION
        ************************************************************
        -       This program shows how to access inputs to the main
                function
                -       In a program invoked from a Command Line Prompt
        -       The inputs are
```

```
                         -     Specified after the name of the program
                         -     Stored in the array argv
*/
#include <iostream>
#include <string>
using namespace std;

int main( int argc, char *argv[] )
{
       //»»»»»»»»»»»»»»»»»»»»»»»»»»»»»»»»»»»»»»»»»»»»»»»»»»»»»»»»»»»»»»»
       cout << argv[0]<< " ";
       cout << argv[1]<< " ";
       cout << argv[2]<< " ";


       //»»»»»»»»»»»»»»»»»»»»»»»»»»»»»»»»»»»»»»»»»»»»»»»»»»»»»»»»»»»»»»»
       cout << endl<< endl<< "==========="
           << endl
           << "END...C++ Demo: Command Line commands to Main"
           << " "
           << endl;
       while(true){}
       return 0;

}//END...Function MAIN
```

```
Path> Main_CommandLine Hello Goodbye
1st element, argv[0] = Main_CommandLine
2nd element, argv[0] = Hello
2nd element, argv[0] = Goodbye


===========
END...C++ Demo: Command Line commands to Main
```

Program Listing/Output 25.1: Main_CommandLine_Client.cpp

25.3 WORKING WITH TEXT FILES

25.3.1 READING FROM A TEXT FILE

❖ When reading from an external file i.e. bringing data **in**to your program, the input stream is required

> ➤ The ANSI C++ library has the class ifstream for this; its member functions include:

- ifstream   The input to this constructor is the path of the file being read
- isopen   Returns TRUE if the file is open, false if not
- good   Returns TRUE if the file is workable i.e. not corrupt
- close   Closes the file

  o It is essential not to forget this

  o If the file is left open, no other program will be able to use it until this program has been ended

  o This is similar to freeing up heap memory when finished with it

❖ In order to test this

  ➢ *Create a text file, save it with the name* exampleFile.txt *in a path of your choice, such as* C:\Users\YourName\Desktop\

  ➢ *In this file, place some text, any text, then save it; the following is suggested text*

```
Hello, how are you
Bonjour, comment allez vous
Guten Tag, wie geht es Ihnen
```

```
/*     *****************************************************************
       READING FROM A TEXT FILE
       *****************************************************************
       -     This program reads data from a .txt file, using the ifstream
             library in C++
       -     The Path & Filename are specified
*/
#include <iostream>
#include <fstream>
#include <string>
using namespace std;

int main( )
{
        //»»»»»»»»»»»»»»»»»»»»»»»»»»»»»»»»»»»»»»»»»»»»»»»»»»»»»»»»»»»»
        string line_Current;
        ifstream myfile_Reading("C:\\Users\\YourName\\Desktop\\
                        exampleFile.txt");
        if (myfile_Reading.is_open())
        {
                while ( myfile_Reading.good() )
                {
                        getline (myfile_Reading, line_Current );
                        cout    << line_Current<< endl;
```

```
                }
                myfile_Reading.close();
        }
    else
        {
            cout    << "Cannot open specified file; "
                    << "please check the path and try again"
                    << endl;
        }

        //»»»»»»»»»»»»»»»»»»»»»»»»»»»»»»»»»»»»»»»»»»»»»»»»»»»»»»»»
        cout    << endl<< endl<< "==========="
                << endl
                << "END...C++ Demo: Reading from a Text File"
                << " "
                << endl;
        while(true){}
        return 0;

}//END...Function MAIN
```

```
Hello, how are you
Bonjour, comment allez vous
Guten Tag, wie geht es Ihnen


===========
END...C++ Demo: Reading from a Text File
```

Contents of the File exampleFile.txt

```
Hello, how are you
Bonjour, comment allez vous
Guten Tag, wie geht es Ihnen
```

Program Listing/Output 25.2: Read_TextFile_Client.cpp

❖ When the file cannot be found, on this occasion, a conditional statement
   gives feedback

❖ An improvement would be to throw an exception at this stage (section 24),
   since this

   ➢ Is a very common error
   ➢ More than one thing will often be required when it occurs

EXERCISE

❖ *Modify the program in Program Listing/Output 1.2 so that it throws an exception when the file cannot be found*

25.3.2 WRITING TO A TEXT FILE

❖ When writing to an external file i.e. passing data **out** of your program, the output stream is required

  ➢ The ANSI C++ library has the class `ostream` for this; its member functions and operators include:

  • `open`   Creates the file and its associated path

    ○ If the directories do not exist yet, they will be created
    ○ If no path has been given, the environment chooses a default location

      ▪ In MS Visual Studio, when using a Win32 Console Application, it is:
      `PathOfProject\ProjectName\ProjectName`

  • `<<`   This overloaded operator ensures that any data which follows it is fed into the stream

    ○ A common usage is feeding data into `cout`, to be displayed on the console window

  • `close` Closes the file

    ○ Again, it is essential not to forget this

❖ The following program shows how to write data to a text file:

```
/*      ************************************************************
        WRITING TO TEXT FILES
        ************************************************************
        -     This program writes data to a .txt file, using the ostream
              library in C++
        -     The Path & Filename are specified
*/
#include <iostream>
#include <fstream>
#include <string>
using namespace std;

int main( )
{
        //»»»»»»»»»»»»»»»»»»»»»»»»»»»»»»»»»»»»»»»»»»»»»»»»»»»»»»»»»»»»»»»
        ofstream myFile_Write;
```

```
            double london_TravelCardPrice = 8.75;

            myFile_Write.open("C:\\Users\\YourName\\Desktop\\
                              exampleFile_Write.txt");
            myFile_Write<< "Writing this to a file.\n";
            myFile_Write<< "Hello, how are you?"<< endl;
            myFile_Write<< "London is the Capital of the UK"<< endl;
            myFile_Write<< "\tThe price of a daily travelcard here is "
                                << london_TravelCardPrice
                                << endl;
            myFile_Write<< "Washington DC is the Capital of the USA"
                                << endl;
            myFile_Write.close();

            //»»»»»»»»»»»»»»»»»»»»»»»»»»»»»»»»»»»»»»»»»»»»»»»»»»»»»»»»
            cout << endl<< endl<< "==========="
                 << endl
                 << "END...C++ Demo: Write text data to a text file, with"
                 << "the specified path"
                 << " "
                 << endl;
            while(true){}
            return 0;

}//END...Function MAIN
```

```
===========
END...C++ Demo: Write text data to a text file, with the specified
path
```

Contents of the File exampleFile_Write.txt

```
Writing this to a file.
Hello, how are you?
London is the Capital of the UK
        The price of a daily travelcard here is 8.75
Washington DC is the Capital of the USA
```

Program Listing/Output 25.3: Read_TextFile_Client.cpp

25.3.2.1 Overwriting Existing Data

❖ An issue with using the open member function in its default form is that every time a file is opened, anything previously there is erased

❖ This is similar to runtime memory:

➢ Whenever anything new is written to a variable, any existing value is overwritten

➢ In fact, this is the way memory is "cleared"

❖ In many cases, this is useful

➢ The following program shows how this replacement takes place when the same file is opened more than once

➢ Note that only the data from the last opening is contained in the file; the rest is indeed lost

```cpp
/*      ****************************************************************
        WRITING to Text Files - OVERWRITING
        ****************************************************************
        -    This program writes data to a .txt file, using the ostream
             library in C++
        -    It then OVERWRITES the data in that file by recalling the
             open member function in its DEFAULT state
        -    The Path & Filename are specified
*/
#include <iostream>
#include <fstream>
#include <string>
using namespace std;

int main( )
{
    //»»»»»»»»»»»»»»»»»»»»»»»»»»»»»»»»»»»»»»»»»»»»»»»»»»»»»»»»»»»»»
    ofstream my_Txt_File_Write_Overwrite;
    my_Txt_File_Write_Overwrite.open
                    ("C:\\Users\\YourName\\Desktop\\
                        exampleFile_Write_Overwrite.txt");
    my_Txt_File_Write_Overwrite<< "Writing this to a file.\n";
    my_Txt_File_Write_Overwrite<< "Hello, how are you?"<< endl;
    my_Txt_File_Write_Overwrite<< "London is the Capital of the UK";
    my_Txt_File_Write_Overwrite<< "Washington DC is the Capital of
                        the USA";
    my_Txt_File_Write_Overwrite.close();

    // ALL The previous text gets overwritten when the file is opened
    //      again

    my_Txt_File_Write_Overwrite.open
                    ("C:\\Users\\YourName\\Desktop\\
                        exampleFile_Write_Overwrite.txt");
    my_Txt_File_Write_Overwrite
            << "Bonjour, comment allez vous?" << endl;
    my_Txt_File_Write_Overwrite
            << "Paris, c'est la Capital de France"<< endl;
    my_Txt_File_Write_Overwrite.close();
```

```
//»»»»»»»»»»»»»»»»»»»»»»»»»»»»»»»»»»»»»»»»»»»»»»»»»»»»»»»»»»»»»»»
cout    << endl<< endl<< "==========="
        << endl
        << "END...C++ Demo: Write text data to a text file, with "
        << "specified path, then OVERWRITE"
        << " "
        << endl;
    while(true){}
    return 0;

}//END...Function MAIN
```

```
===========
END...C++ Demo: Write text data to a text file, with specified path,
    then OVERWRITE
```

Contents of the File exampleFile_Write_Overwrite.txt

```
Bonjour, comment allez vous?
Paris, c'est la Capital de France
```

Program Listing/Output 25.4: Write_Text File_Overwrite_Client.cpp

25.3.2.2 Appending a Text File

❖ In order to ensure the file is appended as opposed to overwritten, the `ios::app` flag needs to be activated

➢ This is specified when calling the open member function, as follows

❖ The following program is similar in content to Program Listing/Output 1.4, the difference is now it appends the data:

```
/*      ************************************************************
        WRITING to Text Files - APPENDING
        ************************************************************
        -     This program writes data to a .txt file, using the ostream
              library in C++
        -     It then APPEND the data in that file by recalling the
              open member function
              -     with its Append Flag "ios::app" set
        -     The Path & Filename are specified
*/
#include <iostream>
#include <fstream>
#include <string>
```

```
using namespace std;

int main( )
{
    //»»»»»»»»»»»»»»»»»»»»»»»»»»»»»»»»»»»»»»»»»»»»»»»»»»»»»»»»»»»»
    ofstream my_Txt_File_Write_Append;
    my_Txt_File_Write_Append.open
            ("C:\\Users\\YourName\\Desktop\\
            exampleFile_Write_Append.txt" );
    my_Txt_File_Write_Append<< "Writing this to a file.\n";
    my_Txt_File_Write_Append<< "Hello, how are you?"<< endl;
    my_Txt_File_Write_Append<< "London is the Capital of the UK"
                                         << endl;
    my_Txt_File_Write_Append<< "Washington DC is the Capital of the
                            USA"
                                         << endl<< endl;
    my_Txt_File_Write_Append.close();

    my_Txt_File_Write_Append.open
            ("C:\\Users\\YourName\\Desktop\\
             exampleFile_Write_Append.txt", ios::out | ios::app );
    my_Txt_File_Write_Append<< "Bonjour, comment allez vous?"
                                << endl;
    my_Txt_File_Write_Append<< "Paris, c'est la Capital de France"
                                << endl;
    my_Txt_File_Write_Append.close();

    //»»»»»»»»»»»»»»»»»»»»»»»»»»»»»»»»»»»»»»»»»»»»»»»»»»»»»»»»»»»»»'
    cout      << endl<< endl<< "==========="
              << endl
              << "END...C++ Demo: Write text data to a text file, with"
              << "specified path, then APPEND"
              << " "
              << endl;
    while(true){}
    return 0;

}//END...Function MAIN
```

```
===========
END...C++ Demo: Write text data to a text file, with specified path,
then APPEND
```

Contents of the File exampleFile_Write_Append.txt

```
Writing this to a file.
Hello, how are you?
London is the Capital of the UK
Washington DC is the Capital of the USA
```

```
Bonjour, comment allez vous?
Paris, c'est la Capital de France
```

25.3.3 CONVERTING FROM STRING CHARACTERS

❖ When data is read in from a text file, it is of type string

❖ In order to convert it, **parsing** has to take place

25.3.4 FURTHER TEXT FILES

25.3.4.1 CSV (Comma Separated Values) Files

❖ .csv files are special text files where items are separated by commas

❖ They are useful because most IT programs, such as spreadsheets and databases recognise them i.e.

➢ Excel will display data items

- separated by commas over different lines
- in separate cells in different rows

❖ The following program shows how to write data to a .csv file:

```
/*      ************************************************************
        WRITE TO CSV FILE
        ************************************************************
        -       This program writes data to a csv file, using the ostream
                library in C++
        -       The Path & Filename are specified
*/
#include <iostream>
#include <fstream>
#include <string>
using namespace std;

int main( )
{
    //»»»»»»»»»»»»»»»»»»»»»»»»»»»»»»»»»»»»»»»»»»»»»»»»»»»»»»»»»»»»»»»»»
    ofstream my_CSV_File_Write;
    my_CSV_File_Write.open ("C:\\Users\\YourName\\Desktop\\
                        exampleFile_Write.csv");
    my_CSV_File_Write<< "1, 2, 3\n";
    my_CSV_File_Write<< "4, 5, 6"<< endl;
    my_CSV_File_Write<< "7, 8, 9"<< endl;
    my_CSV_File_Write<< "155, 234, 643"<< endl;
    my_CSV_File_Write.close();

    //»»»»»»»»»»»»»»»»»»»»»»»»»»»»»»»»»»»»»»»»»»»»»»»»»»»»»»»»»»»»»»»»»
```

```
    cout        << endl<< endl<< "==========="
                << endl
                << "END...C++ Demo: Write text data to a csv file, with "
                << "the specified path"
                << " "
                << endl;
    while(true){}
    return 0;

}//END...Function MAIN
```

```
===========
END...C++ Demo: Write text data to a csv file, with the specified path
```

```
Contents of the File exampleFile_Write.csv displayed in a Text Editor

1, 2, 3
4, 5, 6
7, 8, 9
155, 234, 643
```

Contents of the File exampleFile_Write.csv displayed in a Spreadsheet – MS Excel

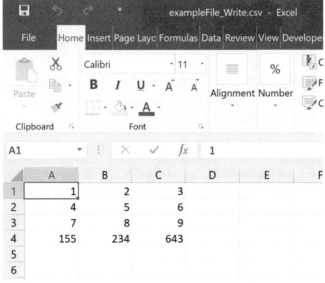

Program Listing/Output 25.6: Write_CSVFile_Client.cpp

25.4 WORKING WITH BINARY FILES

❖ Binary files contain data, in binary form, for a computer to process

> ➤ They are rarely readable by humans
> ➤ The data can be varied, and can include
>   • Numerical data
>   • Images
>   • Multimedia files:
>   ○ Audio files
>   ○ Video files
> ➤ Object files are often referred to as "binaries" since they contain binary data for the computer to read

❖ When you write a variable `int myInt_A = 5` to a file

> ➤ If it is a text file, the value 5 will be converted to type string
> ➤ If it is a binary file, the value 5 will maintain its type int

❖ This part looks at saving, firstly Primitive Variables, then Objects

❖ In each case, the `reinterpret_cast` is used

> ➤ This is used to retain the **value** of the data
> ➤ In order to place the data in the file, type `.dat` for instance, it needs to be cast into a different type
> ➤ When reading from the file, it needs to be cast back to the original type
> ➤ The `static_cast`, section Error! Reference source not found., is used to retain the **address** of the data

25.4.1 SAVING & RETRIEVING PRIMITIVE VARIABLES

❖ This section uses the data type int to show how primitive variables can be saved to the hard drive

> ➤ It confirms that the procedure is identical whether saving
>   • A single instance
>   • A data structure, demonstrated here using an array

❖ In order to write the data to a `.dat` file

> ➤ An object of type `ofstream` needs to be created, pointing to the specified file

➤ The data is written, via a reinterpret_cast, to this object, for saving in the file

❖ To read this data

➤ An object of type `ifstream` needs to be created, pointing to the specified file

➤ The data is read, via a reinterpret_cast, from this object

❖ The following program shows this in action

```
/*      *************************************************************
        Binary File: WRITE/READ PRIMITIVE VARIABLES
        *************************************************************
*/
#include <iostream>
#include <fstream>
#include <string>
using namespace std;

int main()
{
    //»»»»»»»»»»»»»»»»»»»»»»»»»»»»»»»»»»»»»»»»»»»»»»»»»»»»»»»»»»»»»»»
    cout << "===============================================" << endl;
    cout << "Binary File: Write/Read Array of Integers" << endl;
    cout << "===============================================" << endl;

    //»»»»»»»»»»»»»»»»»»»»»»»»»»»»»»»»»»»»»»»»»»»»»»»»»»»»»»»»»»»»»»»
    //          Creating the Data & WRITING to the Hard Drive
    //
    //-----------------------------------------------------------
    int arrInt_WriteToHD[] = { 8, 32, 9, 5, 7, 172, 8192, 12, 16 };
    cout << endl << "----------------" << endl;
    cout << "Original Data, to be Saved to the Hard Drive:" << endl;
    for ( int i = 0;
               i < ((sizeof(arrInt_WriteToHD)) / (sizeof(int)));
               ++i )
    {
           cout << arrInt_WriteToHD[i] << " ";
    }

    //-----------------------------------------------------------
    //          WRITING Data, myIntArr, to the Hard Drive
    //          -          Create Output Stream
    //
    ofstream this_ofstream("binaryData_Ints.dat", ios::binary);
    this_ofstream.write(reinterpret_cast<char *>(arrInt_WriteToHD),
                                        sizeof(arrInt_WriteToHD));
    //-----------------------------------------------------------
    this_ofstream.close();
```

```
//»»»»»»»»»»»»»»»»»»»»»»»»»»»»»»»»»»»»»»»»»»»»»»»»»»»»»»»»»»»»
//          READING Data from the Hard Drive
//        -           Create Input Stream
//
int arrInt_ReadFromHD[] = { 0, 0, 0, 0, 0, 0, 0, 0, 0 };

ifstream this_ifstream("binaryData_Ints.dat", ios::binary);
this_ifstream.read(reinterpret_cast<char *>(arrInt_ReadFromHD),
                                  sizeof(arrInt_ReadFromHD));

cout << endl << "----------------" << endl;
cout << "Data Retrieved from Saved File:" << endl;
for(  int i = 0;
          i < ((sizeof(arrInt_ReadFromHD)) / (sizeof(int)));
          ++i )
{
        cout << arrInt_ReadFromHD[i] << " ";
}
//----------------------------------------------------------------
this_ifstream.close();

//»»»»»»»»»»»»»»»»»»»»»»»»»»»»»»»»»»»»»»»»»»»»»»»»»»»»»»»»»»»»
cout << endl << endl << "===========" << endl
        << "END...C++ Demo: Binary Files using Ints"
        << endl;
while (true) {}
return 0;

}//END...MAIN Function
```

```
=================================================
Binary File: Write/Read Array of Integers
=================================================


----------------
Original Data, to be Saved to the Hard Drive:
8 32 9 5 7 172 8192 12 16
----------------
Data Retrieved from Saved File:
8 32 9 5 7 172 8192 12 16


===========
END...C++ Demo: Binary Files using Ints
```

Program Listing/Output 25.7: BinaryFile_IntArray_Client.cpp

25.4.2 SAVING & RETRIEVING OBJECTS

❖ Object Oriented data can be saved to the Hard Drive in a similar means to
   primitive variables

❖ The following sections show how object data can be saved to and retrieved from the Hard Drive

➤ Firstly, a single object is shown, then an array of objects

➤ The objects are of the same Employee type, used since section Error! Reference source not found.

● For clarity, the listings have been shown after the listing of the "client" program, in each case

25.4.2.1 Working with Single Objects

❖ Like with primitive variables, in order to write the single object data to a `.dat` file

➤ An object of type `ofstream` needs to be created, pointing to the specified file

➤ The data is written, via a `reinterpret_cast`, to this object, for saving in the file

❖ To read this data

➤ An object of type `ifstream` needs to be created, pointing to the specified file

➤ The data is read, via a reinterpret_cast, from this object

❖ The following program shows this in action

```
/*      ************************************************************
        Binary File: WRITE/READ SINGLE OBJECTS
        ************************************************************
*/
#include <fstream>
#include <iostream>
#include <string>
using namespace std;

#include "Employee.h"

int main()
{
    cout << "=============================================" << endl;
    cout << "Binary File: Write/Read Single Objects" << endl;
    cout << "=============================================" << endl;

    //»»»»»»»»»»»»»»»»»»»»»»»»»»»»»»»»»»»»»»»»»»»»»»»»»»»»»»»»»»»»»»»»
    //          WRITING a Single Object to Disk
    //
    cout << "===================================" << endl;
```

```
    cout << "Original Data, Writing to Hard Drive" << endl;
    cout << "---------------------------------------" << endl;

    Employee joeBloggs("Joseph Bloggs", "PH 45 67 34 D", 17000.0);
    joeBloggs.greet_AllAttributes();

    ofstream file_WritingEmployeeObject("Employee_Objs.dat",
                                        ios::binary);
    file_WritingEmployeeObject.write( reinterpret_cast<char *>
                                           (&joeBloggs),
                                           sizeof(joeBloggs) );

    // Forgetting to Close will cause Errors...
    file_WritingEmployeeObject.close();

    //»»»»»»»»»»»»»»»»»»»»»»»»»»»»»»»»»»»»»»»»»»»»»»»»»»»»»»»»»
    //        READING a Single Object from Disk
    //
    cout << "=========================================" << endl;
    cout << "Reading in Employee Object from Hard Disk" << endl;
    cout << "-----------------------------------------" << endl;
    Employee temp_Employee;
    ifstream file_ReadingEmployeeObject("Employee_Objs.dat",

    ios::binary);
    file_ReadingEmployeeObject.read( reinterpret_cast<char *>
                                         (&temp_Employee),
                                         sizeof(temp_Employee));

    temp_Employee.greet_AllAttributes();

    file_ReadingEmployeeObject.close();

    //»»»»»»»»»»»»»»»»»»»»»»»»»»»»»»»»»»»»»»»»»»»»»»»»»»»»»»»»»
    cout << endl << endl << "===========" << endl
         << "END...C++ Demo: Binary File: Write/Read Single Objects"
         << endl;
    while (true) {}
    return 0;

}// END...MAIN Function
```

```
=================================================
Binary File: Write/Read Single Objects
=================================================
=======================================
Original Data, Writing to Hard Drive
---------------------------------------

---------------------
Hello, I am an employee, my name is Joseph Bloggs,
my NI Number is PH 45 67 34 D,
```

```
my Gross Salary is 17000
my tax Rate is 0.3%
so the amount I have to pay in tax is 5100
thus my Net Salary is 11900
--------------------
============================================
Reading in Employee Object from Hard Disk
--------------------------------------------


--------------------
Hello, I am an employee, my name is Joseph Bloggs,
my NI Number is PH 45 67 34 D,
my Gross Salary is 17000
my tax Rate is 0.3%
so the amount I have to pay in tax is 5100
thus my Net Salary is 11900
--------------------



===========
END...C++ Demo: Binary File: Write/Read Single Objects
```

Program Listing/Output 25.8: BinaryFile_SingleObjects_Client.cpp

```cpp
#ifndef EMPLOYEE_H
#define EMPLOYEE_H

#include <iostream>
#include <string>
using namespace std;

const double TAXRATE_THRESHOLD_LOWER = 20000.0;
const double TAXRATE_THRESHOLD_INTERMEDIATE = 25000.0;


const double TAXRATE_RATE_LOWER = 0.3;
const double TAXRATE_RATE_INTERMEDIATE = 0.4;
const double TAXRATE_RATE_HIGHER = 0.5;


// Class DECLARATION
class Employee
{
                // DATA MEMBERS
        private:
                string name;
                string nI_Number;
                double salary_Gross;
                double tax_Rate;
                double tax_Amount;
                double salary_Net;
```

```
                // ---------- STATIC Data Member(s) ------------
                static int num_Employees;

                // MEMBER FUNCTIONS
        public:
                // ---------- CONSTRUCTOR(S) ------------
                Employee();
                Employee(string);
                Employee(string, string);
                Employee(string name_IN,
                         string nI_Number,
                         double salary_Gross_IN);

                // ---------- DESTRUCTOR ------------
                ~Employee();

                // ---------- MUTATORS ------------
                double get_salary_Gross() const;
                void set_salary_Gross(double salary_Gross_IN);

                // ---------- STATIC MEMBER FUNCTION(S) ------------
                // ...also a Mutator
                static int get_num_Employees();

                // ---------- "NORMAL Member Functions " ------------
                void greet();
                void greet_Name();
                void greet_Name_NINumber();
                void greet_Name_SalaryGross();
                void greet_AllAttributes();

};//END...Class Declaration Employee

#endif
```

Program Listing/Output 25.9: Employee.h

```
#include "Employee.h"

// ---------- STATIC Data Member(s) ------------
int Employee::num_Employees = 0;

// ---------- CONSTRUCTOR(S) ------------
Employee::Employee()
{
}
Employee::Employee(string name_IN)
{
        name = name_IN;

        // Keeping track of the number of objects
        ++num_Employees;
```

```
}//END...Constructor Employee( string name_IN )


Employee::Employee(string name_IN, string nI_Number_IN)
{
        name = name_IN;
        nI_Number = nI_Number_IN;

        // Keeping track of the number of objects
        ++num_Employees;

}//END...Constructor Employee( string name_IN )

Employee::Employee(string name_IN,
        string nI_Number_IN,
        double salary_Gross_IN)
{
        name = name_IN;
        nI_Number = nI_Number_IN;
        salary_Gross = salary_Gross_IN;
        //----------------------------------------------------
        //          Calculating the Tax Rate
        //
        if (salary_Gross < TAXRATE_THRESHOLD_LOWER)
        {
                tax_Rate = TAXRATE_RATE_LOWER;
        }
        else
        {
                if (salary_Gross < TAXRATE_THRESHOLD_INTERMEDIATE)
                {
                        tax_Rate = TAXRATE_RATE_INTERMEDIATE;
                }
                else
                {
                        tax_Rate = TAXRATE_RATE_HIGHER;
                }
        }

        //----------------------------------------------------
        //          Calculating the Tax Amount & Net Salary
        //
        tax_Amount = tax_Rate * salary_Gross;
        salary_Net = salary_Gross - tax_Amount;


        // Incrementing the Number of Objects/Employees

        // Keeping track of the number of objects
        ++num_Employees;

}/* END...Constructor Employee( string name_IN,
```

```
  string nI_Num_IN,
 double salary_Gross_IN )
 */

 // ---------- DESTRUCTOR ------------
Employee::~Employee()
{
        cout << "Goodbye, I, " << name
                << " am  going out of scope now."
                << endl;

        // Keeping track of the number of objects
        --num_Employees;

}// END...Destructor ~Employee

 // ---------- MUTATORS ------------
 // const keyword is required here as well
double Employee::get_salary_Gross() const
{
        // tax_Rate = 0.6;
        // ILLEGAL - const member function cannot
        //                       modify any data members

        return salary_Gross;

}//END...Mutator get_salary_Gross


void Employee::set_salary_Gross(double salary_Gross_IN)
{
        salary_Gross = salary_Gross_IN;

        //----------------------------------------------------
        //       Calculating the NEW Tax Rate
        //
        if (salary_Gross < TAXRATE_THRESHOLD_LOWER)
        {
                tax_Rate = TAXRATE_RATE_LOWER;
        }
        else
        {
                if (salary_Gross < TAXRATE_THRESHOLD_INTERMEDIATE)
                {
                        tax_Rate = TAXRATE_RATE_INTERMEDIATE;
                }
                else
                {
                        tax_Rate = TAXRATE_RATE_HIGHER;
                }
        }

        //----------------------------------------------------
```

```
        //          Calculating the NEW Tax Amount & Net Salary
        //
        tax_Amount = tax_Rate * salary_Gross;
        salary_Net = salary_Gross - tax_Amount;

}//END...Mutator set_salary_Gross


 // ---------- STATIC MEMBER FUNCTION(S) ------------
 // ...also a Mutator
int Employee::get_num_Employees()
{
        return num_Employees;

}//END...Mutator get_num_Employees


 // ---------- "NORMAL Member Functions " ------------
void Employee::greet()
{
        cout << "Hello, I'm an Employee" << endl;

}//END...Member Function greet

void Employee::greet_Name()
{
        cout << "Hello, I'm an Employee, "
                << "my name is " << name
                << endl;

}//END...Member Function greet_Name

void Employee::greet_Name_NINumber()
{
        cout << "Hello, I'm an Employee, "
                << "my name is " << name
                << " and my NI Number is " << nI_Number
                << endl;

}//END...Member Function greet_Name_NINumber


void Employee::greet_Name_SalaryGross()
{
    cout << "Hello, I'm an Employee, my name is " << name
            << ", my Gross Salary is " << salary_Gross
            << endl;

}//END...Member Function greet_Name_SalaryGross()


void Employee::greet_AllAttributes()
{
```

```
    cout << endl
           << "---------------------"
           << endl
           << "Hello, I am an employee, my name is " << name << ", "
           << endl
           << "my NI Number is " << nI_Number << ","
           << endl
           << "my Gross Salary is " << salary_Gross
           << endl
           << "my tax Rate is " << tax_Rate << "%"
           << endl
           << "so the amount I have to pay in tax is " << tax_Amount
           << endl
           << "thus my Net Salary is " << salary_Net
           << endl
           << "---------------------" << endl;
}// END...MF greet_AllAttributes()
```

Program Listing/Output 25.10: Employee.cpp

25.4.2.2 Working with an Array of Objects

❖ Saving a data structure of objects is similar to saving a data structure of primitive variables

➢ Note that an array is used here only for demonstration purposes – it is not the best data structure for storing objects

➢ An STL data structure, such as a Vector or Deque, is much more sophisticated

• These are covered in Cybernetics in C++ - Expert

❖ In order to write the an array of objects to a .dat file

➢ An object of type ofstream needs to be created, pointing to the specified file

➢ The data is written, via a reinterpret_cast, to this object, for saving in the file

❖ To read this data

➢ An object of type ifstream needs to be created, pointing to the specified file

➢ The data is read, via a reinterpret_cast, from this object

❖ The following program shows this in action:

```
/*      ************************************************************
        Binary File: WRITE/READ ARRAY OF OBJECTS
        ************************************************************
*/
#include <fstream>
#include <iostream>
#include <string>
using namespace std;

#include "Employee.h"

int main()
{
    cout << "===============================================" << endl;
    cout << "Binary File: Write/Read Array of Objects" << endl;
    cout << "===============================================" << endl;

    //»»»»»»»»»»»»»»»»»»»»»»»»»»»»»»»»»»»»»»»»»»»»»»»»»»»»»»»»»»»»
    //          WRITING Array of Objects to Disk
    //
    cout << "=====================================" << endl;
    cout << "Original Data, Writing to Hard Drive" << endl;
    cout << "-------------------------------------" << endl;

    Employee arrEmployee_WriteToHD[] =
    { Employee("Joseph Bloggs", "PH 45 67 34 D", 17000.0),
            Employee("John Smith", "GH 67 34 67 D", 18000.0),
            Employee("Amanda Taylor", "SG 36 78 24 F", 22000.0)
    };

    for (int i = 0;
            i < ((sizeof(arrEmployee_WriteToHD)) / (sizeof(Employee)));
            ++i)
    {
            arrEmployee_WriteToHD[i].greet_AllAttributes();
    }

    //----------------------------------------------------------------
    ofstream this_ofstream("Employee_ArrObjs.dat", ios::binary);
    this_ofstream.write( reinterpret_cast<char *>
                                    (arrEmployee_WriteToHD),
                            sizeof(arrEmployee_WriteToHD) );

    //----------------------------------------------------------------
    this_ofstream.close();

    //»»»»»»»»»»»»»»»»»»»»»»»»»»»»»»»»»»»»»»»»»»»»»»»»»»»»»»»»»»»»
    //          READING Multiple Objects from Disk
    //
```

```
cout << "=========================================" << endl;
cout << "Reading in Employee Array from Hard Disk" << endl;
cout << "-----------------------------------------" << endl;

//»»»»»»»»»»»»»»»»»»»»»»»»»»»»»»»»»»»»»»»»»»»»»»»»»»»»»»»»
//          READING Data from the Hard Drive
//       -            Create Input Stream
//
Employee arrEmployee_ReadFromHD[3];

ifstream this_ifstream("Employee_ArrObjs.dat", ios::binary);
this_ifstream.read( reinterpret_cast<char *>
                                 (arrEmployee_ReadFromHD),
                          sizeof(arrEmployee_ReadFromHD) );

for(int i = 0;
     i < ((sizeof(arrEmployee_ReadFromHD)) / (sizeof(Employee)));
     ++i)
{
     arrEmployee_ReadFromHD[i].greet_AllAttributes();
}

//»»»»»»»»»»»»»»»»»»»»»»»»»»»»»»»»»»»»»»»»»»»»»»»»»»»»»»»»
this_ifstream.close();

//»»»»»»»»»»»»»»»»»»»»»»»»»»»»»»»»»»»»»»»»»»»»»»»»»»»»»»»»
cout  << endl << endl << "===========" << endl
      <<"END...C++ Demo: Binary File, Write/Read Array of Objects"
      << " "
      << endl;
while (true) {}
return 0;

}// END...MAIN Function
```

```
=================================================
Binary File: Write/Read Array of Objects
=================================================
=================================================
Original Data, Writing to Hard Drive
-------------------------------------------

---------------------
Hello, I am an employee, my name is Joseph Bloggs,
my NI Number is PH 45 67 34 D,
my Gross Salary is 17000
my tax Rate is 0.3%
so the amount I have to pay in tax is 5100
thus my Net Salary is 11900
---------------------

---------------------
```

```
Hello, I am an employee, my name is John Smith,
my NI Number is GH 67 34 67 D,
my Gross Salary is 18000
my tax Rate is 0.3%
so the amount I have to pay in tax is 5400
thus my Net Salary is 12600
---------------------


---------------------
Hello, I am an employee, my name is Amanda Taylor,
my NI Number is SG 36 78 24 F,
my Gross Salary is 22000
my tax Rate is 0.4%
so the amount I have to pay in tax is 8800
thus my Net Salary is 13200
---------------------
================================================
Reading in Employee Array from Hard Disk
--------------------------------------------


---------------------
Hello, I am an employee, my name is Joseph Bloggs,
my NI Number is PH 45 67 34 D,
my Gross Salary is 17000
my tax Rate is 0.3%
so the amount I have to pay in tax is 5100
thus my Net Salary is 11900
---------------------


---------------------
Hello, I am an employee, my name is John Smith,
my NI Number is GH 67 34 67 D,
my Gross Salary is 18000
my tax Rate is 0.3%
so the amount I have to pay in tax is 5400
thus my Net Salary is 12600
---------------------


---------------------
Hello, I am an employee, my name is Amanda Taylor,
my NI Number is SG 36 78 24 F,
my Gross Salary is 22000
my tax Rate is 0.4%
so the amount I have to pay in tax is 8800
thus my Net Salary is 13200
---------------------


===========
END...C++ Demo: Binary File, Write/Read Array of Objects
```

Program Listing/Output 25.11: BinaryFile_ArrayObjects_Client.cpp

```
#ifndef EMPLOYEE_H
#define EMPLOYEE_H

#include <iostream>
#include <string>
using namespace std;

const double TAXRATE_THRESHOLD_LOWER = 20000.0;
const double TAXRATE_THRESHOLD_INTERMEDIATE = 25000.0;


const double TAXRATE_RATE_LOWER = 0.3;
const double TAXRATE_RATE_INTERMEDIATE = 0.4;
const double TAXRATE_RATE_HIGHER = 0.5;


// Class DECLARATION
class Employee
{
        // DATA MEMBERS
        private:
                string name;
                string nI_Number;
                double salary_Gross;
                double tax_Rate;
                double tax_Amount;
                double salary_Net;

                // ---------- STATIC Data Member(s) ------------
                static int num_Employees;

                // MEMBER FUNCTIONS
        public:
                // ---------- CONSTRUCTOR(S) ------------
                Employee();
                Employee(string);
                Employee(string, string);
                Employee(string name_IN,
                        string nI_Number,
                        double salary_Gross_IN);

                // ---------- DESTRUCTOR ------------
                ~Employee();

                // ---------- MUTATORS ------------
                double get_salary_Gross() const;
                void set_salary_Gross(double salary_Gross_IN);

                // ---------- STATIC MEMBER FUNCTION(S) ------------
```

```
                         // ...also a Mutator
                         static int get_num_Employees();

                         // ---------- "NORMAL Member Functions " ------------
                         void greet();
                         void greet_Name();
                         void greet_Name_NINumber();
                         void greet_Name_SalaryGross();
                         void greet_AllAttributes();

};//END...Class Declaration Employee

#endif
```

Program Listing/Output 25.12: Employee.h

```
#include "Employee.h"

// ---------- STATIC Data Member(s) ------------
int Employee::num_Employees = 0;

// ---------- CONSTRUCTOR(S) ------------
Employee::Employee()
{
}
Employee::Employee(string name_IN)
{
        name = name_IN;

        // Keeping track of the number of objects
        ++num_Employees;

}//END...Constructor Employee( string name_IN )


Employee::Employee(string name_IN, string nI_Number_IN)
{
        name = name_IN;
        nI_Number = nI_Number_IN;

        // Keeping track of the number of objects
        ++num_Employees;

}//END...Constructor Employee( string name_IN )

Employee::Employee(string name_IN,
        string nI_Number_IN,
        double salary_Gross_IN)
{
        name = name_IN;
        nI_Number = nI_Number_IN;
        salary_Gross = salary_Gross_IN;
```

```
        //------------------------------------------------------
        //          Calculating the Tax Rate
        //
        if (salary_Gross < TAXRATE_THRESHOLD_LOWER)
        {
                tax_Rate = TAXRATE_RATE_LOWER;
        }
        else
        {
                if (salary_Gross < TAXRATE_THRESHOLD_INTERMEDIATE)
                {
                        tax_Rate = TAXRATE_RATE_INTERMEDIATE;
                }
                else
                {
                        tax_Rate = TAXRATE_RATE_HIGHER;
                }
        }

        //------------------------------------------------------
        //          Calculating the Tax Amount & Net Salary
        //
        tax_Amount = tax_Rate * salary_Gross;
        salary_Net = salary_Gross - tax_Amount;


        // Incrementing the Number of Objects/Employees

        // Keeping track of the number of objects
        ++num_Employees;
}/* END...Constructor Employee( string name_IN,
 string nI_Num_IN,
 double salary_Gross_IN )
 */

 // ---------- DESTRUCTOR ------------
Employee::~Employee()
{
        cout << "Goodbye, I, " << name
                << " am  going out of scope now."
                << endl;

        // Keeping track of the number of objects
        --num_Employees;

}// END...Destructor ~Employee

 // ---------- MUTATORS ------------
 // const keyword is required here as well
double Employee::get_salary_Gross() const
{
        // tax_Rate = 0.6;
```

```
        // ILLEGAL - const member function cannot
        //                      modify any data members

        return salary_Gross;

}//END...Mutator get_salary_Gross


void Employee::set_salary_Gross(double salary_Gross_IN)
{
        salary_Gross = salary_Gross_IN;

        //---------------------------------------------------
        //        Calculating the NEW Tax Rate
        //
        if (salary_Gross < TAXRATE_THRESHOLD_LOWER)
        {
                tax_Rate = TAXRATE_RATE_LOWER;
        }
        else
        {
                if (salary_Gross < TAXRATE_THRESHOLD_INTERMEDIATE)
                {
                        tax_Rate = TAXRATE_RATE_INTERMEDIATE;
                }
                else
                {
                        tax_Rate = TAXRATE_RATE_HIGHER;
                }
        }

        //---------------------------------------------------
        //        Calculating the NEW Tax Amount & Net Salary
        //
        tax_Amount = tax_Rate * salary_Gross;
        salary_Net = salary_Gross - tax_Amount;

}//END...Mutator set_salary_Gross


 // ---------- STATIC MEMBER FUNCTION(S) ------------
 // ...also a Mutator
int Employee::get_num_Employees()
{
        return num_Employees;

}//END...Mutator get_num_Employees


 // ---------- "NORMAL Member Functions " ------------
void Employee::greet()
{
        cout << "Hello, I'm an Employee" << endl;
```

```
}//END...Member Function greet

void Employee::greet_Name()
{
    cout << "Hello, I'm an Employee, "
            << "my name is " << name
            << endl;

}//END...Member Function greet_Name

void Employee::greet_Name_NINumber()
{
    cout << "Hello, I'm an Employee, "
            << "my name is " << name
            << " and my NI Number is " << nI_Number
            << endl;

}//END...Member Function greet_Name_NINumber


void Employee::greet_Name_SalaryGross()
{
    cout << "Hello, I'm an Employee, my name is " << name
            << ", my Gross Salary is " << salary_Gross
            << endl;

}//END...Member Function greet_Name_SalaryGross()


void Employee::greet_AllAttributes()
{
    cout << endl
            << "---------------------"
            << endl
            << "Hello, I am an employee, my name is " << name << ", "
            << endl
            << "my NI Number is " << nI_Number << ","
            << endl
            << "my Gross Salary is " << salary_Gross
            << endl
            << "my tax Rate is " << tax_Rate << "%"
            << endl
            << "so the amount I have to pay in tax is " << tax_Amount
            << endl
            << "thus my Net Salary is " << salary_Net
            << endl
            << "---------------------" << endl;
}// END...Member Function greet_AllAttributes()
```

Program Listing/Output 25.13: Employee.cpp

25.5 FURTHER FILE CATEGORIES

❖ Text files and Binary files are amongst the simplest and most commonly used files

➤ Their simplicity is an asset in that the raw data from them can be read and, if required, parsed directly

➤ There is no danger of further control characters being read in

❖ For several commercial applications, including financial and internet based services, specialist activities need to be carried out, which can be provided by more sophisticated IT applications, including

➤ Spreadsheets

➤ Databases

These will be covered in Cybernetics in C++ - Expert

GENERIC PROGRAMMING

❖ Generic Programming is a powerful paradigm in software, which brings in "Generic Thinking"

❖ While procedural and object oriented programming have made significant contributions to the modern world, the fact remains, even they alone will have problems coping with modern day demands

❖ Generic Programming comprises being able to separate the algorithmic thinking from the types

❖ C++ supports generics primarily using the template, and is widely considered the strongest support in any language in modern times

❖ Several programming languages now support generics

➢ Java
➢ .NET languages, including
• C#
• Visual Basic.NET

❖ The versatility of Generic programming makes it ideal for libraries; examples include:

➢ The STL Standard Template Library
➢ Boost
➢ OpenGL (Open Graphics Library)

❖ Generic programming is not only an essential paradigm in its own right, but one that underpins several others:

➢ Metaprogramming: The practice of creating one program which generates/manipulates other programs, has the template at its core
• It is a major means by which modern day programs can be several million lines long
• The notion, programmers are expected to write 10 original lines of code a day remains

559

26. WHAT IS GENERIC PROGRAMMING?

26.1 GENERIC THINKING

❖ Generic thinking also brings in Abstraction: It comprises taking a **broad** look at activities, often very different ones, and focusing on what tasks **they have in common**

> The objective then is to perform the tasks in one way which can be reused

> Lack of Generic thinking often has catastrophic consequences

- At best, it will mean wasting a lot of time and effort on repeating tasks
- The frequent worst will mean mistakes will be duplicated
  - ○ Copying from a separate entity will mean everything which was specific will need to be eliminated:
    - ▪ this is subject to human error and consumes time and energy

❖ The fundamental tool used in generic thinking is the **template**

> **Mantra of Templates:** Do the hard work once, reuse it for the rest of your life

> Generic thinking using templates is practised abundantly outside the programming world:

26.1.1 GENERIC THINKING OUTSIDE THE PROGRAMMING WORLD

❖ Graphics Templates & Stencils

> A draughtsperson will often work on several different technical drawings such as architectural drawings, electronic circuits, aeronautical diagrams

> There will be several repetitive tasks such as drawing a 3mm curve

> A template which enables drawing this needs to be used

- Stencils in the pre computer era
- Shape templates in CAD tools nowadays

> In modern times, individual shapes in CAD tools can each be thought of as mini-templates

❖ Music

The same chord templates are frequently used by considerably different songs

➢ Guitar chords template: G major: G, C, D is used by
  - The Times they are A Changin' and Blowin' in the Wind - Bob Dylan
  - Cry, Cry, Cry and Ring of Fire - Johnny Cash
  - Brown Eyed Girl - Van Morrison
  - I'm a Believer - the Monkees
  - Where have all the Flowers Gone - written by Pete Seeger, performed by many artists including Joan Baez and Peter, Paul & Mary
  - Grandma's Feather Bed - John Denver
  - Coffee House (Indian Bengali song) - Manna Dey

➢ Guitar chords template: A major: A, E, further chords such as F#m, C#m
  - Help! - The Beatles
  - Pretty Woman - Roy Orbison
  - The Road to Amarillo - composed by Neil Sedaka, performed by various artists, especially Tony Christie
  - I want it that way - Backstreet Boys
  - Blue Suede Shoes - Carl Perkins, performed by several further artists especially Elvis

➢ Guitar chords template: Em, Am,
  - Losing my Religion - REM
  - Venus - Shocking Blue
  - Caballo Viejo - Simón Díaz (Venezuelan folk song)
    ○ Bamboleo - Gipsy Kings
  - Koto Dur Aar Koto Dur (Indian Bengali song) - Film Marutirtha Hinglaj

➢ Guitar chords template: C, Am, F, G
  - Let it Be - The Beatles
  - Blue Moon - various artists inc Elvis
  - Rawhide
  - Right here waiting for you - Richard Marx

➢ Guitar chords template: C, Am, F, D, E7
  - House of the Rising Sun - performed by various artists including The Animals & Bob Dylan
  - Hotel California - the Eagles

❖ Management

  ➢ Templates are widely used to analyse ideas, situations and case studies

- SWOT analysis template: Strengths, Weaknesses, Opportunities, Threats
- PEST analysis template: Political, Economic, Social, Technological
  - PESTLE: adds Legal, Environmental to PEST

❖ Law

➢ Templates are widely used to analyse cases
- CIRIP Conclusion, Issue, Rule, Interweaving, Policy
- CIRIC Conclusion, Issue, Rule, Interweaving, Conclusion

❖ IT

➢ Most tasks in IT comprise several repetitive tasks

	DOCUMENT	**TEMPLATE**
• MS Word	`.doc(x)`	`.dot(x)`
• MS Excel	`.xls(x)`	`.xlt(x)`
• MS PowerPoint	`.ppt(x)`	`.pot(x)`
• Dreamweaver	`.html, .htm`	`.dwt`

➢ Rough Guide: When you find yourself needing to do something twice or more, this is the cue that a template may well be called for

26.1.2 GENERIC THINKING INSIDE THE PROGRAMMING WORLD

"Generic programming is about abstracting and classifying algorithms and data structures. It gets its inspiration from Knuth and not from type theory. Its goal is the incremental construction of systematic catalogues of useful, efficient and abstract algorithms and data structures."

Alexander Stepanov,
Pioneer of Generic Programming

"Lift algorithms and data structures from concrete examples to their most general and abstract form."

Bjarne Stroustrup

❖ In software engineering, generic programming is a methodology which focuses on algorithms

❖ It comprises 2 major building blocks, PA

➤ **P**   Polymorphism

➤ **A**   Abstraction

You will note immediately that both of these are also building blocks of Object Oriented Programming

➤ This section will look at

- The similarities and differences of Abstraction in OOP and Abstraction in Generic Programming, then at
- Polymorphism in each paradigm

❖ Generic thinking has consequently had an immense influence in not only the software world but the modern world in general

➤ The introduction of generic programming is a relatively recent initiative, started in 1994

26.1.2.1 Problems: Lack of Generics

❖ Consider the following simple but illustrative problem: We want to compute the greater of two inputs

❖ We could achieve this with a function, as shown below:

```
//      Function: Output the larger of 2 numbers
//      -      If equal, output the first
//
int find_GreaterOfTwo( int myInt_In_A, int myInt_In_B )
{
    if( myInt_In_A >= myInt_In_B )
    {
        return myInt_In_A;
    }
    return myInt_In_B;
}
```

Program Listing/Output 26.1: Function: find_GreaterOfTwo

❖ The issue with functions is that specifying the TYPES of the input(s) and output are prerequisites

➤ In the above example, we chose the type `int`

❖ We now want to compute the greater of two numbers which are of different

➤ Primitive Types
- Doubles
- Floats

- Chars
- long ints
- ...in fact, all appropriate primitive types

➢ ...if applicable, user defined types

❖ The algorithm for each of these will be identical

❖ However, in C programming, the only strategy would be to copy and paste the code, then modify it for different types

➢ This is wasteful of time and effort

➢ Any mistakes in the original will be directly manifest in all functions thereafter

- When such mistakes are discovered, there is no quick way to correct them for all functions

❖ i.e. We have now discovered repetitive requirements in our program

➢ This is the cue that a **template** is required

➢ Virtually all programs will have several repetitive requirements

27. FUNCTION TEMPLATES

27.1 CREATING FUNCTION TEMPLATES

❖ The following program shows how a Function Template can be created for finding the greater of two inputs

➢ Note that the entire template, including the part that was the "function definition" is placed inside a header file

➢ This is a common practice, you will see why in the next section

```
template < typename T >
T find_LargerOfTwo( T var_IN_A, T var_IN_B )
{
        if( var_IN_A > var_IN_B )
        {
                return var_IN_A;
        }
        return var_IN_B;

}//END...Function Template find_LargerOfTwo
```

Program Listing/Output 27.1: File find_LargerOfTwo.h - Function Template

❖ The keyword `template` is used to tell the compiler that this is a function template, not a function

❖ The important point is that this is to work with different types

> Thus the generic type "T" has been used to represent this

  • This is very commonly used in the C++ community

> In order to tell the compiler that this is the generic type, one of two keywords can be used

  • `class`

    ○ However, since this is usually confused with the object oriented concept class, this is avoided in this book

  • `typename`

    ○ This is used throughout this book

❖ Adapting the generic way of thinking,

> The question "What is repetitive i.e. what actions will **all** types require?" is asked

> These are then implemented in the program, using some chosen character(s), "T" being a common choice, instead of a specific data type, whenever a data type is needed

  • Inputs

  • Output type

27.2 INSTANTIATING FUNCTION TEMPLATES

❖ When using the template parameter, 2 actions are required, which subtly look like 1 to the programmer:

> The template needs to be instantiated

> This "instantiated function" needs to be invoked

❖ Both of these can be accomplished with a "call" with variables of a certain type

> For example, if we want to use our template with integers, we need the command:

```
myInt_Bigger_A_B =
find_LargerOfTwo<int>( myInt_A, myInt_B );
```

where `myInt_A`, `myInt_B` and `myInt_Bigger_A_B` are previously defined integers

➤ Note that here, specifying the type `<int>` is good practice but not compulsory

➤ However, it is compulsory throughout the creation and usage of class templates (section 28), so it is best adhered to

❖ The following program shows how the function template is used for different types:

```cpp
#include <iostream>
using namespace std;

#include "find_LargerOfTwo.h"

int main()
{
        //»»»»»»»»»»»»»»»»»»»»»»»»»»»»»»»»»»»»»»»»»»»»»»»»»»»»»»»»»»»»»»
        int myInt_A = 5, myInt_B = 8, myInt_Bigger_A_B = 0;

        double myDbl_A = 3.7, myDbl_B = 7.4, myDbl_Bigger_A_B = 0.0;

        signed int mySignedInt_A = -5, mySignedInt_B = 7,
                                      mySignedInt_Bigger_A_B = 0;

        char myChr_A = 'p', myChr_B = 'r', myChr_Bigger_A_B = ' ';

        int myInt_Bigger_12_14 = 0;

        //»»»»»»»»»»»»»»»»»»»»»»»»»»»»»»»»»»»»»»»»»»»»»»»»»»»»»»»»»»»»»»
        // "Calling" our Function Template
        //
        myInt_Bigger_A_B = find_LargerOfTwo<int>( myInt_A, myInt_B );

        myDbl_Bigger_A_B = find_LargerOfTwo<double>( myDbl_A, myDbl_B );

        mySignedInt_Bigger_A_B =
                find_LargerOfTwo<signed int>( mySignedInt_A, mySignedInt_B
        );

        myChr_Bigger_A_B = find_LargerOfTwo<char>( myChr_A, myChr_B );

        //»»»»»»»»»»»»»»»»»»»»»»»»»»»»»»»»»»»»»»»»»»»»»»»»»»»»»»»»»»»»»»
        cout    << "The larger of "<< myInt_A
                << " and "<< myInt_B
                << " is "<< myInt_Bigger_A_B
                << endl;

        cout    << "The larger of "<< myDbl_A
                << " and "<< myDbl_B
```

```
                    << " is "<< myDbl_Bigger_A_B
                    << endl;

      cout    << "The larger of "<< mySignedInt_A
              << " and "<< mySignedInt_B
              << " is "<< mySignedInt_Bigger_A_B
              << endl;

      cout    << "The larger of "<< myChr_A
              << " and "<< myChr_B
              << " is "<< myChr_Bigger_A_B
              << endl;

       cout   << "The larger of "<< 12
              << " and "<< 14
              << " is "<< find_LargerOfTwo<int>( 12, 14 )
              << endl;

      //»»»»»»»»»»»»»»»»»»»»»»»»»»»»»»»»»»»»»»»»»»»»»»»»»»»»»»»»»»»
      cout << endl<< "===================="
           << endl<< "END...C++ Generics: Function Templates"<< endl;
      while(true){}
      return 0;

}//END...Function main
```

```
The larger of 5 and 8 is 8
The larger of 3.7 and 7.4 is 7.4
The larger of -5 and 7 is 7
The larger of p and r is r
The larger of 12 and 14 is 14

====================
END...C++ Generics: Function Templates
```

Program Listing/Output 27.2: Function_Template_Demo_Client.cpp: Using the Function Template

27.2.1 COMPILER GENERATED CODE

❖ When the function template is "invoked" with a new data type, we say that it has been instantiated

❖ Thus the lines in the above program which invoke the template will generate the following code:

C++ STATEMENT	CODE GENERATED BY THE COMPILER
find_LargerOfTwo<int> (myInt_A, myInt_B);	```int find_LargerOfTwo
(int var_IN_A, int var_IN_B)
{
if(var_IN_A > var_IN_B)
{
return var_IN_A;
}
return var_IN_B;
}```
find_LargerOfTwo<double> (myDbl_A, myDbl_B);	```double find_LargerOfTwo
        (double var_IN_A,
          double var_IN_B)
{
    if(var_IN_A > var_IN_B)
    {
            return var_IN_A;
    }
    return var_IN_B;

}``` |
| find_LargerOfTwo<signed int>
  (mySignedInt_A, mySignedInt_B); | ```signed int find_LargerOfTwo
        (signed int var_IN_A,
          signed int var_IN_B)
{
    if(var_IN_A > var_IN_B)
    {
            return var_IN_A;
    }
    return var_IN_B;
}``` |
| find_LargerOfTwo<char>
    (myChr_A, myChr_B); | ```char find_LargerOfTwo
        (char var_IN_A,
          char var_IN_B)
{
    if(var_IN_A > var_IN_B)
    {
            return var_IN_A;
    }
    return var_IN_B;
}``` |
| find_LargerOfTwo<int>(12, 14) | Nothing - the function for the type int has already been generated |
| **Program Listing/Output 27.3**
Client Code "invoking" the Function Template | **Program Listing/Output 27.4**
Code generated by the Compiler |

❖ Thus at runtime, it will be as if the code was written by a human

27.2.1.1 Compiler Intelligence

❖ It is the **compiler** which has generated this code

❖ Up until now, the compiler was seen as an important but unintelligent tool

- ➢ It performed the roles of 2Ps, the Pigeon and the Parrot
- ➢ This is the case throughout the Procedural, Memory Management and Object Oriented paradigms

❖ Now, in the Generic paradigm, the compiler is an intelligent tool, which generates code

- ➢ This phenomenon is the basis of several further paradigms of computer science, including Metaprogramming

27.3 POLYMORPHISM IN GENERIC PROGRAMMING

❖ Looking again at Program Listing/Output, notice that

- ➢ The template is a boilerplate
- ➢ which binds to different TYPES

❖ Generic programming thus exhibits POLYMORPHISM

❖ However, this is a different sort of polymorphism from that encountered in Object Oriented Programming

❖ Generic Polymorphism is

- ➢ Unbounded (noninvasive/nonintrusive): The interface of the types participating in the polymorphism are not predetermined
- ➢ Static: The binding takes place at compile time

❖ OOP Polymorphism is

- ➢ Bounded (invasive/intrusive): The interface of the types participating in the polymorphism are predetermined
- ➢ Dynamic: The binding takes place at runtime

28. CLASS TEMPLATES

❖ The principle of creating and using Class Templates is similar to that of function templates

❖ Now, entire classes are created

- ➢ Instead of coding "concrete types" for the
  - • Data Members

- Member Functions
  - Input types
  - Output type

(a) generic type(s) is/are used

28.1 CREATING CLASS TEMPLATES

❖ The example used here is to create a template to implement the special purpose data structure, the " Stack "

➤ The idea is that this has a LIFO Last In First Out system

```cpp
#include "assert.h"

const int SIZE_STACK = 100;

//»»»»»»»»»»»»»»»»»»»»»»»»»»»»»»»»»»»»»»»»»»»»»»»»»»»»»»»
template < typename dataTypeT >
class Stack
{
      private:
            //=====================================================
            int count;
            dataTypeT stackArray[SIZE_STACK];

      public:
            //=====================================================
            // Initialise the Stack
            Stack()
            {
                  count = 0;
            }

            //=====================================================
            // Member Functions - Functionalities of the Stack
            //
            void push( dataTypeT item_In ); // Insertion
            dataTypeT pop();    // Removal
};

//»»»»»»»»»»»»»»»»»»»»»»»»»»»»»»»»»»»»»»»»»»»»»»»»»»»»»»»
//   push - Inserting data on to the Stack
//
template <typename dataTypeT>
void Stack<dataTypeT>::push( dataTypeT item_In)
```

```
{
    assert( count >= 0 );
    assert( count < sizeof(stackArray) / sizeof( stackArray[0] ) );
    stackArray[count] = item_In;

    ++count;

}//END...Member Function Template push


//»»»»»»»»»»»»»»»»»»»»»»»»»»»»»»»»»»»»»»»»»»»»»»»»»»»»»»»»»»»»»»
//    pop - Removing data from the Stack and returning it
//
template <typename dataTypeT>
dataTypeT Stack<dataTypeT>::pop(  )
{
    --count;
    assert( count >= 0 );
    assert( count < sizeof(stackArray) / sizeof( stackArray[0] ) );

    return stackArray[count];

}//END...Member Function Template push
```

Program Listing/Output 28.1: Stack.h - Class Template for implementing the Stack data structure

❖ Once again,

> ➤ All the data and functionality which **all** types have in common need to be implemented
> ➤ Whenever types need to be mentioned, the generic name "dataTypeT" is used

❖ Note that when implementing each individual member function

> ➤ The fact that it is a template, this time a member function template
> ➤ Which generic data type it will be using

   must be (re)stated

❖ NOTE: Treat this section as reference;

> ➤ It will not be worth your while developing template code for data structures which already exist in established libraries
> ➤ Using templates for data structures from libraries the STL and Boost is covered in detail in Cybernetics in C++ - Expert.

28.2 USING CLASS TEMPLATES

```cpp
// ClassTemplate_Stack.cpp : main project file.
#include <iostream>
using namespace std;

#include "Stack.h"

int main( )
{
    //»»»»»»»»»»»»»»»»»»»»»»»»»»»»»»»»»»»»»»»»»»»»»»»»»»»»»»»»»»»»»»»
    cout    << endl
            << "========================================="<< endl
            << "Stack of INTEGERS"<< endl
            << "========================================="<< endl;
    Stack<int> myIntStack_A;
    myIntStack_A.push(1);
    myIntStack_A.push(2);
    myIntStack_A.push(3);
    myIntStack_A.push(4);

    cout    << "Popping myIntStack, latest element: "
            << myIntStack_A.pop()
            << endl;

    cout    << "Popping myIntStack, latest element: "
            << myIntStack_A.pop()
            << endl;

    cout    << "Popping myIntStack, latest element: "
            << myIntStack_A.pop()
            << endl;

    cout    << "Popping myIntStack, latest element: "
            << myIntStack_A.pop()
            << endl;


    cout    << endl
            << "========================================="<< endl
            << "Stack of DOUBLES"<< endl
            << "========================================="<< endl;
    Stack<double> myDblStack_A;
    myDblStack_A.push(9.23);
    myDblStack_A.push(2.235);
    myDblStack_A.push(25.1236);
    myDblStack_A.push(129.3);

    cout    << "Popping myDblStack_A, latest element: "
            << myDblStack_A.pop()
            << endl;
```

```
   cout    << "Popping myDblStack_A, latest element: "
           << myDblStack_A.pop()
           << endl;

   cout    << "Popping myDblStack_A, latest element: "
           << myDblStack_A.pop()
           << endl;

   cout    << "Popping myDblStack_A, latest element: "
           << myDblStack_A.pop()
           << endl;


   cout    << endl
           << "============================================="<< endl
           << "Stack of CHARS"<< endl
           << "============================================="<< endl;
   Stack<char> myCharStack_A;
   myCharStack_A.push('e');
   myCharStack_A.push('s');
   myCharStack_A.push('a');
   myCharStack_A.push('c');

   cout    << "Popping myCharStack_A, latest element: "
           << myCharStack_A.pop()
           << endl;

   cout    << "Popping myCharStack_A, latest element: "
           << myCharStack_A.pop()
           << endl;

   cout    << "Popping myCharStack_A, latest element: "
           << myCharStack_A.pop()
           << endl;

   cout    << "Popping myCharStack_A, latest element: "
           << myCharStack_A.pop()
           << endl;

   //»»»»»»»»»»»»»»»»»»»»»»»»»»»»»»»»»»»»»»»»»»»»»»»»»»»»»»»»
   cout    << endl<< "===================="
           << endl<< "END...Demo C++ Class Templates Demo"
           << endl;
   while(true){}
   return 0;

}//END...Function main
```

```
=============================================
Stack of INTEGERS
```

```
==========================================
Popping myIntStack, latest element: 4
Popping myIntStack, latest element: 3
Popping myIntStack, latest element: 2
Popping myIntStack, latest element: 1


==========================================
Stack of DOUBLES
==========================================
Popping myDblStack_A, latest element: 129.3
Popping myDblStack_A, latest element: 25.1236
Popping myDblStack_A, latest element: 2.235
Popping myDblStack_A, latest element: 9.23


==========================================
Stack of CHARS
==========================================
Popping myCharStack_A, latest element: c
Popping myCharStack_A, latest element: a
Popping myCharStack_A, latest element: s
Popping myCharStack_A, latest element: e


====================
END...Demo C++ Class Templates Demo
```

Program Listing/Output   28.2: ClassTemplateDemo_Stack_Client.cpp Using the Class Template Stack.h

28.3 TEMPLATE PARAMETERS & ARGUMENTS

❖ Revision: What are Parameters and Arguments?

➢ Parameters are placeholders which will be replaced by an "actual value" at a certain time

● These are often called "Formal Parameters"

➢ Arguments - these are the actual values

● These are often called "Actual Parameters"

➢ For templates (and macros), the replacement takes place at compile time

➢ For functions, the replacement takes place at runtime

❖ When a template parameter is introduced using the keyword `typename` or `class`, it is called a Type Parameter

➢ These are by far the most common, and usually go without say

28.3.1 NON-TYPE TEMPLATE PARAMETERS

❖ Non-Type Template Parameters are when the parameters represent constant values of a certain type, which can be determined at compile or link time

❖ The type (i.e. the type of the forthcoming value) is restricted to one of the following:

➢ An Integer or Enumeration

➢ A Pointer

➢ A Reference

```
template<typename dataTypeT
    typename dataTypeT::Allocator* Allocator
    >
 class Linked_List
```

❖ The first `typename dataTypeT` is a type parameter

❖ The second, which is followed by a name with the scope resolution operator, is a non-type parameter

❖ Non-type parameters are always r-values, never l-values

28.3.2 TEMPLATE TEMPLATE PARAMETERS

❖ Template Template Parameters are when the parameters represent class types

```
template < template<typename dataType T> class myClass >
void myFunction( myClass<int>* ptr_IN );
```

❖ Note that ADTs, including structs and unions, cannot be used

28.3.3 DEFAULT TEMPLATE ARGUMENTS

❖ In several situations, it is useful to have (a) default type(s) established:

➢ The syntax will be as follows:

```
template < typename T1 = int, typename T1 = double,
typename T3 >
```

➢ This way, when a type has not been specified when instantiating the template, the default will be used

❖ This is particularly useful in situations where type flexibility is useful, but there is one type which will be the most suitable most of the time

> e.g. Salaries will almost always be of type double
>   • However, if a situation arises when integer values can be used, such as a small service being outsourced, using the type int will ensure, amongst other things, half the memory space is used

28.4 CLASS TEMPLATE DESIGN IN UML

❖ UML supports the usage of templates, by denoting the generic type using a rectangle with a dotted line in the top right corner of the class diagram

❖ The following diagram can be used to represent the class template Stack in UML:

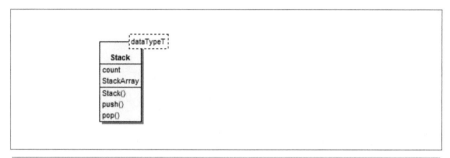

Figure 28.1: UML Diagram for the Stack Template

In BOUML, in order to introduce the generic type, here `dataTypeT`
❖ Right click on the class name in the navigator, then click on Edit

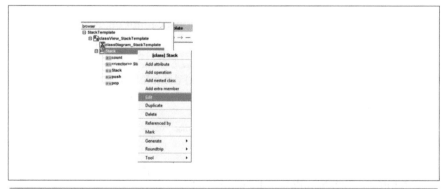

Figure 28.2: Clicking on Edit in the Navigator to bring up the "Class dialogue" dialogue box

❖ Now, in the "Class Dialog" dialogue box which follows,

  ➢ Click on "Parametrized"
  ➢ In the column "Name", type the name of the generic type, in our case
    `dataTypeT`

28.5 BEAUTIES OF GENERIC PROGRAMMING

❖ Generic Programming using templates unleashes a lot of power
❖ New level of software design and thinking

  ➢ By applying abstraction, generic programming enables a new level of thinking algorithmically
    • You can concentrate on the algorithm
    • The template will ensure that details like which types it operates on are taken care of

❖ Debugging

  ➢ When an error is found, only the template needs to be fixed, then a recompilation will regenerate all the required code
  ➢ In C programming, the only possibility is to manually modify each program

❖ Reusability

  ➢ When a specified operation needs to be performed on data structures, a template can be reused for different types
  ➢ For this reason, libraries are mostly template libraries; examples include
    • The C++ Standard Library
      ○ A major component is the STL Standard Template Library
    • Boost
    • OpenGL

❖ Utilising the Compiler

  ➢ It introduces a level of smartness to the compiler not encountered in C programming

❖ Improvement in runtime performance

  ➢ Since it is the compiler which carries performs the code generation,

- A lot of work can be carried out at compile time which would otherwise have to be done at runtime
- Examples include:
  - Error checking
  - The overhead associated with Object Oriented Classes when objects are created and used

➢ Using templates can significantly improve runtime performance; examples include

- Functors (Cybernetics in C++ - Expert)
- STL and Boost data structures and algorithms (Cybernetics in C++ - Expert)

❖ Reflection of the real world

➢ Since templates are used abundantly outside the computer world (section 26.1.1), whenever they need to be implemented in C++, a direct translation is often possible

29. TEMPLATE SPECIALISATION & OVERLOADING

29.1 WHAT IS TEMPLATE SPECIALISATION

❖ Generic Programming using templates unleashes a lot of power

➢ It enables thinking algorithmically, then ensuring the algorithm works for different types

➢ It introduces a level of smartness to the compiler not encountered in C programming

❖ However, it may seem like a compromise at first glance

➢ The same algorithm might not work the same way for all types of data

➢ Even if it does, it might not give optimal performance; e.g. a find algorithm will behave differently for different data structures such as arrays and linked lists (Cybernetics in C++ - Expert)

❖ What we want is a generic template which will work for the bulk of cases, yet give specialist performance for a few special cases

➢ This is template specialisation

29.2 FUNCTION TEMPLATE SPECIALISATION

❖ In order to implement Function Template Specialisation, **overloading** is used

> A new function with the same name as the function template, needs to be created

> That way, the program will initially look for any function by this name

> When it finds that there is no function which is appropriate for the given inputs, it will then go for the template

29.2.1 IMPLEMENTING FUNCTION TEMPLATE SPECIALISATION

❖ In our example, we want to be able to compare strings as well

❖ The problem is that

> since strings are implemented as arrays-of-Chars,

> the name of a string variable is a pointer to the first address in memory where the string/array is stored

❖ Thus if we try to use our template function as follows:

```
string surname_A = "Smith";
string surname_B = "Jones";
string string_Bigger_A_B = " ";
string_Bigger_A_B = find_LargerOfTwo<string>( surname_A, surname_B );
```

> We will end up comparing the pointers of their first addresses

❖ Therefore, we create a function, with

> The overloaded name `find_LargerOfTwo`

> Inputs: 2 strings

> Output: 1 string

```
string find_LargerOfTwo( string str_IN_A, string str_IN_B )
{
        if( strcmp( str_IN_A.c_str(), str_IN_B.c_str() ) )
        {
                return str_IN_A;
        }
        return str_IN_B;

}//END...Function Template find_LargerOfTwo
```

Program Listing/Output 29.1: Specialist `Function find_LargerOfTwo`

❖ The comparison is carried out using the Standard Library function `strcmp`

❖ When our program sees the word `"find_GreaterOfTwo"`

  ➤ It first looks at whether there is a simple function by this name
  ➤ If not, it then looks for templates

❖ Thus specialisation is achieved using Overloading

❖ The full program is as follows:

```
// FUNCTION TEMPLATE
template < typename T >
T find_LargerOfTwo( T var_IN_A, T var_IN_B )
{
      if( var_IN_A > var_IN_B )
      {
      return var_IN_A;
      }
      return var_IN_B;

}//END...Function Template find_LargerOfTwo
```

Program Listing/Output 29.2: find_LargerOfTwo.h

```
/*      ***************************************************************
        Function Template Specialisation/Overloading Demo
        ***************************************************************

        -       Want a "function" which will return the larger of 2 inputs
                -       We want it to work for several different types:
                        -       ints
                        -       doubles
                        -       unsigned ints
                -       Issue: a function requires the TYPES of the
                        inputs and output to be known
                -       We write a Function TEMPLATE

                -       Generic Programming exhibits POLYMORPHISM
                        -       The Function Template is a Boilerplate
                        -       It binds to different types

                -       OOP vs Generic Polymorphism
                        -       OOP Polymorphism: Binding takes place at
                                RUNTIME
                        -       Generic Polymorphism: Binding takes place at
                                COMPILE TIME
 */

#include <iostream>
using namespace std;
```

```cpp
#include <string>

#include "find_LargerOfTwo.h"

string find_LargerOfTwo( string, string );


int main()
{
        //»»»»»»»»»»»»»»»»»»»»»»»»»»»»»»»»»»»»»»»»»»»»»»»»»»»»»»»»»»»»»»
        int myInt_A = 5, myInt_B = 8, myInt_Bigger_A_B = 0;

        double myDbl_A = 3.7, myDbl_B = 7.4, myDbl_Bigger_A_B = 0.0;

        signed int mySignedInt_A = -5, mySignedInt_B = 7,
                        mySignedInt_Bigger_A_B = 0;

        char myChr_A = 'p', myChr_B = 'r', myChr_Bigger_A_B = ' ';

        int myInt_Bigger_12_14 = 0;

        string surname_A = "Smith", surname_B = "Jones",
                    surname_Later_A_B = " ";

        //»»»»»»»»»»»»»»»»»»»»»»»»»»»»»»»»»»»»»»»»»»»»»»»»»»»»»»»»»»»»»»
        // "Calling" our Function Template
        //
        myInt_Bigger_A_B = find_LargerOfTwo<int>( myInt_A, myInt_B );

        myDbl_Bigger_A_B = find_LargerOfTwo<double>( myDbl_A,
                            myDbl_B );

        mySignedInt_Bigger_A_B =
                find_LargerOfTwo<signed int>( mySignedInt_A,
                    mySignedInt_B );

        myChr_Bigger_A_B = find_LargerOfTwo<char>( myChr_A, myChr_B );

        surname_Later_A_B = find_LargerOfTwo( surname_A, surname_B );

        //»»»»»»»»»»»»»»»»»»»»»»»»»»»»»»»»»»»»»»»»»»»»»»»»»»»»»»»»»»»»»»
        cout    << "The larger of "<< myInt_A
                << " and "<< myInt_B
                << " is "<< myInt_Bigger_A_B
                << endl;

        cout    << "The larger of "<< myDbl_A
                << " and "<< myDbl_B
                << " is "<< myDbl_Bigger_A_B
                << endl;

        cout    << "The larger of "<< mySignedInt_A
                << " and "<< mySignedInt_B
```

```
                    << " is "<< mySignedInt_Bigger_A_B
                    << endl;

        cout    << "The larger of "<< myChr_A
                    << " and "<< myChr_B
                    << " is "<< myChr_Bigger_A_B
                    << endl;

        cout    << "The larger of "<< 12
                    << " and "<< 14
                    << " is "<< find_LargerOfTwo<int>( 12, 14 )
                    << endl;

        cout    << "The later of "<< surname_A
                    << " and "<< surname_B
                    << " is "<< surname_Later_A_B
                    << endl;

        //»»»»»»»»»»»»»»»»»»»»»»»»»»»»»»»»»»»»»»»»»»»»»»»»»»»»»»»»»
        cout    << endl<< "===================="<< endl
                    << "END...C++ Generics:"
                    << " Function Template Specialisation and Overloading"
                    << endl;
        while(true){}
        return 0;

}//END...Function main

//+++++++++++++++++++++++++++++++++++++++++++++++++++++++++++++++++
string find_LargerOfTwo( string str_IN_A, string str_IN_B )
{
        if( strcmp( str_IN_A.c_str(), str_IN_B.c_str() ) )
        {
                return str_IN_A;
        }
        return str_IN_B;

}//END...Function Template find_LargerOfTwo
```

```
 The larger of 5 and 8 is 8
 The larger of 3.7 and 7.4 is 7.4
 The larger of -5 and 7 is 7
 The larger of p and r is r
 The larger of 12 and 14 is 14
 The later of Smith and Jones is Smith

 ====================
 END...C++ Generics: Function Template Specialisation and Overloading
```

Program Listing/Output 29.3: Function_Template_Specialisation_Demo_Client.cpp

29.3 CLASS TEMPLATE SPECIALISATION

❖ Class Templates can be specialised in a similar but not quite the same the same manner as Function Templates

➢ A class name cannot be "overloaded"

❖ Instead, two forms of specialisation are possible:

➢ Explicit Specialisation

➢ Implicit Specialisation

❖ When a class template is to be specialised, each member function must also be specialised

➢ While it is possible to specialise an individual member function

• doing so will mean the rest of the template cannot be specialised

❖ The specialisation must start with the qualifier template ◇

29.3.1 EXPLICIT SPECIALISATION

❖ The following program shows Explicit Specialisation being used:

```cpp
#include <iostream>
using namespace std;


//      ################################################################
//      "USUAL/GENERIC" CLASS i.e. NO Specialisation
template<typename T = double, int i = 7>
class myClass
{
     public:
          myClass();
          int myInt_A;
};

// CONSTRUCTOR: Non Specialist Class
template<typename T, int i>
myClass<T, i>::myClass()
     : myInt_A(i)
{
     cout    << "Generic Template i.e. No Specialisation, "
          << " myInt_A = "<< myInt_A
          << endl;
}

//################################################################
// EXPLICIT SPECIALISATION 1
```

```
template<>
class myClass<>
{
      public:
            myClass();
};

// CONSTRUCTOR: Specialist Class 1
myClass<>::myClass()
{
      cout << "Explicit specialization (1) called, "
            << "Parameters: Default"
            << endl;
}

//###############################################################
// EXPLICIT SPECIALISATION 2
template<>
class myClass<double, 10>
{
      public:
            myClass();
};

// CONSTRUCTOR: Specialist Class 2
myClass<float, 9>::myClass()
{
      cout    << "Explicit specialization (2) called, "
            << "Parameters: <float, 9>"<< endl;
}

int main()
{
      cout   << endl<< "=============="<< endl;
      cout   << "Instantiating with myClass<int,5> x"<< endl;
      myClass<int,5> x;


      cout   << endl<< "=============="<< endl;
      cout   << "Instantiating with myClass<> y"<< endl;
      myClass<> y;

      cout << endl<< "=============="<< endl;
      cout << "Instantiating with myClass<float, 9> z"<< endl;
      myClass<float, 9> z;

      cout << endl<< "=============="<< endl;
      cout << "Instantiating with myClass<float, 10> z"<< endl;
      myClass<float, 10> a;

      //»»»»»»»»»»»»»»»»»»»»»»»»»»»»»»»»»»»»»»»»»»»»»»»»»»»»»»
      cout << endl<< "===================="
            << endl<< "END...C++ Generics: Class Templates, "
```

```
                          << "EXPLICIT/FULL Specialisation"<< endl;
      while(true){}
      return 0;

}//END...Function main
```

```
==============
Instantiating with myClass<int,5> x
Generic Template i.e. No Specialisation,  myInt_A = 5

==============
Instantiating with myClass<> y
Explicit specialization (1) called, Parameters: Default

==============
Instantiating with myClass<float, 9> z
Explicit specialization (2) called, Parameters: <float, 10>

==============
Instantiating with myClass<float, 10> z
Generic Template i.e. No Specialisation,  myInt_A = 10

===================
END...C++ Generics: Class Templates, EXPLICIT/FULL Specialisation
```

Program Listing/Output 29.4: ExplicitSpecialisation_Client.cpp

29.3.2 PARTIAL SPECIALISATION

❖ Partial specialisation is an extension of explicit specialisation in that it involves template arguments as well as template parameters

❖ Partial specialisation is a step further from explicit, in that it instead of fully substituting the template parameters, alternative implementations are given which enable the further specialisations

❖ The following program demonstrates this

```
#include <iostream>
using namespace std;

//##############################################################
template<typename T, typename U, int I>
class myClass
{
      public:
            void myMemFunc()
            {
```

```cpp
                        cout << "Generic template invoked"
                                << endl;
                }
};

//################################################################
template<typename T, int I>
class myClass<T*, T, I>
{
        public:
                void myMemFunc()
                {
                        cout << "Partial specialization 1 invoked"
                                << endl;
                }
};

//################################################################
template<typename T, typename U, int I>
class myClass<T*, U, I>
{
        public:
                void myMemFunc()
                {
                        cout << "Partial specialization 2 invoked"
                                << endl;
                }
};

//################################################################
template<typename T>
class myClass<int, T*, 10>
{
        public:
                void myMemFunc()
                {
                        cout << "Partial specialization 3 invoked"
                                << endl;
                }
};

//################################################################
template<typename T, typename U, int I>
class myClass<T, U*, I>
{
        public:
                void myMemFunc()
                {
                        cout << "Partial specialization 4 invoked"
                                << endl;
                }
};
```

```
int main()
{
        //»»»»»»»»»»»»»»»»»»»»»»»»»»»»»»»»»»»»»»»»»»»»»»»»»»»»»»»»»»»»»»
        // "Invoking the various Partial Specialisations"
        //
        myClass<int, int, 7> myObj_A;
        myObj_A.myMemFunc();

        myClass<double*, double, 9> myObj_B;
        myObj_B.myMemFunc();

        myClass<double*, char, 10> myObj_C;
        myObj_C.myMemFunc();

        myClass<int, double*, 10> myObj_D;
        myObj_D.myMemFunc();

        myClass<double, float*, 9> myObj_E;
        myObj_E.myMemFunc();

        // myClass<int, int*, 10> myObj_E;
                // ERROR: AMBIGUITY: Both Specialisations 3 and 4 match
     this

        //»»»»»»»»»»»»»»»»»»»»»»»»»»»»»»»»»»»»»»»»»»»»»»»»»»»»»»»»»»»»»»
        cout    << endl<< "===================="
                << endl<< "END...C++ Generics: Class Templates, "
                << "PARTIAL Specialisation"<< endl;
        while(true){}
        return 0;

}//END...Function main
```

```
Generic template invoked
Partial specialization 1 invoked
Partial specialization 2 invoked
Partial specialization 3 invoked
Partial specialization 4 invoked
====================
END...C++ Generics: Class Templates, PARTIAL Specialisation
```

Program Listing/Output 29.5: PartialSpecialisation_Client.cpp

GETTING IT TOGETHER: PROGRAMMING IN C++

❖ The aim of this section is to serve as both revision and reference material
❖ Throughout this book, several concepts have been encountered, which have been related, but by their nature, have had to be covered separately

➢ This is the part where they are compared and contrasted
➢ When there have been ambiguities, especially with duplicate keywords and symbols, the aim is to clarify matters here

30. DUPLICATE KEYWORDS

30.1 THE KEYWORD "STATIC"

❖ Often considered the joint most confusing single aspect of C++, the keyword `static` has no less than 5 different meanings, only some of which are related

	PARADIGM	MEANING
❖ Static Global Variable	Procedural	Accessible only within that file
❖ Static Global Function	Procedural	Accessible only within that file
❖ Static Local Variable	Procedural & Object Oriented	Retains its value between function calls i.e. does not go out of scope throughout the duration of the program
❖ Static Data Member	Object Oriented	Class specific data member i.e. all objects of that class share the same value e.g. number of objects created
❖ Static Member Function	Object Oriented	A member function which can be called via the Class name, as well as via an object The only data members it can access are static ones

30.2 THE KEYWORD "VIRTUAL"

❖ The keyword `virtual` is used on 2 different cases, which are also largely unrelated:

➢ Virtual Class in Multiple Implementation Inheritance

● Used in Diamond Shaped Multiple Inheritance to ensure the overall derived class has one, no duplicate, copy of the data from the overall base class

➢ Virtual Member Function in OOP Polymorphism

● Used to ensure that the Pointer to a Base Class binds to the Derived Class at runtime

30.3 THE KEYWORD "CLASS"

❖ Likewise, the word `class` has 2 different, largely unrelated, meanings in C++

➢ The Object Oriented Data Type `class`
➢ The Generic type `class,` used in Templates

❖ For these reasons, you are often advised to use the keyword `typename` when dealing with templates

31. DUPLICATE SYMBOLS/OPERATORS

31.1 THE ASTERISK *

❖ (Q) What does the ampersand symbol `*` mean?
❖ (A) Depends on the context of its usage; this can be deciphered from the entities around it, judging whether they are data types or variable names

➢ Single asterisk "`*`" with variables on either side: Multiplication
➢ Single asterisk after a data type: "Of Type Pointer-to-DataType" operator for defining a pointer
➢ Single asterisk before a variable name: "Value of the contents of" operator for dereferencing a pointer

31.2 THE AMPERSAND &

❖ (Q) What does the ampersand symbol `&` mean?
❖ (A) Depends on the context of its usage; this can be deciphered from the entities around it, judging whether they are data types or variable names

> Single ampersand & with variables on either side: Bitwise AND
> Double ampersand && Logical AND
> Single ampersand & before a variable name: "Address of" operator for populating a pointer
> Single ampersand & after a data type: "Of Type Reference-to-DataType" operator for defining a reference

31.3 THE MEANINGS OF THE = SYMBOL

❖ One of the most familiar and commonly used symbols in programming is the =

❖ In C++, it has several related but different meanings:

> In all (but one) cases, it is a movement operator, used to move the value(s) of the right hand side into the address of the left hand side
> The nature of this movement depends on the types of data being operated on

OPERATION	MEANING
`a = b;`	Equating primitive variables

`richardGilmour = joeBloggs;`
 	
`After`
`Employee richardGilmour();`
`Employee joeBloggs ();`
	Equating two existing objects
	❖ The assignment operator is called

`Employee susanParker` `    = traceyCarpenter;`	Defining and initialising an object ❖ The Assignment Operator and ❖ Copy Constructor are called
`Employee *ptr_richardGilmour` `    = &richardGilmour;`	Defining and initialising a Pointer-to-anObject ❖ The **address-of** operator is called
`virtual void sound() = 0;`	Declaring `sound()` as a pure virtual function in OOP Polymorphism

❖ An important, but arguably most controversial usage of the = operator, is defining a Virtual Member Function as pure

> The Notation = 0 is used

• This is one of the major criticisms of C++

```
class Animal
{
    public:
            // PURE Virtual Function
            virtual void sound (    ) = 0;

};//END...Class Animal
```

❖ The presence of the pure virtual member function means that

  ➤ sound () will bind to a Derived class, not Base class, at Runtime
  ➤ The Class Animal must be Abstract

  • Any Attempt to instantiate an Object of this type

    Animal fred;

    will result in a compile-time error

31.4 THE PARENTHESES OPERATOR ()

❖ Parentheses are used extensively in C and C++
❖ Their several uses include:

  ➤ In C and C++:

    • Encapsulating an expression

      ○ Definition
      ○ Invocation

    • Initialising a Variable

      ○ Runtime stack
      ○ Runtime Heap

    • Function Operator

      ○ Definition
      ○ Invocation

    • Macro Operator

      ○ Definition
      ○ Invocation

  ➤ In C++:

    • Calling the Constructor of a Class

31.5 THE SCOPE RESOLUTION OPERATOR ::

❖ As we have been seeing, the Scope Resolution Operator :: has at least 4 uses

➤ Defining Members of a class in a Definition

• Member Functions
• Static Data Members

➤ Declaring Variables as part of a Namespace
➤ Calling a Static Member Function via a Class
➤ Resolving Ambiguities in Mulitple Inheritance

❖ It is one of the most important and widely used operators in C++

➤ This is partly why overloading it is disallowed

32. USING CONST

❖ As we have been discussing throughout this book, the qualifier `const` forces an entity to be read only
❖ This is of fundamental importance in all walks of programming, since it is essential to prevent unwanted changes to variables

➤ It is especially important in C++, since Polymorphism in both Object Oriented and Generic programming is all about type flexibility
➤ Using macro definitions, a common practice in C programming, is mostly unsuitable in C++

❖ Several entities can be read only

➤ Global and local variables
➤ Data Members
➤ Function inputs: Formal parameters
➤ The Function output
➤ Pointers
➤ Data being pointed to by pointers
➤ Data being referred to by references
➤ Member functions: they cannot write data to data members

32.1 CONST POINTERS

❖ As we have been discussing, when we deal with pointers, we are primarily dealing with two entities of data:

➤ The pointer itself

➢ The data it points to, the pointee

❖ So when we say "Const Pointer", which of these is forced to become read-only?

❖ In order to address this crucial issue, look at where the keyword `const` is placed in relation to the asterisk

➢ If the `const` keyword is **B**EFORE the asterisk, the data BEING **P**OINTED-**T**O is read only

➢ If the `const` keyword is **A**FTER the asterisk, the **P**OINTER is read only

❖ You can use the abbreviation BP TAP

33. CASTING

❖ As we have been seeing, casting is the practice of converting from one type into another

❖ The following are supported in C++

➢ `static_cast`	Used to convert from- one static type to another at compile time
➢ `const_cast`	Used to convert from into and out of read-only status at compile time
➢ `reinterpret_cast`	Used to convert a pointer from one class type to another, including an unrelated one, at compile time
➢ `dynamic_cast`	Used in OOP polymorphism to convert a pointer from one class type to another in the same inheritance hierarchy at runtime ...covered in Cybernetics in C++ - Expert

SUMMARY & CONCLUSIONS

❖ In this book, we have covered a substantial part of C++ and laid a strong foundation, both broad and deep, for the whole of computer science

❖ We started by looking at Imperative Programming, means of creating structured, organised programs

  ➢ C++ supports this using Data Types, Operators, Control Statements including conditional statements and loops

❖ We next looked at Procedural Programming

  ➢ C++ supports this using Functions, as does C, but the former goes further in that it allows Function Overloading

  ➢ C++ also allows Macros, widely used in C, but best avoided in the former due to the fact that It strongly supports Polymorphism

  ➢ Recursion, where an idea contains itself, was next covered

❖ Memory & Resource Management was covered next, the idea of accessing hardware directly;

  ➢ C++ supports this using
    • Pointers, as is also covered in C
    • References

  ➢ Runtime memory, with the Stack and Heap, is used extensively
    The challenging nature of this paradigm and the power it places in the hands of the programmer, are 2 of the reasons several languages, including Java and Swift largely hide this from the programmer

  ➢ The fact that it is pervasive in C++ is a major reason why It remains the de facto standard when high performance is required

❖ Data Structure programming, using Arrays came next

❖ Object Oriented Programming followed

  ➢ Firstly, how Abstraction is implemented using

- Structs, Unions and Enums in C and C++
- Classes in C++, which also implement Encapsulation
➤ Special members including Constructors, Destructors, Mutators, Static Data Members and Member Functions, Inner Classes and Friend Functions/Classes
➤ Special Operations, including Operator Overloading
➤ Inheritance
  - Including implementation and interface, single and multiple, inheritance with encapsulation qualifiers, and what to use and avoid
➤ Polymorphism, the powerful practice of associating multiple behaviour with a single type was next covered

❖ Generic Programming

➤ The practice of also implementing Abstraction and Polymorphism, this time in different ways, using templates, was the final paradigm covered

❖ The final "Getting it Together" section looked at points which are known to cause confusion

❖ The question now, is what is the way(s) forward

➤ If you have enjoyed all aspects of the course, and the fact that C++ is so powerful in several different paradigms, a further C++ book is advised
  - Cybernetics in C++ - Expert is aimed at this
➤ If you have enjoyed the Procedural and Memory & Resource Management Programming, but not the Object Oriented and Generic
  - C Programming is the way forward
➤ If you have enjoyed the Object Oriented and Generic Programming, but not the Memory & Resource Management
  - There are several programming languages that fall into this category including
    ○ Java
    ○ Swift
    ○ C#.NET
    ○ Web Programming Languages
    ○ Python
  - One category here, which is especially popular in recent times is programming mobile apps for Smartphones and Tablets; these include
    ○ iOS Apps programming using Swift
    ○ Android Apps programming using Java

APPENDIX

❖ Treat these figures as a guideline as opposed to a definitive list

  ➢ Different operating systems and programming environments do vary, along with their versions e.g.

  • PC Operating systems, such as MSDOS, MS Windows 3.1 - 8, Apple Mac, Linux Distros
  • Smart Phone and Tablet Operating Systems, such as Android, Apple iOS, Windows Mobile

  ➢ Significant changes take place whenever products move from one number system to another

  • 16-bit to 32-bit
  • 32-bit to 64-bit

APPENDIX A. C++ PRIMITIVE DATA TYPES

	TYPE	TYPICAL SIZE		COMMON RANGE VALUES		EXPs	
		Bytes	Bits	Lowest	Highest	Lowest	Highest
Character or small Integer (char)	char	1	8	0	255	0	$2^8 - 1$
	unsigned char	1	8	0	255	0	$2^8 - 1$
	signed char	1	8	-128	127	-2^7	$2^8 - 1$
	wide char wchar_t	2	16				
Integer (int)	int	4	32	0	4294967295	0	$2^{32} - 1$
	unsigned int	4	32	0	4294967295	0	$2^{32} - 1$
	signed int	4	32	-2147483648	2147483647	-2^{31}	$2^{31} - 1$
	short int / unsigned short int	2	16	0	65535	0	$2^{16} - 1$
short	signed short int	2	16	-32768	32767	-2^{15}	$2^{15} - 1$
	long int	4	32	0	4294967295	0	$2^{32} - 1$
	unsigned long int	4	32	0	4294967295	0	$2^{32} - 1$
long	signed long int	4	32	-2147483648	2147483647	-2^{31}	$2^{31} - 1$
Floating Point Number (double)	float	4	32			-3.4 E + 38	3.4 E + 38
	double	8	64			-1.8 E + 308	1.8 E + 308
long	long double	8	64			-1.8 E + 308	1.8 E + 308
Binary Digit Value (boolean)	bool	1	8	0	1	0	1

APPENDIX B. C++ OPERATORS

OPERATOR	NAME	INPUT(S)	OPERATION PROCESS	OUTPUT	OVERLOADING POSSIBLE?
+	Addition	2 numerical **values** (any primitive type)	Adds them	The Total	Yes
-	Subtraction	2 numerical **values** (any primitive type)	Subtracts the RHS from the LHS	The Difference	Yes
*	Multiplication	2 numerical **values** (any primitive type)	Multiplies them	The Product	Yes
/	Division	2 numerical **values** (any primitive type)	Divides the LHS (the Numerator) by the RHS (the Denominator) - Be careful the denominator is not zero	The Quotient	Yes
%	Remainder/ Modulo	2 numerical **values** (any primitive type)	Calculates the Remainder when the LHS (the Numerator) is divided by the RHS (the Denominator)	The Remainder	Yes
&	Bitwise AND	2 numerical **values** (any primitive type)	Performs AND on each digit of a Binary Number	1 Numerical Value	Yes
|	Bitwise OR	2 numerical **values** (any primitive type)	Performs OR on each digit of a Binary Number	1 Numerical Value	Yes
~	Bitwise NOT	1 numerical **value**	Performs NOT on each digit of a	1 Numerical Value	Yes

APPENDIX B. C++ OPERATORS

OPERATOR	NAME	INPUT(S)	OPERATION PROCESS	OUTPUT	OVERLOADING POSSIBLE?
		(any primitive type)	Binary Number		
!	NOT	Boolean value	Performs NOT on the input	1 Boolean Value	Yes
=	Assignment Equals	Takes in 2 inputs, returns NO output One address, the l-value, placed on the left One VALUE, the r-value, placed on the right Functionality:	Place the value in the address	None	Yes
<	Less than	Takes in 2 input values, returns boolean output	Compares the values	Returns TRUE if the Left hand value is less than the right hand value Returns FALSE otherwise	Yes
>	Greater than	Takes in 2 input values, returns boolean output Returns TRUE if the Left hand value is greater than the right hand value Returns FALSE otherwise	Compares the values		Yes

APPENDIX B. C++ OPERATORS

OPERATOR	NAME	INPUT(S)	OPERATION PROCESS	OUTPUT	OVERLOADING POSSIBLE?
+=	Addition Assignment	Address l-value, numerical value, r-value	Takes the Value in the Address, Adds to it the r-value, puts this in the location of the l-value	None	Yes
-=	Subtraction Assignment	Address l-value, numerical value, r-value	Takes the Value in the Address, Subtracts from it the r-value,	None	Yes
*=	Multiplication Assignment	Address l-value, numerical value, r-value	Takes the Value in the Address, Mutiplies it by the r-value, puts this in the location of the l-value	None	Yes
/=	Division Assignment	Address l-value, numerical value, r-value	Takes the Value in the Address, Divides it by the r-value, puts this in the location of the l-value	None	Yes
%=	Remainder Assignment	Address l-value, numerical value, r-value	Takes the Value in the Address, Finds the remainder when divided by the r-value, puts this in the location of the l-value	None	Yes
^=	Bitwise XOR Assignment	Address l-value, numerical value, r-value	Takes the Value in the Address, Performs Bitwise XOR with the r-value, puts this in the location of the l-value	None	Yes
&=	Bitwise AND Assignment	Address l-value, numerical value, r-value	Takes the Value in the Address, Performs Bitwise AND with the r-value, puts this in the location of the l-value	None	Yes
|=	Bitwise OR Assignment	Address l-value, numerical value, r-value	Takes the Value in the Address, Performs Bitwise OR with the r-value, puts this in the location of	None	Yes

APPENDIX B. C++ OPERATORS

OPERATOR	NAME	INPUT(S)	OPERATION PROCESS	OUTPUT	OVERLOADING POSSIBLE?
<<	Bitwise Left Shift Inputs to `cout`	2 r-values	the l-value Performs Bitwise Left shift on the left side r-value, by the number of bits in the right side r-value Overloaded to peform Inputs to `cout`	1 r-value	Yes
>>	Bitwise Right Shift Input to `cin`	2 r-values	Performs Bitwise Right shift on the left side r-value, by the number of bits in the right side r-value Overloaded to peform Inputs to `cin`	1 r-value	Yes
>>=	Bitwise Left Shift Assignment	1 l-value, 1 r-value	Takes the Value in the Address, Performs Bitwise Left Shift by the number of bits in the r-value, puts this in the location of the l-value	None	Yes
<<=	Bitwise Right Shift Assignment	1 l-value, 1 r-value	Takes the Value in the Address, Performs Bitwise Right Shift by the number of bits in the r-value, puts this in the location of the l-value	None	Yes
==	Equality	2 r-values	Checks if the two r-values are equal	TRUE if equal FALSE if unequal	Yes
!=	Not equal to	2 r-values	Checks if two values are unequal; returns	Boolean TRUE if unequal FALSE if equal	Yes
<=	Less than or Equal to	Takes in 2 input values, returns boolean output	Compares the 2 values	Returns TRUE if the Left hand value is less than or equal to	Yes

APPENDIX B. C++ OPERATORS

OPERATOR	NAME	INPUT(S)	OPERATION PROCESS	OUTPUT	OVERLOADING POSSIBLE?
>=	Greater than or Equal to	Takes in 2 input values, returns boolean output	Compares the 2 values	Returns TRUE if the Left hand value is greater than or equal to the right hand value Returns FALSE otherwise	Yes
&&	Logical AND	Takes 2 conditions i.e. each return a Boolean Value	Compares the 2 conditions	Returns TRUE if both are TRUE FALSE otherwise	Yes
||	Logical OR	Takes 2 conditions i.e. each return a Boolean Value	Compares the 2 conditions	Returns FALSE if both are FALSE TRUE otherwise	Yes
++	Increment	One Numerical r-value	Adds 1 to the Input post-increment: also holds the old value in a register	New incremented value: r-value	Yes
--	Decrement	One Numerical r-value	Subtracts 1 from the Input post-increment: also holds the old value in a register	New decremented value: r-value	Yes
.	Dot	Multiple uses: Mostly accessing members of an Abstract Data Type			No
. *		Multiple uses			No
: :	Scope	Multiple Uses			No

APPENDIX B. C++ OPERATORS

OPERATOR	NAME	OPERATION INPUT(S)	OPERATION PROCESS	OPERATION OUTPUT	OVERLOADING POSSIBLE?
	Resolution Operator				
`,`	Comma	Separates the formal or actual parameters in a (member) function			Yes
`->`	"Arrow"	Muliple uses, generally Access members of an ADT using a Pointer			Yes
`[]`	Array Brackets/ Square Brackets	Muliple uses, generally Define/Access as an Array/Pointer			Yes
`()`	Parentheses	Multiple Uses			Yes
`new`					No
`delete`					No
`new[]`					No
`delete[]`					No
`operator new`					No
`operator delete`					No
`operator new[]`					No
`operator delete[]`					No
`?:`	Ternary Conditional	The only ternary operator in C++ 3 inputs: ❖ Condition ❖ Action if True ❖ Action if False	Carry out the appropriate action of the conditional statement	None	No
`sizeof`		1 variable	Checks the amount of runtime memory	The Amount of	No

APPENDIX B. C++ OPERATORS

OPERATOR	NAME	INPUT(S)	OPERATION PROCESS	OUTPUT	OVERLOADING POSSIBLE?
typeid		1 variable	it occupies Checks the type ❖ Especially useful in Polymorphism	runtime memory The Type of the Variable	No
static_cast					No
dynamic_cast					No
const_cast					No
reinterpret_cast					No

❖ Note that just because an operator can be overloaded by no means implies that it should be overloaded (see section 21.6)

❖ The following special characters are used as operators as part of a string

- ▲ \\   Literal backslash
- ▲ \"   Double quote
- ▲ \'   Single quote
- ▲ \n   Newline (line feed)
- ▲ \r   Carriage return
- ▲ \b   Backspace
- ▲ \t   Horizontal tab
- ▲ \f   Form feed
- ▲ \a   Alert (bell)
- ▲ \v   Vertical tab
- ▲ \?   Question mark (used to escape trigraphs)
- ▲ \nnn   Character with octal value nnn
- ▲ \xhh   Character with hexadecimal value hh

❖ Treat these listings as a guideline as opposed to a definitive list

➢ Like with Data types, different programming environments do vary, along with their versions

APPENDIX C. RESERVED KEYWORDS IN C++

❖ The following reserved keywords are "inherited" from C programming

```
auto     const      double  float   int       short   struct   unsigned
break    continue   else    for     long      signed  switch   void
case     default    enum    goto    register  sizeof  typedef  volatile
char     do         extern  if      return    static  union    while
```

❖ The following reserved keywords are new to C++:

```
asm          dynamic_cast  namespace  reinterpret_cast  try
Bool         explicit      new        static_cast       typeid
catch        false         operator   template          typename
class        friend        private    this              using
const_cast   inline        public     throw             virtual
delete       mutable       protected  true              wchar_t
```

BIBLIOGRAPHY

[1] The C++ Standard; BSI (The British Standards Institution) (Author), Bjarne Stroustrup (Foreword); 2003 John Wiley & Sons

[2] The C Standard; BSI (The British Standards Institution) (Author), 2003 John Wiley & Sons

[3] The C++ Programming Language, Bjarne Stroustrup, Special Edition, 2000 Addison Wesley, ISBN-13: 978-0321563842, ISBN-10: 0321563840

[4] The Design and Evolution of C++; Author: Bjarne Stroustrup; ISBN-13: 978-0-201-54330-8; ISBN-10: 0-201-54330-3

[5] Programming Principles and Practice Using C++; Author: Bjarne Stroustrup; ISBN-13: 978-0-321-54372-1; ISBN-10: 0-321-54372-6

[6] The Annotated C++ Reference Manual; Authors: Bjarne Stroustrup, Margaret A.Ellis; ISBN: 0-201-51459-1

[7] A Tour of C++, Author: Bjarne Stroustrup, ISBN-10: 0321958314; ISBN-13: 978-0321958310

[8] Exceptional C++; Author: Herb Sutter; ISBN-13: 978-0-201-61562-3; ISBN-10: 0-201-61562-2; Addison-Wesley 2000

[9] More Exceptional C++; Author: Herb Sutter, ISBN-13: 978-0-201-70434-1, ISBN-10: 0-201-70434-X; Addison-Wesley 2002

[10] Exceptional C++ Style; Author: Herb Sutter; ISBN: 0-201-76042-8; Addison-Wesley 2005

[11] C++ Coding Standards; Author: Herb Sutter, Andrei Alexandrescu; ISBN: 0-321-11358-6; Addison-Wesley 2005

[12] Essential C++; Author: Stanley B. Lippman; ISBN-13: 978-0-201-48518-9; ISBN-10: 0-201-48518-4; Addison-Wesley 2000

[13] Applied C++; Author: Philip Romanik, Amy Muntz; ISBN: 0-321-10894-9; Addison-Wesley 2003

[14] Accelerated C++; Authors: Andrew Koenig, Barbera E.Moo; ISBN-13: 978-0-201-70353-5; ISBN-10: 0-201-70353-X; Addison-Wesley 2000

[15] Scott Meyers's Effective Series; Effective C++; Scott Meyers; Author: Scott Meyers; ISBN-13: 978-0-321-33487-9; ISBN-10: 0-321-33487-6

[16] Inside the C++ Object Model; Author: Stanley B. Lippman; ISBN-13: 978-0-201-83454-3; ISBN-10: 0-201-83454-5

[17] C++ Primer; Authors: Stanley B. Lippman, Josée Lajoie, Barbara E. Moo; ISBN-10: 0-321-71411-3; ISBN-13: 978-0-321-71411-4

[18] Advanced C++, Programming Styles and Idioms; Author: David Coplien; ISBN 0-201-54855-0

[19] Expert C++; Author: Herb Schildt; ISBN-10: 0078822092; ISBN-13: 978-0078822094; Publisher: McGraw Hill; Copyright: January 1996

[20] The Art of C++; Author: Herb Schildt; ISBN-13: 978-0072255126; ISBN-10: 0072255129; Publisher: McGrawHill; 2004

[21] C++ the Complete Reference; Author: Herb Schildt; ISBN-10: 0071634800; ISBN-13: 978-0071634809; Publisher: McGrawHill; May 22, 2015 (5th Edition)

[22] C++ for Programmers; Paul J Deitel, Harvey M Deitel, The Deitel Series, Prentice-Hall 2009

[23] C++ Unleashed; Jesse Liberty; Sams 1998

[24] Practical C++, Steve Oualline, O'Reilly 2002

[25] C++ The Core Language; Doug Brown, Gregory Satir; O'Reilly 1995

[26] C++ Cookbook; D. Ryan Stephens, Christopher Diggins, Jonathan Turkanis, Jeff Cogswell; O'Reilly 2005

[27] Expert C Programming, Deep C Secrets, Peter van der Linden, Addison-Wesley 1994

[28] The C Programming Language; Brian W Kernighan, Dennis M Ritchie, Prentice-Hall 2e 1988

[29] C, A Reference Manual; Samuel P. Harbison, Guy L. Steele Jr; Prentice-Hall, 5e 2002

[30] Elements of Programming; Authors: Alexander Stepanov, Paul McJones; ISBN-10: 0-321-63537-X; ISBN-13: 978-0-321-63537-X; Addison-Wesley 2009

[31] Programming Pearls; Author: Jon Bentley; ISBN-10: 0-201-65788-0; ISBN-13: 978-0-201-65788-3; 2e Addison-Wesley 2000

INDEX

ABOUT THE AUTHOR

Avi Bhattacharyya is a Software Engineer, Entrepreneur and Philanthropist based in London, UK. He is the Founder and CEO of Shomiron, clients including Royalty, Government Organisations and International Technical and Financial companies.

He holds the MEng in Electrical/Electronic Engineering from Imperial College, London, and 5 National Engineering Awards, including 1st Place in the 1994 Young Electronic Designer Awards, presented by HRH The Duke of York.

He does extensive charity work throughout the world, including providing Scholarships. He is also interested in calligraphy, country music, and endurance athletics, the latter including the London Marathon.